Rudolf Nieuwenhuys

Chemoarchitecture of the Brain

With 58 Figures

Springer-Verlag
Berlin Heidelberg NewYork Tokyo

RUDOLF NIEUWENHUYS, M.D.
Professor of Neuroanatomy, Department of Anatomy,
University of Nijmegen, The Netherlands

ISBN 3-540-15349-7 Springer-Verlag Berlin Heidelberg New York Tokyo
ISBN 0-387-15349-7 Springer-Verlag New York Heidelberg Berlin Tokyo

Library of Congress Cataloging-in-Publication Data
Nieuwenhuys, R., 1927, Chemoarchitecture of the brain.
Bibliography: p. 199. Includes index. 1. Brain chemistry. 2. Cytoarchitectonics. 3. Histochemistry. I. Title. [DNLM: 1. Brain–anatomy & histology. 2. Brain Chemistry. 3. Brain Mapping. 4. Neuroregulators. 5. Peptides. WL 300 N682c]
QP376.N53 1985 612'.825 85-17258

Printed in Germany.

Reproduction of the figures: Gustav Dreher GmbH, Stuttgart

Typesetting, printing, and bookbinding by Universitätsdruckerei H. Stürtz AG, Würzburg

2121/3130-543210

For Letty

Preface

The purpose of this book is threefold: (1) to present a short but comprehensive survey of the localization of a number of transmitters and other neuroactive principles in the mammalian central nervous system, (2) to provide some comments on the relation between "classical" neuroanatomy and "chemical" neuroanatomy, and (3) to suggest that by a synthesis of these two approaches "new" entities in the brain can be delineated.

The history of the origin of this volume is somwhat unusual and may be worth relating. In a previous book, namely Nieuwenhuys, Voogd and Van Huijzen: *The Human Central Nervous System* (second edition, 1981), there is a brief chapter on monoamine neuron systems. The authors agreed that in the third edition of the work that chapter should be considerably extended so as to cover not the monoamines alone, but all important transmitter-specified neuronal populations. It was my task to write that chapter, so I started working on it, beginning with acetylcholine. As is common practice among the three of us, I forwarded my products at regular intervals to Jan Voogd and Chris van Huijzen. Their comments were in general positive, but after a few months they, and I too, became rather concerned about the rate at which the chapter was growing, and somewhat later its length had surpassed that of the entire text of the work mentioned. It was then agreed that my product would be published as a review paper, and that only the most salient data would be included in *The Human Central Nervous System*. Still later, the total body of text and illustrations had grown well beyond the limits of even a very extensive review, and I realized that the script had spontaneously taken on the size and format of a book. I wrote a letter to Mr. B. Lewerich, of Springer-Verlag, advising him that there would be some delay in the production of the third edition of *The Human Central Nervous System*, that I happened to have generated an entirely different book on a brand of knowledge usually referred to as 'chemical neuroanatomy', and that I should like to publish this work, as previous ones, with Springer. Mr. Lewerich's reply was quite characteristic: 'I must admit that, in spite of seven years of cooperation, you still succeed in surprising me; I had not the slightest idea about your plans so far.' Publishing and neuroanatomy are apparently equally adventurous professions! Mr. Lewerich continued: 'Springer-Verlag will publish this book as a separate monograph.'

During the writing of this book I received the encouragement, advice and help of numerous persons; hence, my gratitude is manifold.

Jan Voogd and Chris van Huijzen generously accepted my excuses for the delay in the production of my share of the third edition of our joint book, as did Springer-Verlag.

Dr. L.H. Bannister very kindly read the entire manuscript. He carefully scrutinized the English and made numerous valuable suggestions. Without his encouragement and most generous help this book would have been impossible.

Drs. A.R. Cools, C. Jaeger and C. Nicholson have read through large parts of the manuscript and given valuable advice concerning special subjects about which they have first-hand knowledge.

Dr. J.W.F.M. van Nispen collected the data on the primary structure of the various neuropeptides and provided information on chemical matters.
Dr. J.G. Veening made some valuable suggestions on neuroethological aspects.
Drs. G. Paxinos and H.W.M. Steinbusch made available manuscripts not yet published.
Dr. P. van Domburg meticulously collected and processed part of the data upon which Chap. 5 is based.
Mr. F. Geurts assisted in preparing the chemical illustrations.
Chris van Huijzen remained what he has been continuously during the past 15 years: my mentor and advisor in the art of preparing clear illustrations. Moreover, he designed the cover and prepared Figs. 48 and 49.
Mr. W.P.J. Maas skilfully executed all of the remaining drawings, and Mr. J. Konings drew the chemical formulae.
Professor A. Hopf and Dr. L. Price kindly provided me with some photomicrographs of neuropathological cases.
The invaluable secretarial assistance afforded by Anneke Siebring and Mia Smeekens is especially acknowledged.

Finally, I extend my most sincere thanks to Springer-Verlag and its staff – especially Mrs. Th. Deigmöller, Mrs. D. Großhans, Mrs. U. Pfaff and Mrs. M. Schäfer, for their help during the publication of this book.

Summer 1985 R. NIEUWENHUYS

Contents

CHAPTER 1

Introduction 1

CHAPTER 2

Plan and Program 6

CHAPTER 3

Survey of Chemically Defined Cell Groups and Pathways 7

Acetylcholine 7

Monoamines 11
Dopamine 14
Noradrenaline 19
Adrenaline 30
Serotonin 33
Histamine 41

Amino Acids 44
γ-Aminobutyric Acid 44
Glutamate and Aspartate 48
Glycine 51
Taurine 51

Gut-Brain Peptides 52
Substance P 52
Vasoactive Intestinal Polypeptide 57
Cholecystokinin 60
Neurotensin 65

Hypophysiotropic Peptides 70
Corticotropin-Releasing Factor 70
Luteinizing Hormone-Releasing Hormone 74
Somatostatin 77
Thyrotropin-Releasing Hormone 82

Neurohypophyseal Peptides 84
Vasopressin and Oxytocin . . . 84

Pro-Opiomelanocortin Derivatives 87
Corticotropin, or Adrenocorticotropic Hormone (ACTH) . . . 89
β-Lipotropin 91
α-Melanocyte-Stimulating Hormone 92
β-Melanocyte-Stimulating Hormone 92
γ-Melanocyte-Stimulating Hormone 93
β-Endorphin 93
Comment on the Functional Significance of POMC Derivatives 94

Enkephalins 97

Dynorphins 104

Angiotensin II 108

CHAPTER 4

Conclusions and Comments 114

CHAPTER 5

'New' Entities in the Central Nervous System: The [Paracrine?] Core and Its Adjuncts 177

CHAPTER 6

Classical Neuroanatomy, 'Chemical' Neuroanatomy and Neurobiology: Some Concluding Remarks 194

References 199

Subject Index 235

CHAPTER 1

Introduction

Although the theory that neurons are true secreting cells which act upon one another by the passage of chemical substances was enunciated at the beginning of the 1900s (Scott 1905), it was only half a century later that the significance of humoral transmission for the processing of information in the central nervous system became fully appreciated. It was then established that the transfer of neural impulses occurs principally at morphologically differentiated contact sites, the synapses, and that the chemical intermezzo in the otherwise electrical flow of signals, though causing some delay, still falls within the millisecond range. It appeared, moreover, that the influence exerted by a presynaptic element via a chemical mediator, or neurotransmitter, could be either excitatory or inhibitory (e.g. Anderson et al. 1964; Eccles 1957, 1969).

The electron microscope and the micro-electrode have been instrumental in these developments. The synapses were found to include, in addition to certain membrane specializations, clusters of small vesicles lying near the presynaptic membrane (Palade and Palay 1954; de Robertis and Bennett 1954, 1955). It became widely accepted that these vesicles contain the transmitter substance and release their content into the synaptic cleft (Whittaker 1965; de Robertis 1966, 1967; Palay 1967; Katz 1969; Andres and von Dühring 1976). It has been possible in a few instances to 'capture' synaptic vesicles in this act of exocytosis (e.g. Heuser et al. 1979). It should be added that a rule or principle first formulated by Dale (1935) for the peripheral nervous system, which may be epitomized as follows: "each neuron releases one and the same neurotransmitter at all its (synaptic) terminals", was generally considered to be valid for the central nervous system as well. Evidence was adduced indicating that the neutrotransmitter molecules, once released into the synaptic cleft, reach and are recognized by specific receptors embedded in the membrane of the postsynaptic element, and that this interaction with the receptors elicits the opening of particular ionic channels and the displacement of certain kinds of ions. It was held that at all of the synaptic terminals of a given neuron, the transmitter substance opens just one type of ionic channel, characterizing either excitatory or inhibitory synapses (e.g. Eccles 1969). The microelectrode enabled the neurophysiologist to visualize and to register the electric potentials correlated with the displacement of ions referred to above (cf. Hubbard et al. 1969). Initially, the number of known central neurotransmitters was extremely small, including hardly more than acetylcholine and noradrenaline (e.g. Paton 1958), and actually, the idea prevailed that the brain could manage with one excitatory and one inhibitory transmitter.

During and after the period just sketched there were two notable developments: (a) the number of known or putative neurotransmitters steadily increased, and (b) the views on the modes in which central neurons may be influenced widened considerably.

As regards neurotransmitters, at the end of the 1960s the following substances were generally considered to be acting as such in the central nervous system: acetylcholine (ACh), noradrenaline, dopamine, adrenaline and histamine. In the late 1960s and the early

1970s came an appreciation that in addition to their metabolic role, certain amino acids, such as γ-aminobutyric acid (GABA), glutamic acid, aspartic acid and glycine, might serve as neurotransmitters (Graham et al. 1967; Johnson and Aprison 1971; Curtis and Johnston 1974). During the past decade there has been a dramatic explosion in the number of possible neurotransmitters, with the increasing recognition that various peptides are neuronally localized in the central nervous system and may well be neurotransmitters (cf. Guillemin 1978; Emson 1979a; Snyder 1980; Hökfelt et al. 1980a; Krieger and Martin 1981a, b; Krieger 1983). Thus, it appeared that the so-called posterior lobe peptides, derived from neurosecretory neurons in the supraoptic and paraventricular nuclei, give rise to extra-hypothalamic projections, the terminals of which do not differ in any fundamental way from the classical transmitter-containing synapses in the brain (Buijs and Swaab 1979). The original concept that other peptidergic hormones, originating from the mediobasal hypothalamus, act solely via the hypothalamo-hypophyseal portal system on the anterior pituitary to regulate the release of the appropriate trophic hormones had to be abandoned. Several of these 'hypophyseotropic factors' (Scharrer 1970), including growth hormone release-inhibiting hormone (somatostatin, SST), thyrotropin releasing hormone (TRH) and luteinizing hormone-releasing hormone (LHRH), have a wide distribution throughout the brain. Moreover, there is strong evidence indicating that several anterior pituitary peptides derived from the precursor pro-opiomelanocortin (POMC) – such as corticotropin (ACTH), α-melanocyte-stimulating hormone (α-MSH) and β-endorphin – are also synthesized in the brain. Finally – this brief summing up is by no means exhaustive – there are a number of peptides which occur in the gastrointestinal tract as well as in the central nervous system. Some of these, such as vasoactive intestinal polypeptide (VIP) and cholecystokinin (CCK), were originally demonstrated to occur in secretory elements of the gastrointestinal tract. Others of these so-called gut brain peptides, such as substance P and neurotensin, were initially isolated and characterized in the brain prior to their description in the gastrointestinal tract. The rapid rate of growth in this field may be illustrated by the fact that in 1980, Snyder mentioned the presence of 20 peptides in the mammalian brain, whereras three years later, Iversen (1983) and Krieger (1983) listed 33 and 38 neuropeptides respectively. Several experts have expressed the opinion that the isolation and characterization of neuropeptides is presumably still in its initial phase. Thus, Sternberger wrote in 1980: 'Only a small proportion of all neuropeptides have been identified so far', and Snyder (1980), in the same year: 'It would not be surprising if the known peptide transmitters represent only 10 percent or less of the total, whose number then may exceed 200'. In the meantime it should be kept in mind that the terms 'neuropeptide' and 'peptidergic neurotransmitter' are not synonymous.

As regards the widening of views on the modes in which central neurons may be influenced, I should like to mention the following aspects:

1. Although there are neurotransmitters that are mainly excitatory in action (e.g. aspartic acid, glutamic acid) and others that are mainly inhibitory (e.g. GABA, glycine), none of the known transmitters can be defined in their own right as either excitatory or inhibitory (and thus as instrumental in opening just one type of ionic channel). The effect elicited by a neurotransmitter depends not only on its chemical structure, but also on the nature of the receptor with which it combines. There is evidence suggesting the existence of both excitatory and inhibitory receptors for the same transmitter (e.g. acetylcholine, noradrenaline) on neurons in the mammalian brain, and in invertebrates, even co-existence of 'opposite receptors' for the same transmitter on the same neuronal membrane has been described (for review see Uchizono 1975; Szabadi 1978).

2. The 'conventional' synaptic transmission is rapid in onset and termination, having a total maximal duration of 30 ms. The process is essentially restricted in time to the duration of transmitter release, which in turn is determined by the duration of the presynaptic action potential. That the time course of action of the transmitter does not exceed the time of its release is due to the availability of mechanisms to rapidly terminate the action of the transmitter, the main mechanisms being (a) inactivation of the transmitter through enzymatic degradation and (b) inactivation through re-uptake by presynaptic terminals or by glial cells. However, evidence has gradually accumulated that some neurotransmitters may elicit transmission processes of much longer duration. To mention a few examples: (a) It is known that dopamine may slowly depolarize the membrane of its target neurons (Bernardi et al. 1978; Libet 1979). (b) Depending on the type of receptor with which it binds, acetylcholine may elicit a rapid or a slow excitatory postsynaptic potential (Brown 1983). (c) Peptidergic neurotransmitters may provide relatively long-lasting signals. One reason for this is that the released peptides are not re-accumulated into the nerve terminal (Emson 1979a).

3. The binding of certain neurotransmitters to certain classes of receptors does not lead to the opening of ionic channels (and to the well-known post-synaptic potentials correlated with the ensuing ion displacement), but rather to the activation of much longer lasting intraneuronal processes. The receptors operative in this type of process are functionally related to the membrane-bound enzyme adenylate cyclase. Activation of this enzyme results in the amplified production of adenosine 3′, 5′ monophosphate, or cyclic AMP. Thus, the neurotransmitter does not enter the cell; in its place cyclic AMP takes over the messenger function within the cell, which is why it is often designated as the 'second messenger'. The enhanced production of cyclic AMP leads to cascade reactions, which ultimately may result in such processes as activation of the production of enzymes which catalyze the synthesis of neurotransmitters or even a genomic effect, involving RNA and protein synthesis (Nathanson 1977; Nathanson and Greengard 1977; Greengard 1978, 1979).

4. In 'classical' synaptic transmission the messenger substance is delivered across a very narrow gap, separating the specialized pre- and postsynaptic parts of the two neurons involved. However, in the mammalian central nervous system, numerous thin, unmyelinated, varicose fibres occur. Many of the varicosities in these fibres are densely filled with typical 'synaptic' vesicles but do not exhibit the characteristic presynaptic membrane specializations. Moreover, most of these varicosities lack a close and distinct apposition to postsynaptic membrane specializations, for example, a local membrane thickening or a subsynaptic web. These findings have led to the hypothesis that the synaptic vesicles present in these varicosities release their contents in the extracellular space, and that by doing so they are able to influence large numbers of receptors located at relatively great distances from the site of release (Palay and Chan-Palay 1974: 'synapses à distance'; Chan-Palay 1977, 1978; Descarries et al. 1975, 1977; Beaudet and Descarries 1978; Léger and Descarries 1978; Pickel et al. 1976, 1977). This hypothesis seems plausible, because analogous processes have long been known to occur in the peripheral autonomic system. It should be noted, however, that no one has so far succeeded in directly visualizing the exocytosis of vesicle contents from such varicosities, so at least some of these structures may not be functional in chemical neurotransmission.

5. Certain neurochemicals not directly involved in the process of synaptic transmission in the strict sense are able to influence this process in various ways. These 'neuromodulators' (see Florey 1967; Barchas et al. 1978; Weight 1979; Kupfermann 1979; Rotsztejn 1980) may be active at the presynaptic site, e.g. by affecting the amount of transmitter released and the time course of

transmitter release, as well as on the postsynaptic side, e.g. by regulating the sensitivity of the receptors. Certain properties are attributed to neuromodulators: (a) they are not only liberated by neurons, but may also be released by glial cells, including ependymal elements, neurosecretory cells and gland cells; (b) the effect of their action usually lasts longer than that of conventional neurotransmitters; (c) most important – they change the capacity of their target cells to respond to a given stimulus, or in other words, they set the levels of excitability at the synaptic junctions. Kupfermann (1979) emphasized that the neuromodulatory influences increase the complexity of the processing of information at the level of individual neurons. He suggested the possibility that each class of synaptic input to a cell could be selectively depressed or enhanced by a corresponding modulatory input; he also proposed that the long duration of modulatory activities and their ability to alter the effects of other synaptic events are properties ideally suited to the control of behavioural phenomena such as learning, motivational state, arousal and sensitization. In practice it is often difficult to make a sharp distinction between neurotransmitters and neuromodulators, and there is evidence suggesting that, for instance, the monoamines noradrenaline and serotonin act either as neurotransmitters or as neuromodulators, according to the site at which the action takes place.

6. We know, thanks largely to the investigations of Tomas Hökfelt and his associates, that in many neurons more than one neuroactive principle is present (for reviews see Hökfelt et al. 1980a; Lundberg and Hökfelt 1983). Two or more neuropeptides may coexist within one and the same neuron, and the combination of a neuropeptide with one of the monoamines is also frequently encountered. Although the functional significance of the phenomenon just indicated remains to be explored, it will be clear that the co-existence and co-release of different neurochemicals potentially greatly increases the number of different chemical signals that a neuron can use in communicating with other neurons (Iversen 1983). The problems involved are complex and cannot be discussed in depth within the frame of the present book. However, the following aspects and possible implications may be briefly mentioned: (a) Hökfelt et al. (1980a) envisioned that a given set of neurons, all containing the 'classical' transmitter A, could be subdivided into a number of subsets, each containing, in addition to A, a different neuropeptide. The specific peptide of each would then confer the ability to convey differentiated messages. (b) The parasympathetic fibres which innervate the sweat glands contain both ACh and VIP. There is evidence indicating that ACh mainly causes secretion, by direct stimulation of the secretory cells in the gland, whereas VIP mainly causes vasodilatation, by relaxing smooth muscle cells around blood vessels (Lundberg et al. 1979). It has now been found that a single impulse preferentially induces a response which is due to the release of ACh, but that upon stimulation with higher frequencies there is an increased functional effect caused by VIP. Lundberg and Hökfelt (1983) suggested that analogous synergistic actions of a classical neurotransmitter and a neuropeptide released by one and the same neuron could also occur in the central nervous system. (c) It is important to note that if a given neuron synthesized and released more than one true neurotransmitter, Dale's principle would be compromised. However, if only one of the substances liberated were a true neurotransmitter, whereas the other(s) exerted (a) neuromodulatory action(s), then Dale's principle could still be considered valid. (d) On the postsynaptic side the co-release of more than one neuroactive principle may lead to a variety of processes; for instance, (i) different terminals of a single axon can innervate different cell types, defined on the basis of their complement of (postsynaptic) receptors (Swanson 1983); (ii) the substances liberated can act on different receptors of the same neurons; (iii) the substances can interact at the same postsynaptic receptors.

7. In spite of the presence of a blood-brain

barrier, certain sets of central neurons are important targets for circulating hormones, produced either by the pituitary or by the peripheral endocrine glands (cf. Stumpf 1970, 1975; Stumpf and Sar 1975; Morell and Pfaff 1978; Héritage et al. 1976, 1980; Rees et al. 1980; Felten and Crutcher 1979; Felten and Sladek 1983). By these sets of neurons, endocrine signals are converted into neural activities. The receptors for most of the hormones are probably situated at the external surface of their target neurons, but it is known that steroid hormones are able to pass through the cell membrane to encounter receptors that are located in the cytoplasm of the target cells. The mapping of the location of steroid hormones in the brain has grown into a neuroanatomical subdiscipline, which has been designated by Stumpf (1975) as 'hormone-architectonics'.

From the foregoing it may be concluded that in the central nervous system, modes of chemical transfer of information occur which diverge markedly from classical synaptic transmission. Chemical signalling appears to be very complex in the central nervous system, and this complexity has led to the introduction of a host of terms such as 'classical or genuine neurotransmitters', 'putative, possible or suspected neurotransmitters', 'neuromodulators', 'neuromediators', 'chemical messengers', 'neuroactive principles', 'neurochemicals' and even (infelicitously) 'neuroregulators'. The significance of several of these terms has been indicated in the preceding pages, and no attempt will be made here to further define all of them (for a critical review, see Van Dongen 1981a).

It is against the background of the developments outlined above that the inception and rise of what has become known as 'chemical neuroanatomy' (Emson 1983) should be appreciated. Taking the general validity of Dale's principle for granted, the initial question was, clearly and succinctly: What is the neurotransmitter of this particular neuron or set of neurons? In this way it was hoped to determine, as it was sometimes put, the 'chemical fingerprint' of neurons. It will be clear that this seemingly simple question has led us into labyrinths of much greater complexity than those of fingerprints. The initial impulse in this area was given by the pioneering work of Dahlström and Fuxe (1964, 1965), who charted the distribution of monoaminergic cells and fibres in the central nervous system of the rat, using the formaldehyde-induced fluorescence technique developed by Falck and Hillarp (Falck 1962; Falck et al. 1962), and by that of Shute and Lewis (Shute and Lewis 1967; Lewis and Shute 1967), who used acetylcholinesterase staining as a potential marker for cholinergic neurons in combination with lesions in the same species. The introduction of the powerful technique of immunohistochemistry (Coons 1958; Sternberger 1979) for the study of the localization of neurotransmitters and their synthesizing enzymes has provided an enormous impetus to the field of chemical neuroanatomy. The dramatic increase in the number of neurotransmitters and possible neurotransmitters during the past several years – already alluded to – is a direct consequence of this technical development.

CHAPTER 2

Plan and Program

Although the mapping of neuroactive principles with the aid of histochemical techniques is still in full progress, a tentative interim summation of what has been accomplished so far may be appropriate. This book is intended to serve that purpose. It (a) presents concise descriptions of the localization of a number of (putative) neuromediators in the mammalian brain, based on the literature, (b) provides pictorial surveys of the localization of most of these substances, and (c) attaches some comments, conclusions and speculations to the data gathered. The following substances will be dealt with:

1. Acetylcholine
2. The monoamines: dopamine, noradrenaline, adrenaline, serotonin and histamine
3. The amino acids: γ-aminobutyric acid, glutamate, aspartate, glycine and taurine
4. The gut-brain peptides: substance P, vasoactive intestinal polypeptide, cholecystokinin and neurotensin
5. The hypophysiotropic peptides: corticotropin-releasing factor, luteinizing hormone-releasing hormone, somatostatin and thyrotropin-releasing hormone
6. The neurohypophyseal peptides: vasopressin, oxytocin and their related neurophysins
7. The pro-opiomelanocortin dervatives: corticotropin (ACTH), α-melanocyte stimulating hormone and β-endorphin
8. The enkephalins and dynorphins
9. Angiotensin II

Although these substances are usually considered (putative) neurotransmitters or neuromodulators, we should not lose sight of the possibility that at least some of them fulfil roles entirely unrelated to the chemical transfer of information.
Before the descriptions of the localization of the compounds mentioned are presented a few notes on the accompanying illustrations are in order. These pictorial overviews were prepared following a style developed in a previous publication (Nieuwenhuys et al. 1981). They are based on a medial view of the bisected human brain. The various centres and pathways have been projected upon the median plane. Several 'manipulations' have been applied. Thus, the hippocampal formation has been 'unrolled' in order to expose its various parts, and the globus pallidus has been turned 90°.

CHAPTER 3

Survey of Chemically Defined Cell Groups and Pathways

Acetylcholine

Acetylcholine was the first neurotransmitter to be identified. Choline acetyltransferase (ChAT) is the enzyme which is responsible for the synthesis of ACh, whereas the enzyme acetylcholinesterase (AChE) degrades this substance (Fig. 1). In numerous older studies, among which those of Shute and Lewis (Shute and Lewis 1967; Lewis and Shute 1967) especially should be mentioned, it was attempted to study the anatomy of central cholinergic neurons by means of AChE histochemistry. However, it is known that although cholinergic neurons generally show a very high content of AChE, a high activity of this enzyme is in itself not a sufficient characteristic for identifying cholinergic neurons (Lehmann and Fibiger 1979; Butcher and Woolf 1982; Eckenstein and Sofroniew 1983). The successful purification of ChAT and the subsequent procurement of antibodies against this enzyme for use in immunochemistry have recently yielded a reliable means of visualizing cholinergic neurons in the central nervous system (cf. McGeer et al. 1984a; Wainer et al. 1984; Rossier 1984). The following survey of the major central cholinergic neuron populations and their efferent projections (Fig. 2) is based primarily on the studies of Kimura et al. (1981), Fibiger (1982) and Mesulam et al. (1983a, b, 1984a).

The skeletal motoneurons (α as well as γ) and preganglionic autonomic neurons in the *brain stem* and *spinal cord* are cholinergic in nature. The localization of these neurons is well known, and has not been indicated in Fig. 2.

The *auditory receptor cells in the organ of Corti* receive a substantial efferent innervation originating from the periolivary nuclei, small groups of cells surrounding the superior olivary complex. Axons arising from these nuclei, which include both crossed and ipsilateral fibres, reach the cochlea via the fasciculus olivocochlearis, or bundle of Rasmussen (1946, 1953; Moore and Osen 1979;

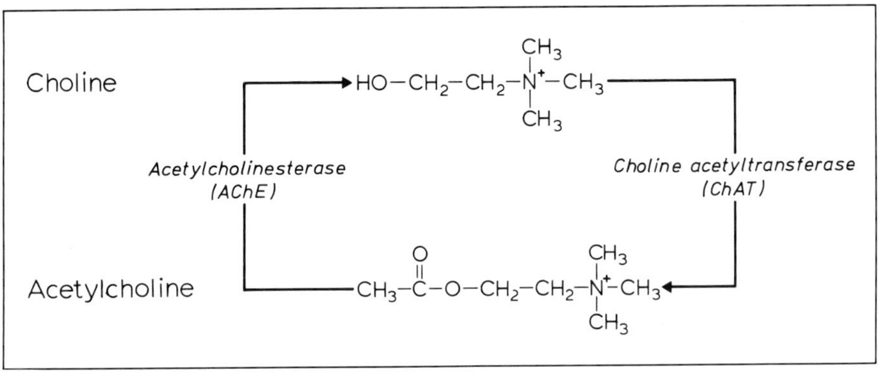

Fig. 1. Synthesis and degradation of acetylcholine

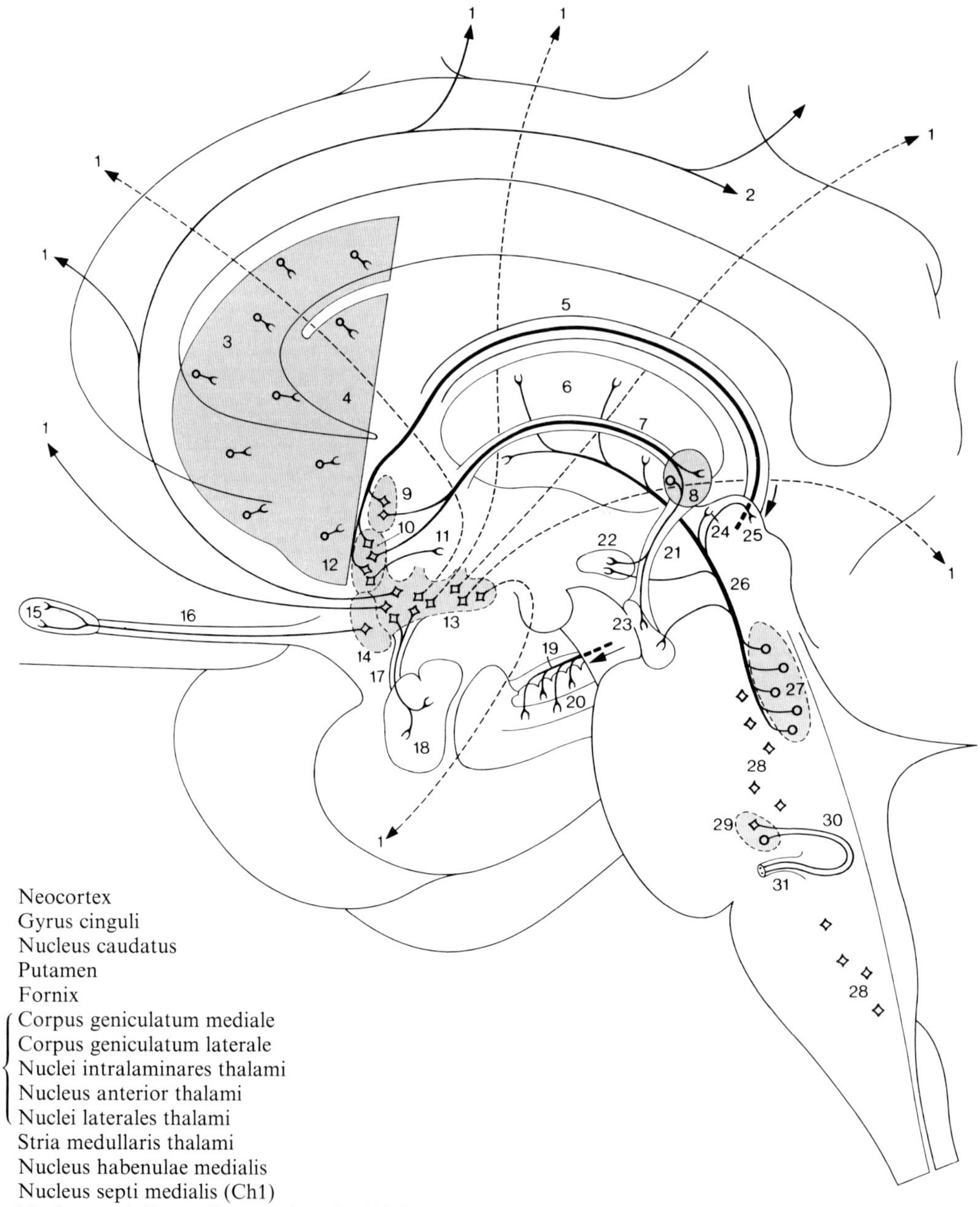

1 Neocortex
2 Gyrus cinguli
3 Nucleus caudatus
4 Putamen
5 Fornix
6 { Corpus geniculatum mediale; Corpus geniculatum laterale; Nuclei intralaminares thalami; Nucleus anterior thalami; Nuclei laterales thalami }
7 Stria medullaris thalami
8 Nucleus habenulae medialis
9 Nucleus septi medialis (Ch1)
10 Nucleus gyri diagonalis, pars dorsalis (Ch2)
11 Area lateralis hypothalami
12 Nucleus accumbens
13 Nucleus basalis of Meynert (Ch4)
14 Nucleus gyri diagonalis, pars ventralis (Ch3)
15 Bulbus olfactorius
16 Tractus olfactorius
17 Fibrae amygdalofugales ventrales
18 Nucleus basalis amygdalae
19 Fimbria hippocampi
20 Hippocampus
21 Tractus habenulointerpeduncularis
22 Area tegmentalis ventralis
23 Nucleus interpeduncularis
24 Area pretectalis
25 Colliculus superior
26 Fasciculus tegmentalis dorsalis (Shute and Lewis)
27 Area tegmentalis dorsolateralis (including nuclei parabrachiales) plus adjacent central grey (Ch5+Ch6)
28 Formatio reticularis medialis
29 Nuclei periolivares
30 Fasciculus olivocochlearis (Rasmussen)
31 Nervus vestibulocochlearis

Fig. 2. Cholinergic cell groups and pathways

White and Warr 1983; Adams 1983). There is strong, though not conclusive evidence that the olivocochlear neurons are cholinergic (Guth et al. 1976). The bundle of Rasmussen forms the final link in a descending corticocochlear projection, by which the brain is able to influence its own auditory input.

The *rhombencephalic medial reticular formation* contains scattered large cholinergic elements (Kimura et al. 1981).

In the *lateral tegmental area of the rostral rhombencephalon* a conspicuous, longitudinally oriented zone of cholinergic neurons is found. This zone does not coincide with a particular cytoarchitectonic unit, but includes parts of the medial and lateral parabrachial nuclei and of the pedunculopontine tegmental nucleus (Armstrong et al. 1983; Mesulam et al. 1983b). Dorsomedially, this area is continuous with a group of cholinergic neurons, most of which are located within the central grey. In rodents these elements lie largely within a cytoarchitectonic entity known as the nucleus laterodorsalis tegmenti (Mesulam et al. 1983b). The complex of cholinergic neurons just outlined gives rise to an ascending fibre system, the dorsal tegmental pathway of Shute and Lewis (1967), which projects upon the superior colliculus, the pretectal area and several parts of the thalamus, including the medial and lateral geniculate bodies, the intralaminar nuclei and the anterior and lateral nuclear groups (Shute and Lewis 1967; Hoover and Jacobowitz 1979; Hoover and Baisden 1980). Fibres originating from the cholinergic neurons in the lateral tegmental zone probably also distribute to the interpeduncular nucleus and the ventral tegmental area of Tsai (Fibiger 1982). Mesulam et al. (1983b, 1984a) designated the cholinergic cells in the lateral rhombencephalic tegmental area as Ch5 and the adjacent sector of the central grey as Ch6. They regarded both groups as forming part of the reticular formation and expressed the opinion that the cholinergic pathway passing from these groups to the thalamus provides an essential component of the ascending reticular activating system.

The *basal forebrain* contains a population of large cholinergic neurons, extending from the septal region rostrally to the subthalamic nucleus caudally. Mesulam et al. (1983a, b, 1984a) subdivided this population into four groups, Ch1–Ch4. (For a charting of these cell masses in the human brain see Perry et al. 1984.)

The Ch1 group corresponds to the medial septal nucleus; about 10% of its neurons are cholinergic.

The Ch2 group corresponds to the vertical limb of the nucleus of the diagonal band of Broca; at least 70% of its neurons are cholinergic.

The Ch3 group most closely corresponds to the horizontal limb of the nucleus of the diagonal band of Broca; only 1% of the neurons of this nucleus are cholinergic.

The Ch4 group, which is very extensive in the human brain, most closely corresponds to the nucleus basalis of Meynert (Meynert 1872; Kölliker 1896). This nucleus lies embedded in the substantia innominata, a vaguely defined area situated ventral to the globus pallidus. At least 90% of the neurons in the nucleus basalis are cholinergic. Large cholinergic neurons situated within the internal and external medullary laminae of the globus pallidus also form part of the Ch4 group.

The four cell groups just discussed give rise to the following cholinergic projections:

1. Fibres originating from Ch1, Ch2 and Ch3 pass via the stria medullaris and the tractus habenulointerpeduncularis to the base of the midbrain, where they terminate in the interpeduncular nucleus and in the area tegmentalis ventralis (Fibiger 1982). Cholinergic fibres contained in the stria medullaris also project heavily to the medial habenular nucleus (Gottesfeld and Jacobowitz 1979), but whether the same or different populations of cholinergic neurons innervate the habenula and the ventromedial mesencephalon has not yet been determined. The cholinergic projection arising from the basal forebrain and passing by way of the tractus habenulointer-

peduncularis to the interpeduncular nucleus is reinforced by fibres originating from cholinergic elements situated in the medial habenular nucleus (Villani et al. 1983).
2. Ch1 and, particularly, Ch2 provide a substantial cholinergic projection to the hippocampus, as has already been suggested by Lewis and Shute (1967). The fibres pass through the fornix and terminate in the cornu ammonis as well as in the fascia dentata.
3. Ch2 also sends fibres to the lateral hypothalamus.
4. Cholinergic neurons situated mainly in Ch3 pass to the olfactory bulb, to terminate in the external layers of that structure.
5. Cholinergic neurons located in the anterolateral portion of Ch4 project, via the ventral amygdalofugal pathway to the amygdaloid complex, mainly to the nucleus basalis.
6. The remaining parts of Ch4 provide a major cholinergic projection to the entire neocortex (Shute and Lewis 1967; Divac 1975; Kievit and Kuypers 1975; Mesulam and Van Hoesen 1976; Bigl et al. 1982; Mesulam et al. 1983a). This projection is organized in a crude topographical fashion, with the more rostral cells projecting to the frontal and parietal cortex, and the more caudal cells innervating the occipital and temporal cortex (Mesulam et al. 1983a; cf. also Mayo et al. 1984), but its precise organization remains to be worked out. Bigl et al. (1982) and Price and Stern (1983) have provided evidence suggesting that the individual cells in the nucleus basalis project to very restricted areas within the cortex. Aston-Jones et al. (1984) have demonstrated that cortically projecting nucleus basalis neurons in the rat have heterogeneous physiological properties. These elements yielded spontaneous discharge rates of 0–40 Hz, a variety of impulse amplitudes and waveforms, and a wide range of conduction latencies from the frontal cortex, yielding fibre conduction velocities of about 0.6–8.0 m/s. These values are within the range of speeds for fine myelinated fibres.

The large population of cholinergic cells situated in the basal forebrain has been implicated in a variety of physiological and behavioural processes. The projection to the olfactory bulb may well play a role comparable to that of the olivocochlear bundle in the auditory system. The pathway to the hippocampus is possibly involved in memory processes, and that to the amygdala may modulate affective behaviour. The nucleus basalis has been suggested to represent a telencephalic extension of the reticular formation of the brain stem (Mesulam and Van Hoesen 1976; Mesulam et al. 1983a). It receives afferents from a variety of cortical and subcortical sources (Norgren 1974; Jones et al. 1976; Leichnetz and Astruc 1977; Saper et al. 1979; Price and Amaral 1981; Mesulam et al. 1983a; Mesulam and Mufson 1984). The cortical input arises from the prepiriform cortex, the orbitofrontal cortex, the anterior insula, the temporal pole, the entorhinal cortex and the medial temporal cortex. Thus, in contrast to its widespread projections to all parts of the neocortex, the nucleus basalis receives reciprocal projections from only a relatively small number of cortical areas. The subcortical afferents of the nucleus basalis originate from the septal nuclei, the nucleus accumbens–ventral pallidum complex, the amygdala, the hypothalamus, the peripeduncular nucleus of the midbrain and the pontine taste area. In light of its afference, Mesulam and co-workers (Mesulam et al. 1983a; Mesulam and Mufson 1984) considered it likely that the Ch4 complex acts as a cholinergic relay between limbic plus paralimbic areas and the entire neocortex in a fashion that may influence complex behaviour according to the prevailing emotional and motivational states. Finally, some recent neuropathological findings deserve a brief mention. It has been reported that in Alzheimer's disease and senile dementia of the Alzheimer type (a) the amounts of ChAT and AChE in the cerebral cortex are considerably reduced (Davies 1979), (b) the neurons in the nucleus basalis of Meynert undergo a profound and selective degeneration (Whitehouse et al. 1981, 1982), (c) there are marked decreases in AChE and ChAT in the substantia innominata (Rossor et al. 1982), and (d) there is a loss of more than 50% of ChAT-containing neurons in the substantia innominata (Nagai et al. 1983a). These findings are of great potential importance for the understanding of the pathophysiology of the disorders mentioned. However, as has been pointed out by Mesulam et al. (1983a), whether the loss of cholinergic neurons in the nucleus basalis (Ch4) causes the loss of cortical cholinergic innervation or instead the primary lesion is in the cortex, the alterations in the nucleus basalis reflecting retrograde degeneration, has not yet been determined.

All of the groups of cholinergic cells discussed so far are composed of long-axoned projection neurons. However, throughout the *neostriatum*, i.e. the nucleus caudatus, the putamen and the nucleus accumbens, cholinergic local circuit neurons are found. These elements are large and comprise approximately 1% of the total population of neostriatal cells (Kimura et al. 1981; Woolf and Butcher 1981; Butcher and Woolf 1982; Fibiger 1982). It is noteworthy that they remain unaffected in Alzheimer's disease (Parent et al. 1984). Mesulam et al. (1984a) reported that in the macaque the density of cholinergic neurons is higher in the ventral striatum (i.e. the nucleus accumbens and the olfactory tubercle) than in the dorsal striatum (i.e. the caudate nucleus and the putamen).

Addenda:

1. Using monoclonal antibodies against ChAT, Sofroniew et al. (1985; see also Cuello and Sofroniew 1984) visualized cholinergic neurons in several additional regions of the central nervous system of the rat, including the cerebral cortex, the hippocampus, the anterior olfactory nucleus, the olfactory tubercle, the arcuate nucleus and layers III–VI of the spinal dorsal horn. The cortical neurons are small and bipolar, and most probably represent local circuit neurons. They occur throughout the neocortical and allocortical parts of the pallium. Most, if not all, of these elements not only are cholinergic, but also show immunoreactivity for VIP (Eckenstein and Baughman 1984). Interestingly, Mesulam et al. (1984a), who also used monoclonal antibodies against ChAT, were unable to visualize cortical ChAT-positive cell bodies in the macaque. They pointed out that if the primate cortex lacked intrinsic cholinergic neurons, then its cholinergic innervation would be even more dependent on the projections from the nucleus basalis.
2. Sugimoto and Hattori (1984) provided experimental evidence suggesting that in the rat, cholinergic neurons situated in the nucleus tegmenti pedunculopontinus pars compacta project to several thalamic centres, including the centrolateral nucleus and the centromedian parafascicular complex, and to the subthalamic nucleus.
3. In patients dying with paralysis agitans, a considerable loss of neurons in the nucleus basalis of Meynert is frequently found (e.g. Arendt et al. 1983). Given the high incidence of senile dementia among Parkinsonians (culminating in the Parkinson–dementia complex of Guam), this finding is not surprising. It seems very likely that there is an interrelationship between the pathological processes underlying the two diseases. However, the nature of this interrelationship remains to be elucidated (see Nakana and Hirano 1983; Dubois et al. 1983; Whitehouse et al. 1983; McGeer 1984).
4. It is worthy of note that in Huntington's chorea, another major degenerative disorder that causes dementia, there is no reduction of cortical ChAT activity and no significant loss of neurons from the nucleus basalis of Meynert (Arendt et al. 1983; Clark et al. 1983).

Monoamines

The biogenic amines include the catecholamines, dopamine, noradrenaline and adrenaline, and the indolamine – serotonin, or 5-hydroxytryptamine. All four of these monoamines have a claim to be regarded as central neurotransmitters. The biosynthesis of monoamines takes place as the stepwise

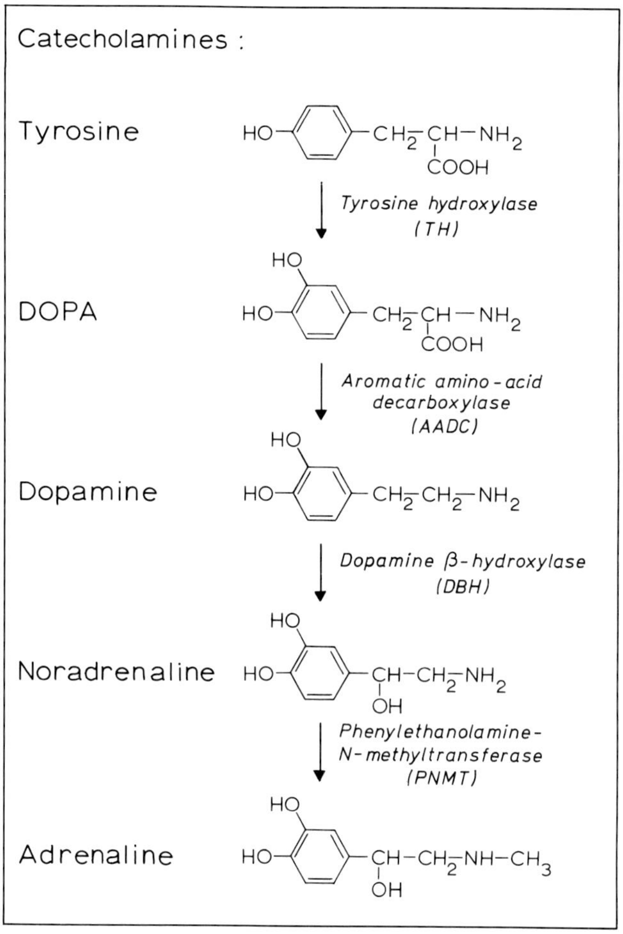

Fig. 3. Biosynthetic pathway of the catecholamines dopamine, noradrenaline and adrenaline

conversion of amino acids in the presence of specific enzymes. The catecholamines are derived from the L-amino acid tyrosine, which is converted into L-dihydroxyphenylalanine (DOPA) by the enzyme tyrosine hydroxylase (TH). A second enzyme, aromatic amino acid decarboxylase (AADC), converts DOPA into dopamine, which may be consecutively converted into noradrenaline and adrenaline, reactions which are specifically catalyzed by the enzymes dopamine β-hydroxylase (DBH) and phenylethanolamine-N-methyl-transferase (PNMT) respectively (Fig. 3). The sequence of reactions by which serotonin is formed involves two steps: the amino acid tryptophan is first converted into 5-hydroxytryptophan by the enzyme tryptophan hydroxylase; then AADC (which also catalyzes the synthesis of dopamine) converts this intermediary amino acid into serotonin or 5-hydroxytryptamine (Fig. 4).

The distribution of monoamine-containing neurons in the central nervous system has been studied in various mammals using fluorescence histochemical techniques based on condensation with formaldehyde (Falck et al. 1962) or glyoxylic acid (Axelsson et al. 1973; Lindvall and Björklund 1974b). The chief limitations of these techniques are that they do not distinguish between the various catecholamines (all fluoresce green) and they have a low sensitivity towards serotonin. More recently, antibodies against the various enzymes involved in catecholamine synthesis have been prepared and antibodies against serotonin (Steinbusch et al. 1978) and dopa-

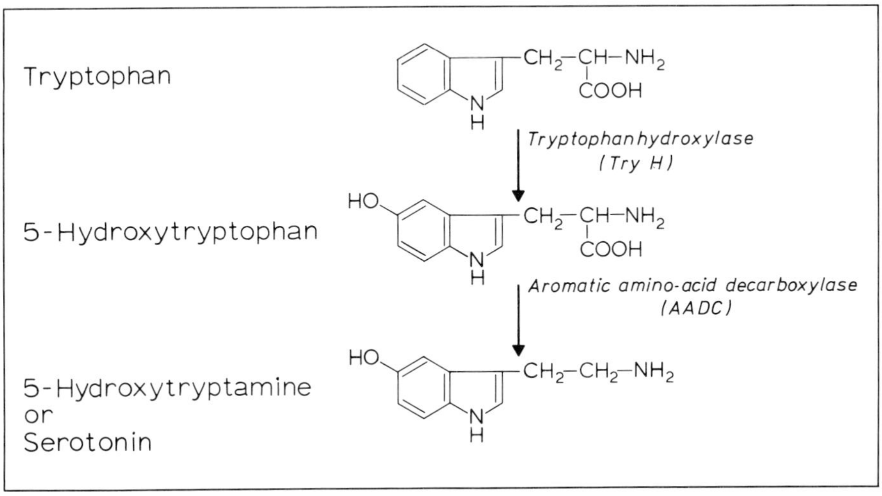

Fig. 4. Biosynthesis of serotonin

mine (Geffard et al. 1984) have also become available, rendering it possible to study the localization of these substances with sensitive immunohistochemical techniques. The data obtained using the classical histofluorescence of monoamines are now being progressively re-examined in the light of the newer immunohistochemical techniques, and further information concerning the organization of the monoaminergic neurons and their projections is being gained by the combination of immunohistochemical techniques with placement of lesions or with retrograde labelling methods.

The literature on the structure and functional significance of the monoamine-containing neuron populations has grown to enormous proportions, and within the framework of this overview no more than a brief survey of the main findings can be presented. For details the reader is referred to the following reviews: Moore and Bloom 1978, 1979; Lindvall and Björklund 1983; Steinbusch and Nieuwenhuys 1983; Felten and Sladek 1983.

The neuron groups which synthesize monoamines are situated mainly in the brain stem and distribute their products be fine-fibred, profusely ramifying projections to a great many regions of the brain and spinal cord. In earlier mapping studies carried out on the brain of the rat, 26 monoaminergic cell groups have been identified (Björklund and Nobin 1973; Dahlström and Fuxe 1964, 1965; Halász et al. 1977; Hökfelt et al. 1974b).

The dopaminergic and noradrenergic cell groups have been collectively designated as A1–A15, the serotoninergic cell groups as B1–B9, and the adrenergic cell groups as C1–C2. This numerical system will be followed here, but the classical neuroanatomical names of these structures will be added wherever possible. Studies on a variety of species, including primates (Garver and Sladek 1975; Schofield and Everitt 1982; Schofield and Dixson 1982; Felten and Sladek 1983) and man (Nobin and Björklund 1973; Olson et al. 1973; Pearson et al. 1983), have shown a remarkable constancy in the organization of the monoamine-containing cell groups among mammals. The distribution of neuromelanin-pigmented neurons in the human brain appears to correspond closely to that of the catecholamine cell groups (Bogerts 1981; Saper and Petito 1982). This is of practical importance, because the study of the pattern of neuromelanin pigmentation may be a rapid and reliable means of assessing the integrity of the catecholamine neuronal populations in neuropathological specimens (Saper and Petito 1982).

In the following survey the dopaminergic, noradrenergic, adrenergic and serotoninergic cell groups and their projections will be dealt with consecutively. Finally, brief consideration will be given to histamine, another biogenic amine. Adequate techniques for studying the distribution of this putative neurotransmitter in the brain have only recently been developed (Watanabe et al. 1983, 1984; Steinbusch and Mulder 1984).

Dopamine

Neurons which synthesize dopamine (Fig. 5) are found in the mesencephalon, the diencephalon and the telencephalon. The mesencephalic dopaminergic cells have been described as forming three cell groups – A8, A9 and A10, but the boundaries between these groups are indistinct (Felten et al. 1974; Hubbard and Di Carlo 1974a; Moore and Bloom 1978; Lindvall and Björklund 1983). The cells of group A8 are found in the lateral tegmentum, just caudal to the level of the red nucleus. This group merges ventromedially with group A9, which is constituted by the compact part of the substantia nigra. Group A10 is an unpaired midline aggregate that is situated medial to the substantia nigra and ventral to the red nucleus. It is limited ventrally by the interpeduncular nucleus. The majority of the cells of group A10 are located within the confines of the ventral tegmental area.

Four different dopaminergic cell groups – A11, A12, A13 and A14 – have been recognized in the diencephalon (Dahlström and Fuxe 1964; Ungerstedt 1971; Björklund and Nobin 1973; Felten et al. 1974; Felten 1976; Felten and Sladek 1983). Group A13 consists mainly of small clusters of cells situated in the caudomedial part of the zona incerta. Group A11 is a diffuse caudal continuation of group A13; its cells are located in the caudal hypothalamus and are scattered principally around the mamillothalamic tract. The compact group A12 is situated largely within the confines of the infundibular nucleus, and cell group A14, which is small and inconspicuous in primates, is found in the periventricular zone of the rostral hypothalamus.

As far as is known, the olfactory bulb is the only telencephalic centre containing dopaminergic neurons. These elements are scattered in the outer zone of the bulb and form part of a set of interneurons, the peri- or juxtaglomerular cells (Halász et al. 1977; Priestley et al. 1979; Fallon and Moore 1978a). They have been collectively designated as group A15 (Halász et al. 1977).

Some findings pertaining to the (possible) presence of additional dopaminergic cell groups should be briefly mentioned. Armstrong et al. (1982) presented immunohistochemical evidence (presence of TH, absence of DBH) suggesting that the dorsal vagal nucleus contains a group of dopaminergic cells. They considered it likely that these elements represented vagal efferent neurons. However, Jaeger et al. (1984) recently reported that these cells lack the enzyme AADC and, hence, are presumably not dopaminergic. Histofluorescence (Hökfelt et al. 1976a; Lindvall and Björklund 1974a) as well as immunohistochemical (Ochi and Shimizu 1978; Miachon et al. 1984) studies have shown that the nucleus raphes dorsalis and its immediate surroundings contain dopaminergic cell bodies. Finally, it is worthy of note that, according to Swanson et al. (1981), the parvocellular portion of the paraventricular nucleus contains significant number of TH-positive, presumably dopaminergic neurons.

The mesencephalic population of dopaminergic cells gives rise to a massive ascending projection which Ungerstedt (1971) divided into two 'systems': the nigrostriatal dopaminergic system, which (as its name implies) originates mainly from the substantia nigra or A9 group, and the mesolimbic dopaminergic system, which arises mainly from the A10 group. In the more recent literature (e.g. Moore and Bloom 1978; Fallon and More 1978b; Lindvall and Björklund 1983) it has been emphasized that the dopaminergic neurons in the midbrain constitute a single group

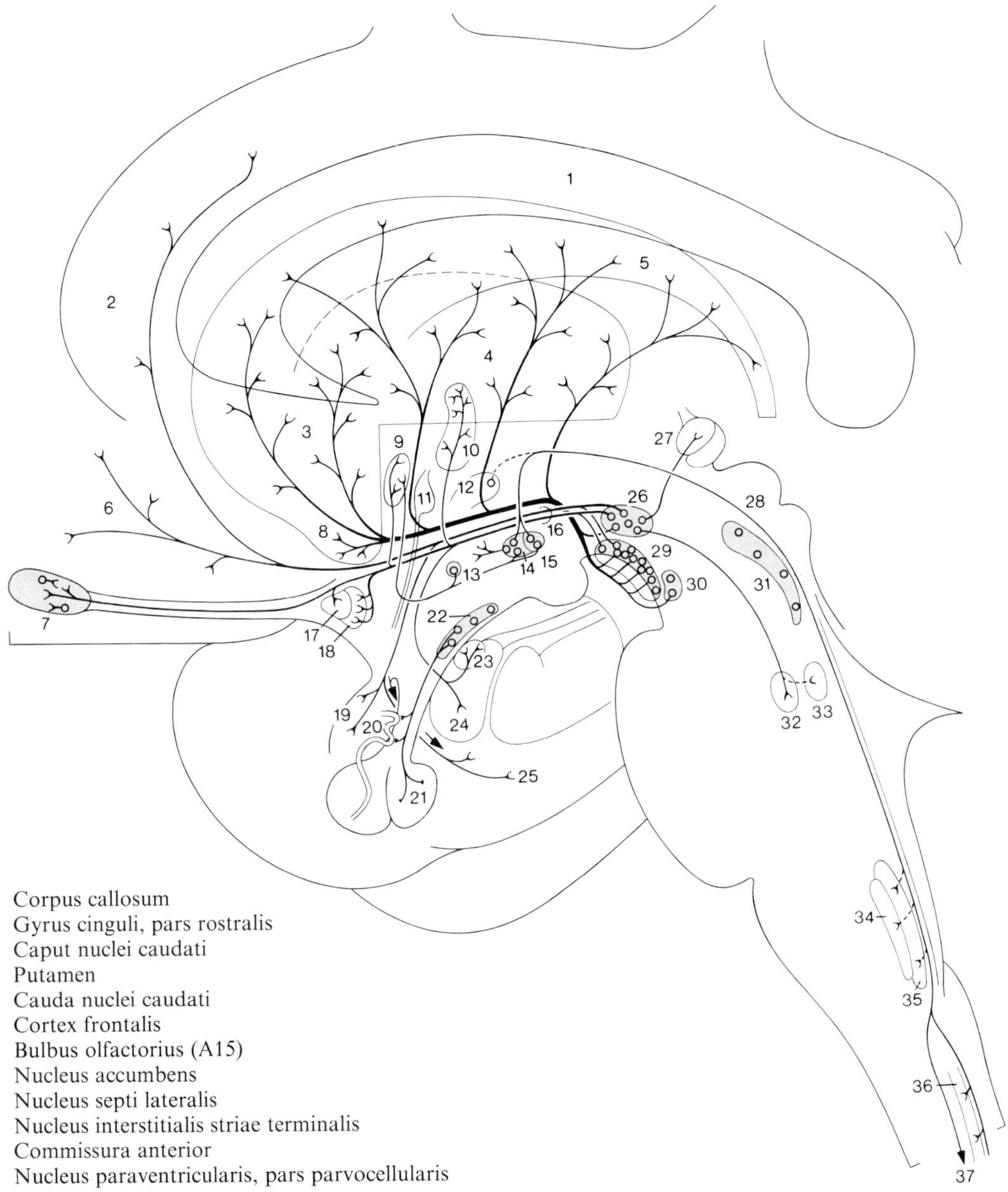

1 Corpus callosum
2 Gyrus cinguli, pars rostralis
3 Caput nuclei caudati
4 Putamen
5 Cauda nuclei caudati
6 Cortex frontalis
7 Bulbus olfactorius (A15)
8 Nucleus accumbens
9 Nucleus septi lateralis
10 Nucleus interstitialis striae terminalis
11 Commissura anterior
12 Nucleus paraventricularis, pars parvocellularis
13 Cell group A14
14 Cell group A13
15 Cell group A11
16 Fasciculus telencephalicus medialis
17 Nucleus olfactorius anterior
18 Substantia perforata anterior
19 Cortex praepiriformis
20 Eminentia mediana
21 Lobus posterior hypophyseos
22 Nucleus infundibularis (A12)
23 Nucleus centralis amygdalae
24 Nucleus basalis amygdalae
25 Cortex entorhinalis
26 Area tegmentalis ventralis (A10)
27 Nucleus habenulae lateralis
28 Fasciculus longitudinalis dorsalis
29 Substantia nigra, pars compacta (A9)
30 Area tegmentalis lateralis (A8)
31 Nucleus raphes dorsalis
32 Locus coeruleus
33 Nucleus parabrachialis lateralis
34 Nucleus dorsalis nervi vagi
35 Nucleus solitarius
36 Substantia gelatinosa
37 Nucleus intermediolateralis

Fig. 5. Dopamine-containing cells and fibres

projecting in at least a crude topographic order to striatal, limbic and cortical telencephalic regions. The entire forebrain projection is, accordingly, now commonly referred to as the 'mesotelencephalic dopaminergic system' (Moore and Bloom 1978; Lindvall and Björklund 1983). Fallon and Moore (1978b) have studied the topographical organization of the mesotelencephalic dopaminergic projection in some detail. According to their findings this projection is organized in three planes, medial-lateral, rostral-caudal and dorsal-ventral. The medial-lateral topography is organized such that the medial sectors of the substantia nigra–ventral tegmental area project to more medially located terminal areas, whereas the lateral sectors of the substantia nigra–ventral tegmental area project to more laterally located areas in the forebrain. Rostrally located mesencephalic dopaminergic cells appeared to project more rostrally, and caudal cells to more caudal areas of the telencephalon. The dorsal-ventral topography appeared to be inverted, in that ventral mesencephalic cells tend to project to more dorsal structures such as the septum, the nucleus accumbens and the neostriatum, whereas dorsal mesencephalic cells tend to project to more ventral structures, such as the olfactory tubercle and the amygdaloid complex. For details on the organization of the dopaminergic projection to the caudatus-putamen complex the reader should also consult the studies of Beckstead et al. (1979) and Veening et al. (1980).

For practical reasons the mesotelencephalic system will be subdivided here into three subsystems, the mesostriatal projection, the mesolimbic projection and the mesocortical projection.

The *mesostriatal projection* originates from the cell groups A8, A9 and A10 (Andén et al. 1964; Lindvall and Björklund 1974a, 1983; Moore and Bloom 1978). Its fibres assemble in the medial tegmentum and enter the diencephalon in the dorsal part of the lateral hypothalamic area. Coursing rostrally through the latter area, most of its fibres diverge dorsally and laterally. After having traversed the internal capsule, these fibres fan out in the caudate nucleus and the putamen, where they form an extremely dense terminal network. The nucleus accumbens, which is often considered a ventromedial extension of the caudate nucleus occupying an intermediate position between the 'extrapyramidal' and 'limbic' systems, is also richly supplied with mesostriatal dopaminergic fibres. The cells projecting to the nucleus accumbens are mainly situated in the A10 group, although some are localized in the medial part of the substantia nigra (Simon et al. 1976; Wang 1981; Swanson 1982). The dopaminergic innervation of the nucleus caudatus and the putamen originates largely from the compact part of the substantia nigra, but the A8 and A10 groups also participate in this innervation. Cell group A8, which may be considered a caudolateral extension of the substantia nigra, has been shown to project to the ventral putamen (Fallon and Moore 1978b; Beckstead et al. 1979). With regard to the contribution of the A10 group to the innervation of the neostriatum, there is no unanimity in the literature. Some groups of investigators (Carter and Fibiger 1977; Fallon and Moore 1978b; Moore and Bloom 1978; Simon et al. 1976) report that this projection is confined to the most rostromedial portion of the caudatus-putamen complex, but another group (Beckstead et al. 1979) presents evidence suggesting that the efferents from the ventral tegmental area are more widely distributed, extending over the entire ventromedial half of the neostriatum. Collaterals of the mesostriatal projection constitute a rather sparse plexus of dopamine-containing terminal axons throughout the globus pallidus (Fallon and Moore 1978b; Lindvall and Björklund 1979).

There is evidence that the dopaminergic mesostriatal projection participates in the regulation of complex behaviour and plays a crucial role in determining the ability of the organism to cope with available exteroceptive sensory information in various ways. This specific capacity has been denoted as 'the ability to arbitrarily switch motor programmes' (see Cools 1980; Cools et al. 1984b; Jaspers et al. 1984).

Parkinson's disease is characterized by a progressive loss of dopaminergic neurons in the substantia nigra, with consequent degeneration of their ascending projections. This degeneration is believed to be responsible for the akinesia and rigidity associated with this disease (Hornykiewicz 1978). The area tegmentalis ventralis, which is rather poorly developed in man, also appears to be affected in Parkinson's disease (Bogerts et al. 1983). In relation to what has been stated above concerning the functions in which the dopaminergic mesostriatal projection is implicated, it is important to note that patients suffering from Parkinson's disease appear to have an impaired ability to arbitrarily switch their behaviour. These patients show a so-called shifting aptitude disorder, which manifests itself both at the motor level and at the cognitive level (Cools et al. 1984a).

The *mesolimbic projection* arises from cell group A10 and from the most medial part of the substantia nigra. Its fibres ascend medial to the mesostriatal projection in the medial forebrain bundle and are distributed to the following telencephalic structures (Ungerstedt 1971; Lindvall 1975; Moore 1978; Fallon et al. 1978; Fallon and Moore 1978a, b; Lindvall and Stenevi 1978; Lindvall and Björklund 1983): bulbus olfactorius (scattered terminals in all layers), anterior olfactory nucleus (a sparse to moderate innervation), olfactory tubercle or anterior perforated substance (a very dense input), lateral septal nucleus, bed nucleus of stria terminalis (a dense plexus, particularly in its dorsal part) and the amygdaloid complex (concentrated in the central and basal nuclei and intercalated cell groups).

There is experimental evidence indicating that activation of the dopaminergic projection from the area tegmentalis ventralis to the nucleus accumbens produces enhanced locomotor activity (Pijnenburg and Van Rossum 1973; Pijnenburg et al. 1976).

It has been suggested that hyperactivity of the area tegmentalis ventralis and its dopaminergic outflow to limbic structures may play an important role in the pathophysiology of schizophrenia (Stevens 1973, 1979). In accordance with this theory, Bird and colleagues (1979) reported that in the brains of patients dying with schizophrenia the dopamine concentrations in the nucleus accumbens and anterior perforated substance are significantly elevated. The curious finding of Reynolds (1983, see also MacKay 1984) that there is gross asymmetry of dopamine concentrations in the amygdalae of post-mortem brains from schizophrenic patients, the left amygdala showing an abnormally high concentration of this catecholamine, should also be mentioned in this context.

A sharp distinction between the *mesolimbic* and the *mesocortical projections* cannot be made, firstly because both projections originate from cell group A10 and from the medial portion of the substantia nigra (Lindvall et al. 1974a; Simon et al. 1976; Carter and Fibiger 1977; Fallon et al. 1978; Fallon and Moore 1978b; Emson and Koob 1978; Lindvall et al. 1978; Beckstead et al. 1979; Simon et al. 1979; Swanson 1982), and secondly, because most of the cortical areas of termination of the latter projection can be considered part of the limbic system. These areas include the medial part of the frontal lobe, the prepiriform and piriform cortices, the entorhinal cortex and the anterior cingulate cortex (Lindvall et al. 1974a, 1978; Hökfelt et al. 1974a; Emson and Koob 1978). In most of these areas the dopaminergic fibres constitute dense and well-delineated fields of termination.

It is important to note that not all fibres which ascend from the compact part of the substantia nigra and the area tegmentalis ventralis to the telencephalon are dopaminergic (Hedreen and Chalmers 1972; Thierry et al. 1980; Guyenet and Crane 1981; Van der Kooy et al. 1981; Swanson 1982). Thus, the mesostriatal projections to caudatus, putamen and nucleus accumbens contain a small proportion of non-dopaminergic fibres. A larger percentage of non-dopaminergic fibres has been found in the mesolimbic projection to the lateral septal nucleus, whereas in the mesocortical projection to the medial frontal cortex only about one-third of the fibres appear to be dopaminergic. It is also worth mentioning that in a certain proportion of the perikarya situated in the ventral tegmental area dopamine co-exists with cholecystokinin, and that some of these cells project to the nucleus accumbens (Hökfelt et al. 1980b; Studler et al. 1984).

Apart from the large mesotelencephalic projection, the mesencephalic dopaminergic neuron population has been reported to give rise to some additional efferent projections. Thus, there is strong evidence that group A10 projects – via the fasciculus retroflexus – to the lateral habenular nucleus (Swanson 1982; Lindvall and Björklund 1983; Skagerberg et al. 1984), that the subthalamic nucleus contains a plexus of catecholaminergic fibres, which is possibly constituted by collaterals of the nigrostriatal bundle (Nobin and Björklund 1973; Lindvall and Björklund 1983), and, finally, that the ventral tegmental area projects bilaterally to the locus coeruleus and possibly to the lateral parabrachial nucleus (Swanson 1982).

The diencephalic dopaminergic cell groups A11–A14 give rise to four efferent systems, the tubero-infundibular projection, the incerto-hypothalamic projection, the hypothalamo-septal projection and the hypothalamo-spinal projection.

The *tubero-infundibular dopaminergic projection* arises mainly from the cells of the A12 group. The axons of these cells pass to the rostroventral part of the infundibulum and terminate in all layers of the median eminence, but are most abundant in the external layer. Many axons continue beyond the median eminence and traverse the pituitary stalk to reach the neurointermediate lobe, where they form a dense plexus (Ungerstedt 1971; Felten 1976; Hökfelt et al. 1978b; Moore and Bloom 1978). According to Björklund et al. (1975), the incerto-hypothalamic projection (to be discussed below) contributes substantially to the dopaminergic innervation of the median eminence. There is evidence suggesting that dopamine released from the tubero-infundibular fibres is transported to the anterior pituitary via the hypothalamo-hypophyseal portal system to inhibit the release of prolactin-secreting cells (Macleod and Lehmeyer 1974; Ben-Jonathan et al. 1977). However, the control of prolactin release appears to be complex, and dopamine is most probably not the only prolactin-inhibiting factor (McCann et al. 1984). The function of the dopaminergic tubero-hypophyseal fibres is unknown.

The *incerto-hypothalamic dopaminergic projection* arises from the complex formed by the A11 and A13 groups. It consists of short, diffusely arranged fibres, which distribute within its areas of origin and in the adjacent dorsal and rostral hypothalamic regions (Björklund et al. 1975).

The *diencephalo-septal dopaminergic projection* consists of fibres which pass from the A11, A13 and A14 groups to the lateral septal nucleus (Lindvall and Stenevi 1978). Thus, this septal centre, which is mainly supplied with dopaminergic fibres originating from the midbrain, receives an additional dopaminergic projection from the diencephalon.

During the past few years experimental evidence has accumulated for the presence of a *dopaminergic hypothalamo-spinal projection* (Björklund and Skagerberg 1979; Hökfelt et al. 1979a; Skagerberg et al. 1982; Lindvall and Björklund 1983). In the rat, this projection originates from perikarya located in the caudal hypothalamus belonging to cell group A11 and passes to the ipsilateral half of the spinal cord, where its fibres descend partly within lamina I of the dorsal horn and the adjoining part of the dorsolateral funiculus and partly along the central canal. The former pathway supplies the lateral parts of the superficial layers of the dorsal horn, whereas the latter innervates the intermediolateral cell column and associated parts in the intermediate zone of the thoracic and upper lumbar segments. The exact course of these descending dopaminergic fibres through the brain stem is not known as yet.

However, it has been suggested (Björklund and Skagerberg 1979; Lindvall and Björklund 1983) that these fibres form part of a dorsal periventricular catecholaminergic bundle, which in turn is a component of the fasciculus longitudinalis dorsalis of Schütz. Blessing and Chalmers (1979) reported that in the rabbit the dopaminergic diencephalospinal fibres originate from the A13, rather than from the A11 group, and it is also worth noting that, according to Swanson et al. (1981), dopaminergic neurons present in the parvocellular part of the paraventricular nucleus project to the region of the dorsal vagal complex and/or to thoracic levels of the spinal cord.

Addendum: Lindvall et al. (1984) provide experimental evidence indicating that in the rat the supraoptic, paraventricular and dorsomedial nuclei of the hypothalamus and the paraventricular nucleus of the thalamus receive a dopaminergic innervation, which most likely originates from the diencephalic A11–A14 cell groups. They suggest that the dopaminergic innervation of hypothalamic neurosecretory nuclei participates in the regulation of oxytocin and vasopressin release from the pituitary.

Noradrenaline

Neurons that synthesize noradrenaline (Figs. 6 and 7) are restricted to the pontine and medullary tegmental regions. Seven noradrenergic cell groups, designated as A1–A7, have been described in rodents (Dahlström and Fuxe 1964). Most of these have also been recognized in primates (Felten et al. 1974; Garver and Sladek 1975; Hubbard and Di Carlo 1973, 1974a; Jacobowitz and MacLean 1978; Schofield and Everitt 1982; Schofield and Dixson 1982; Felten and Sladek 1983), including man (Nobin and Björklund 1973; Olson et al. 1973; Bogerts 1981). Groups A1 and A2 are both situated in the lower part of the medulla oblongata. The cells of group A1 surround the nucleus of the lateral funiculus and extend dorsomedially into the lateral part of the reticular formation; those of group A2 lie in the nucleus solitarius, the dorsal vagal nucleus and the intervening area. Armstrong et al. (1982) reported that numerous neurons of the A2 group are situated in the area postrema. On the basis of its topography the A2 group is also designated as the noradrenergic dorsal medullary cell group. Cells corresponding to group A3 in the rat, lying just dorsal to the inferior olivary complex, have not been observed in primates. Group A4 consists of a band of subependymal neurons which extends along the superior cerebellar peduncle. This group merges rostromedially with the caudal portion of group A6. Group A5 is situated in the caudolateral part of the pontine tegmentum and consists of rather loosely arranged cells lying adjacent to the facial nucleus and the superior olivary complex. Group A6 is a densely packed accumulation of cells within the locus coeruleus, a macroscopically visible blue-black streak of tissue situated in the floor of the fourth ventricle at rostral pontine levels. Evidence suggests that all of the neurons situated in the central part of this structure are noradrenergic (Garver and Sladek 1975; Swanson 1976a). Somewhat schematically it may be said that the noradrenergic cell group situated within the locus coeruleus has three extensions: rostral, ventral and caudolateral. The rostral extension consists of elements lying in the caudolateral part of the mesencephalic central grey (A6cg). The ventral extension is formed by scattered neurons situated within a cytoarchitectonic entity usually referred to as the subcoeruleus area (A6sc). The caudolateral extension of the noradrenergic cell cluster situated within the locus coeruleus is constituted by the A4 group already mentioned. The A4, A6cg and A6sc cell groups are commonly designated together as the (noradrenergic) locus coeruleus complex. The cells of the A7 group are situated in the rostral pontine part of the lateral reticular formation and lie mainly medial to the lateral lemniscus.

The A1, A5 and A7 groups form a caudorostral continuum in primates, which extends throughout the lateral rhombencephalic tegmentum. The caudal part of this (noradrenergic) lateral tegmental complex is connected by strands of noradrenergic cells with the A2 group. Rostrally, a comparable string of cells forms a bridge between the locus coeruleus complex and the A5 and A7 groups. These intervening cells lie partly within the confines of the medial parabrachial and Kölliker-Fuse nuclei (Lackner 1980), these last two centres together constituting the functionally defined 'pneumotaxic centre'.

The noradrenergic cell groups just discussed give rise to extensively branched ascending and descending fibre systems that course the length of the neuraxis and form terminal networks in a variety of grisea. At present we have a fairly complete picture of the total distribution of noradrenergic fibres and terminal fields, but the extent to which individual terminal fields arise from separate neuronal groups is still a matter of dispute. In this condensed and simplified survey a sharp and perhaps too sharp distinction has been made between the projection of the locus coeruleus and those of the remaining noradrenergic cell groups.

Containing about half of the total number of noradrenaline-synthesizing neurons, the locus coeruleus complex (Fig. 6) is quantitatively by far the most important noradrenergic centre of the brain. Its efferents constitute two ascending fibre systems, the large dorsal noradrenergic bundle and the much smaller rostral limb of the dorsal periventricular pathway. Other efferents are distributed to the cerebellum and still others descend to the lower medulla oblongata and to the spinal cord.

The *dorsal noradrenergic bundle* or dorsal tegmental bundle (Lindvall and Björklund 1974a, 1978), which forms part of the major longitudinal catecholamine bundle, described by Jones and Friedman (1983), traverses the midbrain tegmentum in a position ventrolateral to the periaqueductal grey. At the level of the fasciculus retroflexus the bundle arches

Explanations to Figure 6

1 Neocortex
2 Gyrus cinguli
3 Striae longitudinales
4 Corpus callosum
5 Fornix
6 Stria terminalis
7 Nucleus anterior thalami
8 Stria medullaris thalami
9 Thalamus
10 Nucleus interstitialis striae terminalis
11 Lamina medullaris interna
12 Nucleus habenulae lateralis
13 Nucleus habenulae medialis
14 Tractus mamillothalamicus
15 Lamina medullaris externa
16 Corpus geniculatum mediale + laterale
17 Nucleus septi medialis
18 Nucleus paraventricularis, pars parvocellularis
19 Fasciculus telencephalicus medialis
20 Bandeletta diagonalis
21 Bulbus olfactorius
22 Nucleus olfactorius anterior
23 Substantia perforata anterior
24 Nucleus gyri diagonalis
25 Ansa peduncularis + fibrae amygdalofugales ventrales
26 Nucleus centralis amygdalae
27 Nucleus basalis amygdalae
28 Gyrus dentatus
29 Cornu Ammonis
30 Subiculum
31 Gyrus parahippocampalis
32 Tractus habenulointerpeduncularis
33 Fasciculus longitudinalis dorsalis
34 Colliculus superior
35 Colliculus inferior
36 Griseum centrale mesencephali
37 Nucleus raphes dorsalis
38 Nucleus interpeduncularis
39 Cortex cerebelli
40 Locus coeruleus, rostral extension (A6cg)
41 Locus coeruleus (A6)
42 Area subcoerulea (A6sc)
43 Nuclei lemnisci lateralis
44 Locus coeruleus, caudal extension (A4)
45 Brachium conjunctivum
46 Nuclei centrales cerebelli
47 Nuclei pontis
48 Formatio reticularis metencephali
49 Nucleus sensorius principalis nervi trigemini
50 Nucleus cochlearis ventralis
51 Nucleus cochlearis dorsalis
52 Formatio reticularis myelencephali
53 Nucleus solitarius
54 Nucleus dorsalis nervi vagi
55 Nucleus spinalis nervi trigemini
56 Cornu posterius (Laminae IV, V, VI)
57 Cornu anterius

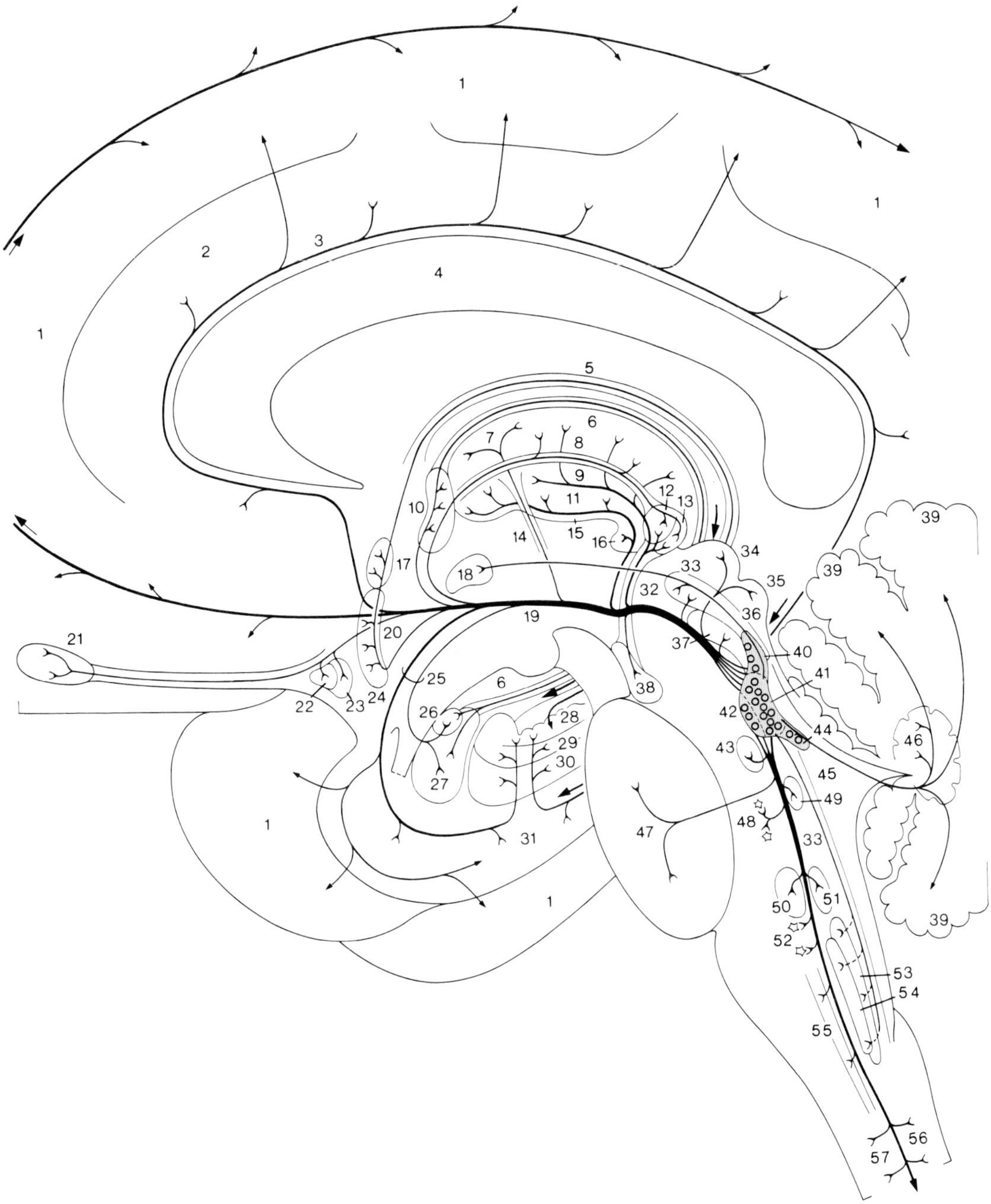

Fig. 6. Noradrenaline-containing cells and fibres: I. The locus coeruleus complex

rostroventrally and attains the hypothalamus, where it joins the dorsal portion of the medial forebrain bundle complex. Within this complex the dorsal noradrenergic bundle proceeds to the septal region. Along its course, numerous smaller and larger branches emerge to innervate a large number of mesencephalic, diencephalic and telencephalic grisea. The mesencephalic areas of termination include the central grey substance, the dorsal raphe nucleus, the superior and inferior colliculi and the interpeduncular nucleus (Levitt and Moore 1979; Marchand et al. 1980). In the rostral mesencephalon a large group of fibres of the dorsal noradrenergic bundle turns dorsally and passes along the fasciculus retroflexus toward the habenular complex. However, a major component of this group does not reach the epithalamus but rather enters the internal medullary lamina of the thalamus and ascends within it. A second group of fibres leaving the dorsal bundle at the level of the fasciculus retroflexus ascends within the external medullary lamina. These two fibre groups together provide most of the very large input of the locus coeruleus to the dorsal thalamus. A more rostral branch of the dorsal noradrenergic bundle, the fibres of which join the stria medullaris, also supplies a number of thalamic centres as well as the epithalamus. Within the latter, both the medial and lateral habenular nuclei receive a noradrenergic input (Moore and Bloom 1979). Practically all areas of the dorsal thalamus receive afferents from the locus coeruleus, most notably, the anterior, ventral and lateral nuclear complexes and the medial and lateral geniculate bodies. The fibres to the anterior nuclei ascend with the mamillothalamic tract (Lindvall et al. 1974b). Ishikawa and Tanaka (1977) found that in the rhesus monkey, noradrenergic fibres originating from the locus coeruleus distribute to most of the thalamus, except for the midline and medial nuclei. The hypothalamus does receive noradrenergic input from the locus coeruleus (Jacobowitz 1975; Jones and Moore 1977; Sawchenko and Swanson 1982a), but most of these fibres reach this part of the brain via the dorsal periventricular system to be discussed below.

The main telencephalic areas of termination of the dorsal noradrenergic bundle and its branches are: (a) the amygdala (mainly the central and basal nuclei; Tohyama et al. 1974; Jones and Moore 1977; Fallon et al. 1978); (b) the olfactory tubercle or anterior perforated substance, the anterior olfactory nucleus and the olfactory bulb (Blackstad et al. 1967; Ungerstedt 1971; Swanson and Hartman 1975; Fallon and Moore 1978a); (c) the nucleus of the diagonal band, the medial septal nucleus and the bed nucleus of the stria terminalis (Moore 1978; Lindvall and Stenevi 1978; Krayniak et al. 1981); (d) the hippocampal formation (gyrus dentatus, cornu Ammonis and subiculum; Blackstad et al. 1967; Ungerstedt 1971; Lindvall et al. 1974a; Jones and Moore 1977; Loy et al. 1980); and (e) the entire neocortex, including the cingulate, retrosplenial and entorhinal cortical areas (Ungerstedt 1971; Gatter and Powell 1977; Lindvall and Björklund 1978; Levitt and Moore 1978; Lindvall et al. 1978). In most of the areas mentioned the afferents from the locus coeruleus are mixed with axons originating from other noradrenergic cell groups. However, the noradrenergic fibres innervating the hippocampus and the neocortex seem to originate exclusively from the locus coeruleus.

The paths along which the efferents from the locus coeruleus ascend to the various telencephalic centres mentioned, via the dorsal noradrenergic bundle, are complex; several of these centres are approached along two or more different channels. The following simplified survey of the organization of these pathways is based mainly on Moore and Bloom (1979). Along its course through the hypothalamus, a large contingent of noradrenergic fibres leaves the medial forebrain bundle laterally to enter the complex formed by the ansa peduncularis and the ventral amygdalofugal pathway. Some of these fibres enter the amygdala, but others innervate the entorhinal cortex and the hippocampal formation. Upon attaining the septal region, the dorsal noradrenergic bundle breaks up into four major fibre groups. The first of these turns medially into the diagonal band of Broca to innervate the nucleus surrounding that fibre system and the medial septal nucleus. Some of the fibres of this group enter the fornix, along which they pass to the hippocampus. The second group enters the stria terminalis and follows this path to the amygdaloid complex. The third group continues in the medial forebrain bundle as it enters the basal telencephalon. A certain proportion of the fibres of this group innervate the olfactory tubercle, the anterior olfactory nucleus and the olfactory bulb, but others continue rostrally in the external capsule. These may well correspond partly to the large fibre contingent which, according to Morrison et al. (1981), continues rostrally from the medial forebrain bundle into the frontal pole of the hemisphere, from where they arch caudally to supply the entire frontal, dorsal and lateral neocortex (see also Shimizu et al. 1974; Tohyama et al. 1974). The fourth group of fibres traverses the rostral septum and encircles the corpus callosum to innervate the neocortex and the hippocampal formation. Ungerstedt (1971) suggested that this group of fibres travels caudally in the cingulum bundle and furnishes laterally directed branches that innervate practically the entire neocortex. Morrison et al. (1981) reported, however, that the fibres arching around the corpus callosum do not follow the cingulum bundle but rather the supracallosal striae, and that these fibres furnish only a marginal innervation of the medial parts of the neocortex. In several earlier studies (e.g. Ungerstedt 1971; Gatter and Powell 1977) the noradrenergic innervation of the neocortex is described as diffuse and uniform throughout. However, more recent investigations have revealed regional variations in pattern and density of cortical noradrenergic innervation, adhering to cytoarchitectonic boundaries (e.g. Morrison et al. 1979, 1982a; Levitt et al. 1984).

It is noteworthy that most of the centres mentioned above receive, in addition to a projection from the ipsilateral locus coeruleus, a limited number of fibres from the contralateral locus coeruleus. These fibres pass from one dorsal noradrenergic bundle to the other via a number of commissures, among which the posterior commissure, the dorsal supraoptic decussation and the anterior commissure may be mentioned (Jones and Moore 1977; Moore and Bloom 1979).

The *rostral limb of the dorsal periventricular pathway* arises mainly from the A6, A6sc, and A6cg cell groups and ascends to the diencephalon within the ventromedial part of the mesencephalic periaqueductal grey, forming part of the fasciculus longitudinalis dorsalis complex (Lindvall and Björklund 1974a; Schofield and Everitt 1982; Tanaka et al. 1982; Felten and Sladek 1983). Rostrally, this pathway continues into the diencephalic periventricular fibre plexus. It has been suggested that its fibres innervate several hypothalamic centres, including the dorsomedial nucleus, the paraventricular nucleus and the supraoptic nucleus (Lindvall and Björklund 1974a; Jones and Moore 1977). However, the experimental studies of Sawchenko and Swanson (1981, 1982a) have revealed that only the parvocellular portion of the paraventricular nucleus receives a substantial projection from the locus coeruleus complex. As described below, the major noradrenergic innervation of the hypothalamus arises from the dorsomedial medullary (A2) and lateral tegmental (A1, A5, A7) cell groups.

The noradrenergic fibres passing to the *cerebellum* follow the superior cerebellar peduncle and terminate in the central nuclei as well as in the cortex. These fibres originate mainly from the locus coeruleus (Ungerstedt 1971; Olson and Fuxe 1971), but some have been reported to arise from the subcoeruleus area (Pasquier et al. 1980).

The descending efferents of the locus coeruleus complex follow two different routes, which may be designated as the caudal limbs of the dorsal periventricular pathway and the dorsal noradrenergic bundle respectively.

The *caudal limb of the dorsal periventricular pathway*, like the rostral limb of the same system, forms part of the fasciculus longitudinalis dorsalis complex (Felten and Sladek 1983). The contribution of the locus coeruleus to this pathway in the rat is confined to a limited number of fibres which arise from the ventral part of that centre and terminate in the nucleus solitarius (Takahashi et al. 1979). However, Westlund and Coulter (1980) traced a considerable number of fibres in the rhesus monkey from the locus coeruleus to the dorsal motor vagus nucleus.

The *caudal limb of the dorsal noradrenergic bundle* forms part of the major longitudinal catecholamine bundle, described by Jones and Friedman (1983). It descends through the lateral part of the rhombencephalic reticular core and continues caudally into the lateral funiculus of the spinal cord (Moore and Bloom 1979; Jones and Friedman 1983). In primates and man this bundle follows over a certain distance the trajectory of the central tegmental tract (Pearson et al. 1983). Noradrenergic fibres originating from the locus coeruleus complex supply, partly directly and partly by way of the bundle just discussed, the following rhombencephalic centres (Levitt and Moore 1979): the nuclei of the lateral lemniscus, the principal sensory trigeminal nucleus, the cochlear nuclei (Kromer and Moore 1976, 1980), the pontine nuclei, the spinal trigeminal nucleus and the entire rhombencephalic reticular formation. The three centres last mentioned also receive, in addition to fibres from the locus coeruleus, input from other noradrenergic cell groups.

The locus coeruleus complex innervates all segments of the spinal cord (Nygren and Olson 1977; Westlund and Coulter 1980). This innervation is bilateral, the decussation occurring at spinal levels (Karoum et al. 1980; Commissiong 1981). The coeruleospinal fibres originate in the ventral part of the locus coeruleus and in the subcoeruleus area (Satoh et al. 1977; Hancock and Fougerousse 1976; Mason and Fibiger 1979; Guyenet 1980; Westlund et al. 1984), descend in the lateral funiculus of the cord (Westlund and Coulter 1980), and terminate in the ventral parts of the dorsal horn (laminae IV, V, VI), the intermediate grey and the ventral horn (Nygren and Olson 1977; Westlund and Coulter 1980). The locus coeruleus complex does not innervate the sympathetic intermediolateral column in the thoracic cord (Nygren and Olson 1977; Commissiong et al. 1978; Westlund and Coulter 1980); however, in the rhesus monkey the parasympathetic preganglionic neurons in the sacral cord have been shown to receive major input from the locus coeruleus (Westlund and Coulter 1980).

From the foregoing survey it appears that the locus coeruleus complex projects widely over vast areas of the neuraxis, from the olfactory bulb to the spinal cord. It was long thought that this projection is diffuse, with little topography. In several earlier publications evidence was presented suggesting that not only the locus coeruleus as a whole, but also its individual neurons innervate widely different regions of the central nervous system via collateral branches (Olson and Fuxe 1971; Ungerstedt 1971; Tohyama et al. 1974; Nygren and Olson 1977). More recently, these findings have been substantiated by the work of several groups of investigators, using multiple fluorescent retrograde tracers (Adèr et al. 1980; Room et al. 1981; Nagai et al. 1981; Steindler 1981). However, it has also been experimentally established that, with respect to their efferent projections, the locus coeruleus neurons show a considerable degree of regional topographic organization. Thus, the study of Mason and Fibiger (1979) has shown that, like the coeruleospinal neurons (see above), the elements projecting to the septum, the thalamus and the hypothalamus are preferentially located in particular areas of the locus coeruleus complex. On the other hand, the projections to the hippocampus, the neocortex, the amygdala and the cerebellum originate, according to those authors, mainly from cells scattered throughout the complex. As regards the noradrenergic projection to the neocortex, Loughlin et al. (1982) presented experimental evidence indicating that the axons of coeruleocortical neurons arborize more extensively in the rostrocaudal direction than in the mediolateral direction, which is in keeping with the pattern of noradrenergic cortical innervation described by Morrison et al. (1981). According to Nagai et al. (1981), the coeruleocortical projection arises from two different types of neurons: a predominant type innervating a restricted region and, less commonly, those projecting widely to various areas of the cerebral cortex. Both types were found intermingled in the A6 and A4 groups. Fallon and Loughlin (1982) also distinguished between locus coeruleus cells with divergent collateral systems and those with more restricted arborizations in a study of the patterns of monoaminergic innervation of the forebrain; moreover, these authors established that the neurons with divergent collateral systems are located mainly in the central zone of the locus coeruleus, whereas in the peripheral zone the less highly collateralized elements prevail. These few examples may suffice to illustrate that, with respect to the organization of its efferent projections, the locus coeruleus is very complex, and that this centre most probably contains a number of subunits with different target areas.

The afferents of the locus coeruleus are multifarious and include fibres from the central nucleus of the amygdala, the preoptic region, the bed nucleus of the stria terminalis, the lateral hypothalamic area, the periaqueductal grey, the ventral tegmental area, the mesencephalic and rhombencephalic reticular formation, the contralateral locus coeruleus, the parabrachial region, most of the raphe nuclei, the vestibular nuclear complex, the deep cerebellar nuclei, the A1, A2, A5, C1 and C2 cell groups and the marginal zone of the spinal dorsal horn (Conrad and Pfaff 1976; Edwards 1975; Pierce et al. 1976; Cedarbaum and Aghajanian 1978; Clavier 1979; Beckstead et al. 1979; Morgane and Jacobs 1979). Interestingly, many of these connections reciprocate efferent projections of the locus coeruleus. To what extent the fibres of these afferent systems are distributed over the entire complex or specifically address neuronal subpopulations within the complex is unknown at present.

A discussion of the vast literature on the possible functional significance of the locus coeruleus complex is beyond the scope of this book. However, some idea of the various theories and concepts concerning the functioning of this intriguing centre may be gained from the following brief notes.

1. It has been found that many of the noradrenergic terminals of locus coeruleus neurons are intimately associated with cerebral arterioles and capillaries, and the suggestion has been made that the locus coeruleus exerts an influence on the regulation of cerebral microcirculation via these structures (Hartman 1973; Raichle et al. 1975; Swanson et al. 1977; Bates et al. 1977). In relation to this and other presumed functions, the locus coeruleus has been characterized as 'a central analogue of a sympathetic ganglion' (Amaral and Sinnamon 1977). However, Grzanna et al. (1978) reported that the vast majority of the noradrenergic fibres in the brain bear no relationship to cerebral blood vessels, and Dahlgren et al. (1981) remained unable to detect any influence of the locus coeruleus on cerebral blood flow.

2. It has repeatedly been reported that stimulation of the locus coeruleus elicits a rise in heart rate and blood pressure. The pathways along which these cardiovascular reactions are effected are not known as yet (Lightman et al. 1984; Gurtu et al. 1984).

3. Redmond and colleagues (Redmond et al. 1977, 1979; Charnay and Redmond 1983) noted that locus coeruleus cells are activated by stressful, threatening stimuli, and that stimulation of the locus coeruleus may produce a syndrome of behaviours characteristic of fear. They proposed that the locus coeruleus functions as part of an 'alarm system'. The cardiovascular effects mentioned under point 2 were included in the reactions caused by this system.

4. Combining the evidence that the locus coeruleus is activated by stressful stimuli with the results of physiological experiments indicating that noradrenaline released from locus

coeruleus terminals inhibits ongoing activity in most target neurons, Amaral and Cinnamon (1977) hypothesized that the locus coeruleus has a 'stress-dampening function'.

5. It has been shown that locus coeruleus neurons are not exclusively activated by noxious or other threatening or stressful stimuli. Foote et al. (1980), for instance, demonstrated that in rats locus coeruleus cells can be activated by the presentation of various non-noxious auditory, visual or somatosensory stimuli, and that in the monkey these cells respond vigorously to complex arousing stimuli such as preferred food. It also appears that the activity of locus coeruleus cells cannot be described simply in terms of excitation or inhibition. Rather, their action selectively enhances or diminishes the effects of the neurotransmitters released by other afferents or by intrinsic neurons in their target areas (see Van Dongen 1981 a). On the basis of this and other information (Mason 1980, 1981; Van Dongen 1980, 1981 b; Foote et al. 1983), the generalization may be made that in the waking state the locus coeruleus exerts an 'attention function', i.e. continuously 'monitors' the environment for important stimuli/events, and prepares the organism to cope with emergency situations.

As regards pathology in the brains of patients dying with Parkinson's disease and Alzheimer's disease, a significant loss of neurons in the locus coeruleus has been found (Van Dongen 1981 b). According to a recent report (Mann et al. 1984), in younger patients with Alzheimer's disease the cell loss is even greater in the locus coeruleus than in the nucleus basalis of Meynert. It is also worthy of note that in certain forms of schizophrenia there may be a disturbance of the locus coeruleus and its outflow. In postmortem examinations of brains of chronic paranoid schizophrenia patients, Farley et al. (1978) found noradrenaline concentrations to be above normal in several limbic forebrain regions, including the ventral part of the septum and the bed nucleus of the stria terminalis, i.e. terminal areas of the locus coeruleus.

Turning now to the efferents of the dorsal medullary (A2) and lateral tegmental (A1, A5, A7) noradrenergic cell groups (Fig. 7), it should be mentioned first of all that, according to the classical mapping study of Ungerstedt (1971), these cell groups collectively give rise to a long, ascending fibre system, the *ventral noradrenergic pathway*. According to his description, this pathway ascends through the reticular zone of the brain stem and continues rostrally, mainly within the medial forebrain bundle. Terminal areas of this pathway were observed in the mesencephalon (the ventrolateral part of the substantia grisea centralis and the reticular formation), the diencephalon (the entire hypothalamus, especially the dorsomedial, periventricular, infundibular, supraoptic and paraventricular nuclei and the internal layer of the median eminence), and in the telencephalon (the preoptic area and the bed nucleus of the stria terminalis). Later studies (Swanson and Hartman 1975; Lindvall and Björklund 1974a, 1978; Moore and Bloom 1979; Jones and Friedman 1983) have shown that the dorsal and ventral noradrenergic bundles as described by Ungerstedt (1971) cannot be sharply separated; they form one complex, which has been termed the 'central tegmental tract' (Lindvall and Björklund 1974a; Swanson et al. 1981; Moore and Bloom 1979) or the 'major longitudinal catecholamine bundle' (Jones and Friedman 1983). In this complex, ascending fibres are mixed with descending ones. The axons of the lateral tegmental group feed into it at successive levels through radially coursing, transverse fibres (Jones and Friedman 1983).

As regards the areas of termination of the dorsal medullary and lateral tegmental noradrenergic cell groups, the findings of Ungerstedt (1971) have been confirmed and extended by several more detailed studies. Thus, it has been reported that fibres arising from these cell groups form a very dense terminal plexus in the ventral part of the bed nucleus of the stria terminalis and also project to the nucleus of the diagonal band and

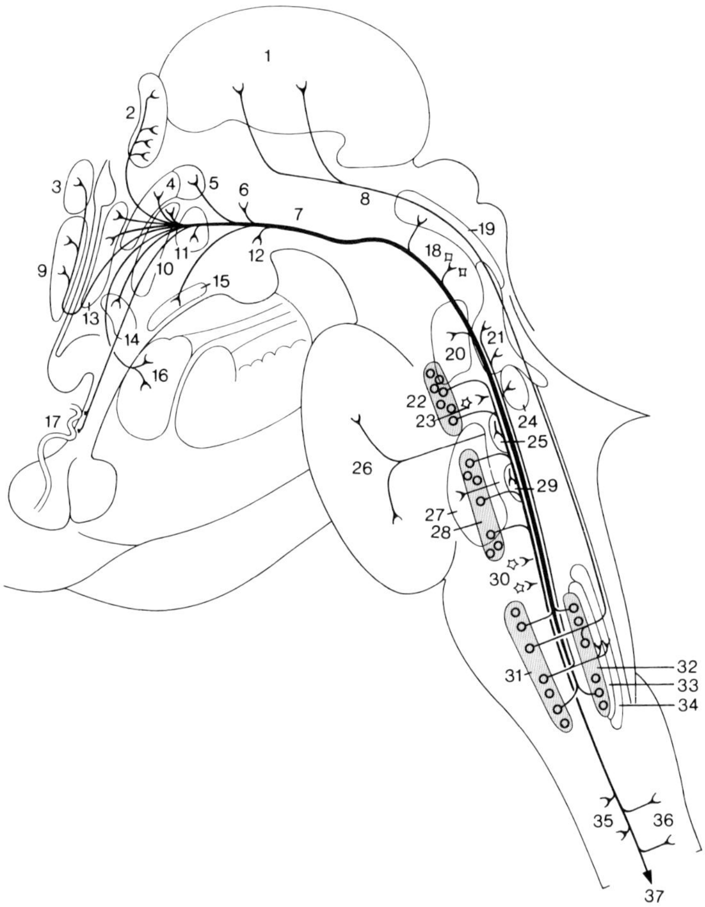

1 Thalamus, periventricular region
2 Nucleus interstitialis striae terminalis
3 Nucleus septi lateralis
4 Nucleus paraventricularis, pars magnocellularis
5 Nucleus paraventricularis, pars parvocellularis
6 Area lateralis hypothalami
7 Fasciculus telencephalicus medialis
8 Fasciculus longitudinalis dorsalis
9 Nucleus gyri diagonalis
10 Nucleus anterior hypothalami
11 Nucleus dorsomedialis
12 Area caudalis hypothalami
13 Nucleus praeopticus medialis
14 Nucleus supraopticus
15 Nucleus infundibularis
16 Corpus amygdaloideum
17 Eminentia mediana
18 Formatio reticularis mesencephali
19 Griseum centrale mesencephali
20 Nucleus centralis superior
21 Locus coeruleus
22 Cell group A7
23 Formatio reticularis metencephali
24 Nuclei parabrachiales
25 Nucleus motorius nervi trigemini
26 Nuclei pontis
27 Nucleus raphes magnus
28 Cell group A5
29 Nucleus nervi facialis
30 Formatio reticularis myelencephali
31 Cell group A1
32 Cell group A2
33 Nucleus dorsalis nervi vagi
34 Nucleus solitarius
35 Substantia grisea centralis
36 Substantia gelatinosa
37 Nucleus intermediolateralis

Fig. 7. Noradrenaline-containing cells and fibres: II. Remaining cell groups

to the lateral septal nucleus (Lindvall and Stenevi 1978; Moore 1978). The amygdaloid complex, particularly the central nucleus, has been shown to receive a non-coerulean noradrenergic projection via the ventral amygdalofugal pathway (Fallon et al. 1978). Speciale et al. (1978) found a considerable decrease of the noradrenaline levels in the nucleus caudatus, the piriform cortex, the bed nucleus of the stria terminalis, the medial preoptic nucleus and the median eminence following lesions in the A5 group, whereas Palkovits et al. (1980) concluded from comparable experiments, supplemented with electron microscopical identification of degenerating terminals, that the medullary noradrenergic cell groups, particularly the A1 group, project to the paraventricular, periventricular, dorsomedial, ventromedial and infundibular nuclei. Finally, Moore and Bloom (1979) reported that the medial preoptic nucleus, the anterior hypothalamic area, the supraoptic nucleus and the paraventricular nucleus are all densely innervated by the lower brainstem cell groups, but the ventromedial nucleus is nearly free of noradrenergic innervation.

Apart from the ventral noradrenergic pathway (Ungerstedt 1971), or central tegmental tract (Lindvall and Björklund 1974a; Swanson et al. 1981; Moore and Bloom 1979), the *dorsal periventricular pathway* may also contain long, ascending noradrenergic fibres. It has been suggested that in the rat a considerable number of fibres originating from the lower medullary cell groups project via this route to the nucleus paraventricularis thalami (Lindvall et al. 1974b; Lindvall and Björklund 1983).

Studies using tracer techniques (Ricardo and Koh 1978; Sakumoto et al. 1978; Berk and Finkelstein 1981; Ottersen 1981; McKellar and Loewy 1982), and particularly analyses in which the use of retrograde tracers was combined with techniques for identification of labelled catecholaminergic cells (Day et al. 1980; Blessing et al. 1982; Ottersen 1981; Sawchenko and Swanson 1981, 1982a), have shown that the non-coerulean noradrenergic projections to basal telencephalic and hypothalamic cell masses originate exclusively from the caudal medullary A1 and A2 groups. Thus, it has been established that the A1 group innervates the bed nucleus of the stria terminalis and the medial preoptic area (McKellar and Loewy 1982); the contralateral amygdala (Ottersen 1981); the anterior, lateral and posterior hypothalamic areas (Sakumoto et al. 1978); the dorsal hypothalamic area, the dorsomedial nucleus and the median eminence (McKellar and Loewy 1982); and the parvocellular and magnocellular portions of the paraventricular nucleus as well as the supraoptic nucleus (Sawchenko and Swanson 1981; 1982a). In the magnocellular part of the paraventricular nucleus and in the supraoptic nucleus the projections from the A1 group appear to terminate preferentially in areas rich in vasopressinergic neurons (Sawchenko and Swanson 1982a). The A2 group has been shown to innervate the medial preoptic area (Ricardo and Koh 1978; Sakumoto et al. 1978; Day et al. 1980; Berk and Finkelstein 1981), the ipsilateral amygdala (Ottersen 1981); the anterior, lateral and posterior hypothalamic areas and the nucleus dorsomedialis (Sakumoto et al. 1978; Berk and Finkelstein 1981); the nucleus infundibularis (Ricardo and Koh 1978); and the parvocellular portion of the paraventricular nucleus (Basbaum et al. 1978). Most of the long, ascending projections from the A1 and A2 cell groups are bilateral with an ipsilateral predominance.

In addition to long, ascending projections, the dorsal medullary and lateral tegmental noradrenergic cell groups give rise to various propriobulbar projections and to fibres descending to the spinal cord. With regard to the *propriobulbar connections*, Levitt and Moore (1979) concluded from biochemical and histofluorescence studies done following locus coeruleus lesions that the dorsal motor nucleus of the vagus, the facial nucleus, the motor trigeminal nucleus, the nucleus solitarius, the rhombencephalic raphe nuclei and the parabrachial nuclei are heavily innervated by the non-coerulean noradrenergic cell groups, whereas the hypoglossal nucleus, the pontine nuclei, the locus coeruleus and the pontine and medullary reticular formation receive a less rich innervation from these sources. They noted that the locus coeruleus complex innervates mainly sensory and association nuclei, whereas the axons of the remaining noradrenergic cell groups are distributed primarily to motor and visceral nuclei. Little is known of the exact localization of the cells of origin of the various projections described by Levitt and Moore (1979). However, there is experimental evidence indicating that the A1 group innervates the nucleus solitarius and other nuclei of the dorsal vagal complex, the locus coeruleus and the parabrachial nuclei (Blessing et al. 1981a; Loewy et al. 1981; Sawchenko and Swanson 1982a), and that the A2 group projects heavily and bilaterally to the nucleus solitarius (Takahashi et al. 1979).

According to the earlier studies of Dahlström and Fuxe (1965) and Lindvall and Björklund (1974a), the *non-coerulean noradrenergic fibres descending to the spinal cord* originate mainly from the lower medullary A1 and A2 groups. However, more recent experimental investigations have revealed that this projection arises largely, if not exclusively, from the pontine A5 and A7 groups (Loewy et al. 1979b; Blessing et al. 1981b; Ross et al. 1981; Sawchenko and Swanson 1982a; Stevens et al. 1982; Lindvall and Björklund 1983; Westlund et al. 1983, 1984). This pontospinal projection presumably descends in the major longitudinal catecholamine bundle of Jones and Friedman (1983) and enters the deeper part of the lateral funiculus of the spinal cord (Satoh et al. 1977; Loewy et al. 1979b). Its fibres terminate in the superficial layers of the dorsal horn, in the area around the central canal and, with great density, in the thoracic sympathetic intermediolateral column (Satoh et al. 1977; Crutcher and Bingham 1978; Loewy et al. 1979b; Westlund et al. 1984).

The caudal medullary and lateral tegmental noradrenergic cell groups and their efferent projections are probably involved in a wide range of visceral functions, including cardiovascular control and respiration. The nucleus solitarius, in which the A2 group is largely embedded, receives via the vagus and glossopharyngeal nerves impulses from, among others, atrial stretch receptors and aortic and carotid baro- and chemoreceptors. These viscerosensory impulses may evoke coordinated autonomic and neuroendocrine responses along various routes. Relying heavily upon the excellent studies of Ricardo and Koh (1978) and Sawchenko and Swanson (1982a), some of these routes may be briefly indicated as follows:

1. A substantial, though non-catecholaminergic pathway passes from the nucleus solitarius to the A1 group, and the A1 and A2 groups both project to the A5 group. Loewy and collaborators (1979a, b) demonstrated that the A5 group projects directly to the sympathetic nucleus intermediolateralis in the thoracic cord, and presented physiological evidence indicating that the A5 group represents a vasomotor centre.
2. Alterations in blood pressure and cardiac frequency, as well as modifications of respiratory function, can be induced by stimulation of the bed nucleus of the stria terminalis, the central amygdaloid nucleus and the medial preoptic area (see Ricardo and Koh 1978). All of these telencephalic centres have been shown to receive a direct projection from the A1 and/or A2 groups.
3. The parvocellular portion of the paraventricular nucleus may well occupy an analogous position at the diencephalic level. This centre, which is known to influence cardiovascular functions, receives direct input from the A1 and A2 groups and projects to various autonomic centres in the lower brain stem and spinal cord (Sawchenko and Swanson 1982a).
4. It has been shown that the peripheral cardiovascular receptors connected to the vagus and glossopharyngeal nerves participate in the mechanism of vasopressin (=antidiuretic hormone) release. The baroreceptors and the atrial stretch receptors inhibit the release mechanism, whereas the chemoreceptors

probably stimulate vasopressin secretion (Ricardo and Koh 1978). The glossopharyngeal and vagus fibres involved terminate in the nucleus solitarius, and the cells which release vasopressin in the posterior pituitary are located in the supraoptic and paraventricular nuclei. Sawchenko and Swanson (1982a) consider it likely that the central route taken by the visceroceptive inputs to the hypothalamus involves (a) the large, non-catecholaminergic pathway which connects the solitary nucleus with the A1 region and (b) the direct projection which they demonstrated to run from the A1 group to the magnocellular parts of the paraventricular nucleus and to the supraoptic nucleus.

Finally, it should be mentioned that the medullary noradrenergic cell groups have been implicated in the control of various anterior pituitary hormones, including growth hormone, luteinizing hormone and ACTH (for reviews see Weiner and Ganong 1978; Moore and Bloom 1979). Given the fact that the activity of the anterior pituitary is regulated by neurosecretory processes in the median eminence, it is noteworthy that noradrenergic fibres originating from the cell groups mentioned are involved in three different projections leading to that structure, one direct and two indirect. The direct projection originates from cell group A1. The indirect projections are constituted by fibres arising from the A1 and A2 groups which terminate in the nucleus infundibularis and in the parvocellular division of the paraventricular nucleus. Both of the centres last mentioned are known to project to the median eminence (Sawchenko and Swanson 1982a).

Adrenaline

Adrenaline (Fig. 8) is the final product of the catecholamine chain, dopamine-noradrenaline-adrenaline (see Fig. 3). The conversion of noradrenaline to adrenaline is specifically catalyzed by the enzyme PNMT. The presence of adrenaline in the central nervous system was first suggested by Marthe Vogt (1954) and later substantiated by a considerable number of biochemical studies (Gunne 1962; McGeer and McGeer 1964; Saavedra et al. 1974; Van der Gugten et al. 1976; Lew et al. 1977; Goldstein et al. 1978, 1980). Immunohistochemical evidence for the existence of adrenaline-containing neurons in the central nervous system has been produced by some groups of investigators (Hökfelt et al. 1974b, 1980d; Goldstein et al. 1978; Howe et al. 1980), all of whom used antibodies against the synthesizing enzyme PNMT.

Three PNMT-containing cell groups have been distinguished, all of which are located in the caudal rhombencephalon. By analogy with the nomenclature of Dahlström and Fuxe (1964) these groups have been designated as C1, C2 and C3 (Hökfelt et al. 1974b; Howe et al. 1980).

Cell group C1 is the largest of the three. It is located in the ventrolateral myelencephalon between the inferior olivary complex and the nucleus funiculi lateralis, and in the rat it contains 69% of the total number of PNMT-positive cells (Howe et al. 1980).

Cell group C2, which like group C1 consists of multipolar, medium-sized perikarya, lies partly within and partly adjacent to the nucleus solitarius. It comprises 22% of the total number of PNMT-containing cells.

Cell group C3, which contains the remaining 9% of the PNMT-positive cells in the rat, is situated between the dorsal raphe region and the initial parts of the intramedullary axons of the hypoglossal nerve. The elements of this small group lie interspersed amongst the bundles of the medial longitudinal fascicle (Howe et al. 1980).

Hökfelt and colleagues (1974b), who first described the cell groups C1 and C2, pointed out that these groups are identical to the rostral parts of the catecholaminergic cell groups A1 and A2 respectively of Dahlström and Fuxe (1964). Howe and co-workers (1980), on the other hand, emphasized that the PNMT-containing cell groups C1 and C2 are situated rostrally to, rather than within the confines of, the cell groups A1 and A2.

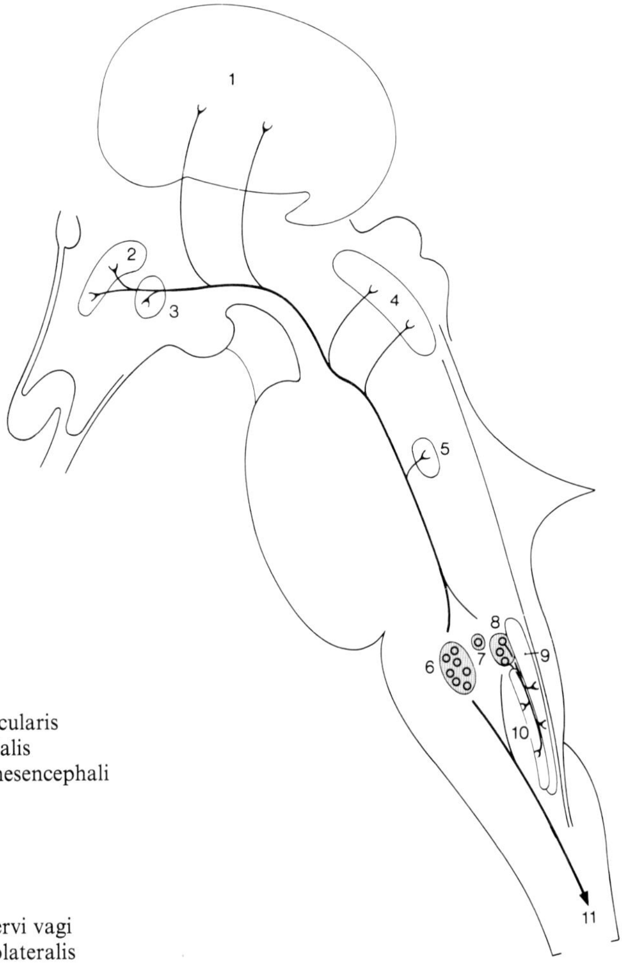

1 Thalamus
2 Nucleus paraventricularis
3 Nucleus dorsomedialis
4 Griseum centrale mesencephali
5 Locus coeruleus
6 Cell group C1
7 Cell group C3
8 Cell group C2
9 Nucleus solitarius
10 Nucleus dorsalis nervi vagi
11 Nucleus intermediolateralis

Fig. 8. Adrenaline-containing cells and fibres

According to their observations, most of the PNMT-positive cells do not show catecholamine histofluorescence. However, they demonstrated by pharmacological manipulation that these cells are capable of synthesizing and storing catecholamines.

From the overall area in which the cell groups C1–C3 are located a bundle of PNMT-containing axons can be traced, ascending through the reticular formation, the area tegmentalis ventralis and the lateral hypothalamic area (Hökfelt et al. 1974b). This bundle closely follows the trajectory of the ventral noradrenergic bundle described by Ungerstedt (1971).

PNMT-containing fields of terminals have been observed in many areas of the central nervous system. The most prominent of these terminal fields include: some thalamic midline nuclei; the dorsomedial and paraventricular hypothalamic nuclei, the latter of which shows a high density of terminals; the ventral part of the periaqueductal grey and the lateral periventricular zone of the rostral rhombencephalon; the ventral part of the locus coeruleus; the nucleus solitarius; the nucleus dorsalis of the vagal nerve; and the nucleus intermediolateralis in the spinal cord. The three nuclei last mentioned all show a high density of PNMT-containing terminals (Hökfelt et al. 1974b). Exact knowledge of which cell groups give rise to which terminal networks is lacking, and therefore the relations shown in Fig. 8 should be considered highly speculative. However, Hökfelt and colleagues (1980d) have presented experimental evidence indicating that the projections to the hypothalamus and the spinal cord originate mainly from cell group C1. The experiments of Ross et al. (1981) have also shown that adrenaline-containing neurons in the C1 group project to the thoracic cord. Sawchenko and Swanson (1982a) considered it likely that the C1 group projects to the supraoptic nucleus and to the magnocellular and parvocellular portions of the paraventricular nucleus, and that the parvocellular portion of the nucleus last mentioned also receives afferents from the C2 group.

The authors who have studied the adrenaline-containing projections with the help of antibodies against PNMT (Hökfelt et al. 1974b; Goldstein et al. 1978) cautioned that the sensitivity of their technique was presumably too low to obtain a complete picture of the distribution of adrenergic axons and terminals. It is interesting to note in this context that some groups of investigators who have analyzed the regional distribution of adrenaline (or PNMT) in the brain with biochemical techniques (Saavedra et al. 1974; Van der Gugten et al. 1976; Lew et al. 1977) detected this substance (or the activity of its synthesizing enzyme) not only in all terminal regions described by Hökfelt and colleagues (1974b), but also in several other brain areas, including the basal ganglia, the nucleus accumbens, the amygdala, the septum, the preoptic region, the habenula, the nucleus infundibularis and the median eminence.

Only little is known concerning the central actions of the adrenergic neurons. The dense innervation of the paraventricular nucleus suggests that adrenaline might be involved in oxytocin and vasopressin secretion, and the innervation of the dorsomedial hypothalamic nucleus indicates a possible influence on food intake. Effects on the regulation of blood pressure and respiration are suggested by the dense adrenergic innervation of the nucleus solitarius, the dorsal motor vagal nucleus and the sympathetic intermedioventral nucleus in the spinal cord.

Physiological experiments (see e.g. Dampney and Moon 1980) have shown that the ventrolateral portion of the medulla oblongata contains a circumscribed and highly sensitive vasopressor area, and Goodchild et al. (1984) recently presented evidence suggesting that this centre corresponds to the adrenergic C1 group. It has also been reported (Goldstein et al. 1978) that in spontaneously hypertensive rats the PNMT activity in the C1 and C2 regions is notably elevated, and that in these animals the adrenaline level in the C2 region is also much higher than in normotensive rats.

Serotonin

Serotonin-containing neurons (see Fig. 9) occur in the mesencephalon, pons and medulla oblongata, but in all of these parts of the brain the elements mentioned are essentially confined to the median and paramedian zones. In marked contrast to most of the catecholaminergic cell groups, the serotoninergic neurons are distributed mainly within specific cytoarchitectonic entities, namely the nuclei of the raphe. However, it is important to note that the groups of serotoninergic neurons, although widely overlapping the raphe nuclei, are by no means congruent with them. It was for this reason that Dahlström and Fuxe (1964) introduced a new classification of the indolamine-containing cells in the brain stem of the rat, distinguishing nine cell groups numbered B1–B9. Most of these groups have also been recognized in other mammals (Felten and Cummings 1979; Takeuchi et al. 1982b; Howe et al. 1983), including primates (Felten et al. 1974; Hubbard and Di Carlo 1974b; Sladek and Walker 1977; Sladek et al. 1982; Schofield and Everitt 1981; Takeuchi et al. 1982c; Felten and Sladek 1983). The most important incongruities between the raphe nuclei and the serotoninergic cell groups are that (a) in many places scattered serotoninergic neurons spread lateral to the raphe nuclei into the adjacent medial reticular formation, and (b) throughout the brain stem, but particularly at pontine levels, serotoninergic elements form wing-like expansions, which extend laterally either in the subventricular area or in the most ventral part of the tegmentum (for details see Steinbusch 1981; Steinbusch and Nieuwenhuys 1983; Takeuchi et al. 1982b, c; Felten and Sladek 1983; Howe et al. 1983). It should also be mentioned that not all of the neurons situated in the raphe nuclei contain serotonin. In fact, in most raphe nuclei only a minority of the neurons are serotoninergic (Wiklund et al. 1980). Within the confines of these nuclei, cells containing the following (putative) neurotransmitters are present: dopamine, noradrenaline, GABA, CCK, L-enkephalin, M-enkephalin, substance P, VIP, SST and TRH (for references, see Steinbusch and Nieuwenhuys 1983; Bowker et al. 1983). In several raphe nuclei, cells containing serotonin as well as a peptide have been observed, e.g. serotonin plus substance P (Hökfelt et al. 1978c; Chan-Palay et al. 1978) and serotonin plus L-enkephalin (Glazer et al. 1981); moreover, evidence has been presented that in some medullary raphe neurons serotonin, co-exists with substance P as well as TRH (Johansson et al. 1981). Hence, the raphe nuclei may be aptly designated as a multiple transmitter complex (Bowker et al. 1983). In the following synopsis of the serotoninergic cell groups the nomenclature of Dahlström and Fuxe (1964) will be employed. The description of the various raphe nuclei follows, as far as possible, those of Taber et al. (1960) and Braak (1970).

Cell group B1 is situated in the ventral part of the medulla oblongata and borders ventrally on the pyramidal tracts. It is limited mainly to the nucleus raphes pallidus, although some of its cells extend laterally in the ventral part of the reticular formation. The rostral part of group B1 is continuous with the caudal part of group B3. Cell group B2 is situated at the same level as group B1, but occupies a more dorsal position. Its cells form two narrow paramedian sheets that coincide with the nucleus raphes obscurus. Cell group B3 is situated in the borderland between the medulla oblongata and the pons. Most of its cells are found within the nucleus raphes magnus, but others constitute a laterally extending band along the fibre bundles of the corpus trapezoideum. Cell group B4 is difficult to delineate as a separate entity. It is reportedly situated at the level of the medial vestibular nucleus lying just dorsal to the nucleus prepositus hypoglossi.

Cell group B5 is rather small and located within the nucleus raphes pontis at the level of the motor nucleus of the fifth nerve. In the upper pons and lower midbrain the data derived from cytoarchitectonic studies are difficult to reconcile with those resulting

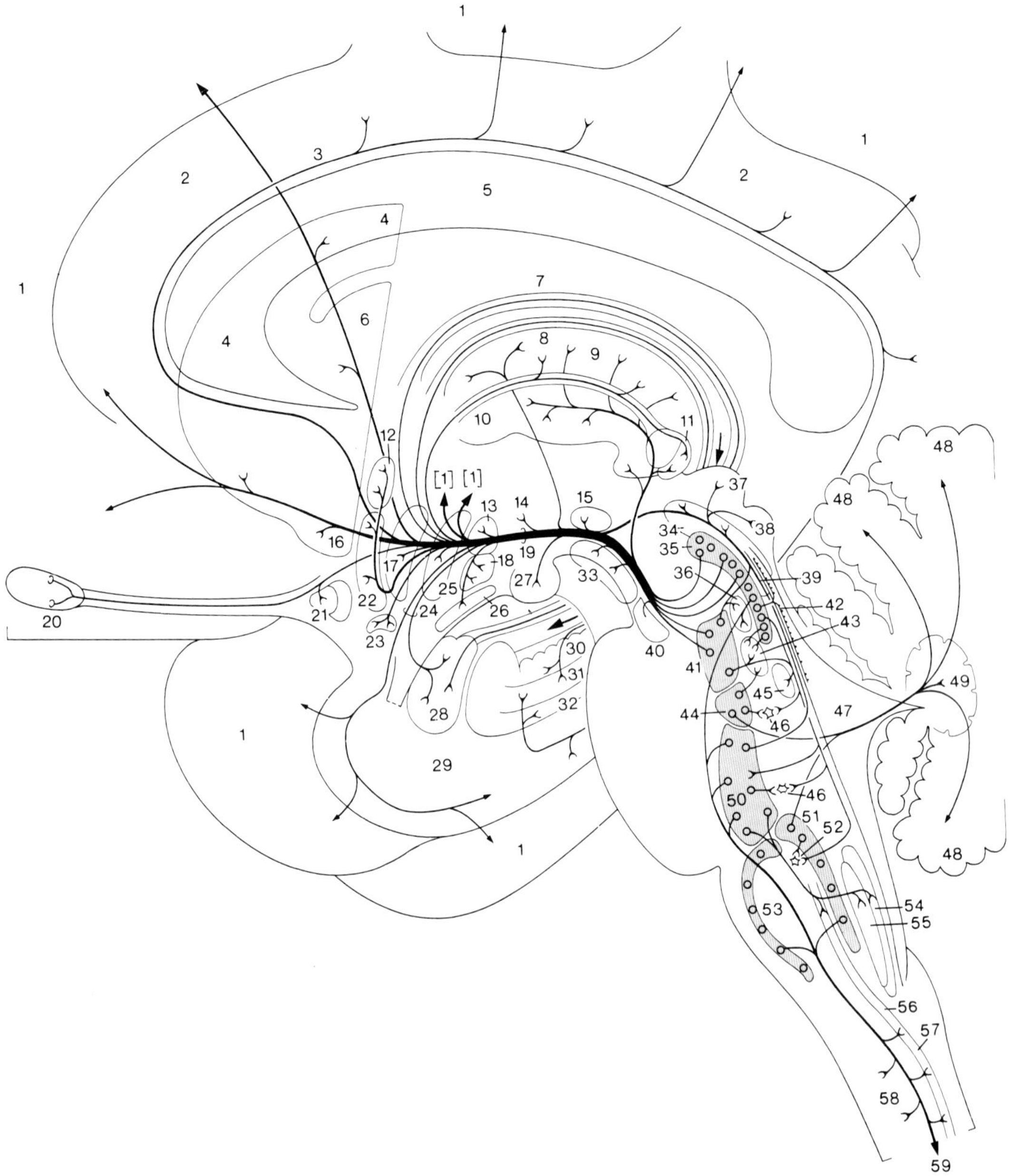

Fig. 9. Serotonin-containing cells and fibres

Explanations to Figure 9

1 Neocortex
2 Gyrus cinguli
3 Striae longitudinales + cingulum
4 Nucleus caudatus
5 Corpus callosum
6 Putamen
7 Fornix
8 Stria terminalis
9 Thalamus
10 Stria medullaris
11 Nucleus habenulae medialis
12 Nucleus septi medialis + lateralis
13 Nucleus dorsomedialis
14 Area lateralis hypothalami
15 Area tegmentalis ventralis
16 Nucleus accumbens
17 Nucleus praeopticus medialis + lateralis
18 Nucleus ventromedialis
19 Fasciculus telencephalicus medialis
20 Bulbus olfactorius
21 Nucleus olfactorius anterior
22 Nucleus gyri diagonalis
23 Nucleus suprachiasmaticus
24 Ansa peduncularis + fibrae amygdalofugales ventrales
25 Nucleus anterior hypothalami
26 Nucleus infundibularis
27 Corpus mamillare
28 Corpus amygdaloideum
29 Gyrus parahippocampalis
30 Gyrus dentatus
31 Cornu Ammonis
32 Subiculum
33 Substantia nigra
34 Griseum centrale mesencephali
35 Nucleus raphes dorsalis (B7)
36 Nucleus tegmentalis dorsalis
37 Colliculus superior
38 Colliculus inferior
39 Fasciculus longitudinalis dorsalis
40 Nucleus interpeduncularis
41 Nucleus centralis superior (B6 + B8)
42 Plexus supraependymalis
43 Locus coeruleus
44 Nucleus raphes pontis (B5)
45 Nuclei parabrachiales
46 Formatio reticularis metencephali
47 Ventriculus quartus
48 Cortex cerebelli
49 Nuclei centrales cerebelli
50 Nucleus raphes magnus (B3)
51 Nucleus raphes obscurus (B2)
52 Formatio reticularis myelencephali
53 Nucleus raphes pallidus (B1)
54 Nucleus solitarius
55 Nucleus dorsalis nervi vagi
56 Nucleus spinalis nervi trigemini
57 Substantia gelatinosa
58 Cornu anterius
59 Nucleus intermediolateralis

from histofluorescence studies on the distribution of serotoninergic cells. However, it seems likely that cell groups B6 and B8 both lie largely within the confines of the superior central nucleus of Bechterew. This nucleus is situated in the upper part of the tegmentum pontis and extends rostrally into the tegmentum of the midbrain. At rostral pontine levels numerous serotoninergic neurons belonging to the B6 and B8 groups extend laterally beyond the boundaries of the raphe nuclei into the lateral tegmentum, where some of them lie intermingled with the noradrenergic cells of the locus coeruleus, the subcoeruleus area and the parabrachial nuclei (Sladek and Walker 1977; Léger et al. 1979; Sladek et al. 1982; Takeuchi et al. 1982b).

The large mesencephalic cell group B7 is localized mainly within the nucleus raphes dorsalis. The latter is situated in and ventral to the periaqueductal grey. It extends from the level of the dorsal tegmental nucleus to the caudal pole of the oculomotor nucleus. The ventral part of the nucleus is situated between the two medial longitudinal fascicles. Cell group B9, finally, is small and occupies a position in the midbrain close to the medial lemniscus.

It has already been mentioned that the raphe nuclei contain considerable numbers of non-serotoninergic neurons. Wiklund et al. (1980) determined the proportions of serotoninergic neurons in the various raphe nuclei of the cat to be as follows: nucleus raphes obscurus, 35%; nucleus raphes pallidus, 50%; nucleus raphes magnus, 15%; nucleus raphes pontis, 10%; nucleus centralis superior, 35% and nucleus raphes dorsalis, 70%. However, Descarries et al. (1982) reported that in the rat only one-third of the neurons in the nucleus raphes dorsalis are serotoninergic.

The serotoninergic neurons in the brain stem are provided with highly branched fibres which innervate virtually the entire central nervous system, thus comprising the most expansive central neuronal network yet described. The axons of the serotonin-containing cells differ in diameter, though most are thin, unmyelinated and varicose. Azmitia and Gannon (1983) studied the serotoninergic fibres in the medial forebrain bundle of the rat and monkey at the ultrastructural level, with an antibody against serotonin. They found that the percentage of serotonin-immunoreactive myelinated axons is much greater in the monkey than in the rat (25.4% versus 0.7% of the total number of serotonin-immunoreactive fibres respectively).

The fibre connections of the serotoninergic cell groups have been studied with a variety of techniques, including fluorescence histochemistry (Felten and Sladek 1983), histofluorescence combined with chemical or surgical lesions (Ungerstedt 1971; Fuxe and Jonsson 1974), radioimmunology combined with surgical lesions (Palkovits et al. 1977), autoradiography following intraventricular administration of tritiated serotonin (Parent et al. 1981), amino acid-autoradiography following chemical lesions (Moore and Halaris 1975), immunohistochemistry (Kojima et al. 1982, 1983; Kojima and Sano 1983) and retrograde transport of HRP in combination with immunohistochemistry (Bowker et al. 1983). In addition, the efferents of the various raphe nuclei have been studied with retrograde (Sakai et al. 1977; Van der Kooy and Kuypers 1979) and anterograde (Conrad et al. 1974; Bobillier et al. 1975, 1976, 1979; Taber Pierce et al. 1976; Basbaum et al. 1978; Azmitia and Segal 1978; Moore et al. 1980; Tohyama et al. 1980) tracer techniques. On the basis of these studies and of additional publications cited below, the following mainly – or at least partly – serotoninergic projections may be distinguished: (a) the ventral ascending pathway, (b) the dorsal ascending pathway, (c) projections to various rhombencephalic centres, (d) the cerebellar pathway, (e) bulbospinal pathways, and (f) the supra-ependymal plexus. These fibre systems will now be briefly discussed.

The large *ventral ascending serotoninergic pathway* arises mainly from the cell groups B6–B8, which include the nucleus raphes dorsalis and the nucleus centralis superior. Its fibres sweep ventrally from these nuclei and then curve rostrally through the ventral tegmentum, after which they enter the medial forebrain bundle in the lateral hypothalamic area. During its course through the midbrain, fibres are given off to the interpeduncular nucleus, the substantia nigra (mainly its compact part), and the area tegmentalis ventralis (Bobillier et al. 1976; Taber Pierce et al. 1976; Palkovits et al. 1977; Fibiger and Miller 1977; Nojyo and Sano 1978; Bobillier et al. 1979). The fibres terminating in the area tegmentalis ventralis originate mainly from the nucleus centralis superior. The major input to the nucleus interpeduncularis and the substantia nigra comes from the nucleus raphes dorsalis. It has been demonstrated that the dorsal raphe neurons which project to the substantia nigra also distribute axonal branches to the caudatus-putamen complex (Van der Kooy and Hattori 1980). In the rostral mesencephalon a number of fibres leave the major group to ascend in and around the habenulo-interpeduncular tract to the thalamus. Some of these fibres terminate in the medial habenular nucleus, but the majority turn rostrally in the internal medullary lamina to innervate various thalamic centres, including several midline nuclei, the nucleus medialis, the nucleus parafascicularis, the anterior and ventral nuclear complexes, the nucleus lateralis dorsalis and the nucleus ventralis corporis geniculati lateralis (Conrad et al. 1974; Bobillier et al. 1976, 1979; Moore et al. 1980; Mantyh and Kemp 1983; Cropper et al. 1984). Some of these centres may receive additional serotoninergic input via the mamillothalamic tract and the stria medullaris (Moore et al. 1980; Bobillier et al. 1979). In the caudal part of the diencephalon fibres are distributed to the posterior hypothalamic area and the corpus mamillare (Bobillier et al. 1976, 1979; Taber Pierce et al. 1976).

Two large fibre contingents leave the ventral ascending serotoninergic pathway along its course through the lateral hypothalamic area, one directed laterally, the other ventromedially. The laterally directed fibres follow the ansa peduncularis–ventral amygdalofugal pathway system and continue through the internal capsule into the amygdala, the striatum and the external capsule to reach the lateral and caudal parts of the neocortex (Fuxe and Jonsson 1974; Bobillier et al. 1976; Moore et al. 1980). The distribution of fibres in the amygdala is complex. Dense innervation has been observed in parts of the anterior amygdaloid area and of the basal, basal posterior, lateral and medial posterior nuclei. As regards the striatum, the nucleus caudatus, the putamen, the nucleus accumbens and the globus pallidus are all supplied by serotoninergic fibres, but in none of these centres does this innervation attain a high density (Steinbusch 1981). The medioventrally directed fibres innervate a large number of preoptic and hypothalamic centres, among which are the dorsomedial, ventromedial and infundibular nuclei, the anterior and lateral hypothalamic areas and the medial preoptic, lateral preoptic and suprachiasmatic nuclei. The ventromedial and suprachiasmatic nuclei are densely innervated (Nojyo and Sano 1978; Van de Kar and Lorens 1979; Steinbusch and Nieuwenhuys 1981).

In front of the hypothalamus the ventral ascending serotoninergic pathway splits up as follows (Azmitia and Segal 1978; Moore et al. 1980; Bobillier et al. 1979; Parent et al. 1981): (a) Some fibres enter the stria medullaris to terminate in the periventricular region of the thalamus and the medial habenular nucleus. (b) Some fibres pass with the stria terminalis to the amygdaloid complex. (c) Some fibres traverse the fornix to the gyrus dentatus and the cornu Ammonis (Moore and Halaris 1975). (d) Some fibres enter the diagonal band of Broca to supply the nucleus of the same name and the medial and lateral septal nuclei (Gall and Moore 1984). (e) Some fibres pass by way of the rostral portion of the external capsule to the rostral part of the nucleus caudatus and to the rostral and lateral parts of the neocortex. (f) A bundle of fibres which traverses the rostral septum enters the cingulum bundle and the supracallosal striae to eventually reach the hippocampus via a caudal approach; during its course fibres are given off to the cingulate and entorhinal cortex and to the adjacent neocortical areas. In the hippocampus the supracallosal projection fans out into the subiculum and the cornu Ammonis. (g) Some fibres proceed rostrally into the stria olfactoria medialis to terminate in the anterior olfactory nucleus and the glomerular layer of the olfactory bulb. (h) Finally, it should be mentioned that serotoninergic fibres ascending with the medial forebrain bundle terminate in the organum vasculosum of the lamina terminalis as well as in the subfornical organ (Moore 1977). There is experimental evidence indicating that the nucleus caudatus and the putamen are supplied by the nucleus raphes dorsalis, whereas the efferents of the nucleus centralis superior are distributed mainly to more medially situated structures, such as the hypothalamus and the hippocampus (Azmitia and Segal 1978; Bobillier et al. 1979; Van de Kar and Lorens 1979). On the basis of anterograde tracer experiments, Tohyama et al. (1980) claimed that the rostral part of the nucleus raphes dorsalis supplies the lateral parts of the cerebral cortex, whereas the caudal part of that nucleus innervates the medial cortical areas. The entire neocortex receives a serotoninergic innervation, which spreads over all cortical layers. In the rat this innervation is relatively uniform (Lidov et al. 1980), but in the monkey different cytoarchitectonic fields may show considerable differences in the density and the laminar distribution pattern of their serotoninergic innervation (Takeuchi and Sano 1983).

The *dorsal ascending serotoninergic pathway* consists of fibres that pass rostrally along the dorsal longitudinal fascicle of Schütz (Bobillier et al. 1976; Moore et al. 1980; Parent et al. 1981) and somewhat more ventrally in the vicinity of the medial longitudinal fascicle (Björklund et al. 1973; Felten and Sladek 1983). Its fibres originate mainly from the nucleus raphes dorsalis but, according to some authors, also from the nucleus centralis superior (Björklund et al. 1973; Felten and Sladek 1983) and the nucleus raphes magnus

(Bobillier et al. 1976). After having distributed fibres to the mesencephalic central grey and the posterior hypothalamic area, most of the fibres of the dorsal pathway enter the medial forebrain bundle to join the ventral pathway. According to Parent et al. (1981), the dorsal ascending serotoninergic pathway (i.e. the periventricular 5-HT system, in their terminology) emits distinct fibre contingents to the inferior and superior colliculi and to the subcommissural organ.

Serotoninergic projections to the various rhombencephalic centres: According to the detailed immunohistochemical mapping study of Steinbusch (1981) in the rat, a great number of rhombencephalic centres receive a serotoninergic innervation. In (parts of) the following cell masses a medium, high, or very high density of serotonin-containing fibres and terminals was observed: nucleus sensorius principalis and nucleus spinalis of the trigeminal nerve, nucleus solitarius, locus coeruleus, nuclei parabrachiales, nucleus tegmenti dorsalis, nucleus praepositus hypoglossi, the pontine and medullary reticular formation, nucleus nervi abducentis, nucleus motorius nervi trigemini, nucleus nervi facialis, nucleus ambiguus and nucleus dorsalis nervi vagi. Little is known of the specific sites of the serotoninergic projections to all of these centres. However, it has been established that the locus coeruleus receives a substantial projection from the nucleus raphes dorsalis (Conrad et al. 1974; Taber Pierce et al. 1976; Bobillier et al. 1976; Sakai et al. 1977; Cedarbaum and Aghajanian 1978; Morgane and Jacobs 1979). The nucleus centralis superior and the nucleus raphes pontis are also reported to project to the locus coeruleus (Léger et al. 1980). The serotoninergic axons in the locus coeruleus are thin and varicose. Pickel et al. (1977) and Léger and Descarries (1978) observed that these axons are frequently in close approximation to the noradrenergic locus coeruleus neurons, but only rarely form specialized synaptic contacts with these elements. The nucleus raphes dorsalis and the nucleus centralis superior also send fibres to the dorsal tegmental nucleus, the tegmental grey, the lower rhombencephalic raphe nuclei and the pontine and medullary reticular formation (Bobillier et al. 1976, 1979; Tohyama et al. 1980; Yezierski et al. 1982). The nucleus raphes pontis and nucleus raphes magnus project likewise to the reticular formation (Bobillier et al. 1976), and the nucleus raphes magnus sends in addition fibres to the nucleus solitarius, the dorsal vagus nucleus and the superficial layers of the spinal trigeminal nucleus (Basbaum et al. 1978). Bobillier et al. (1979) reported that the fibres descending from the nucleus centralis superior are diffusely arranged and follow the trajectories of the dorsal and the medial longitudinal fasciculi.

The *cerebellar serotoninergic pathway* emerges from all raphe nuclei, but particularly from the nuclei raphes pontis and obscurus (Taber Pierce et al. 1977), and passes via the pedunculus cerebellaris medius to the cerebellum, where its fibres are distributed to both the cortex and the central nuclei (Takeuchi et al. 1982a).

The *bulbospinal serotoninergic pathways* are formed by fibres that arise mainly from cell groups B1–B3 (Ungerstedt 1971; Bowker et al. 1981a). In the spinal cord most of these fibres descend immediately below the pia in the lateral funiculi (Bowker et al. 1981b). The bulbospinal serotoninergic fibres project throughout the length of the cord to both the dorsal and ventral horns (Steinbusch 1981). In the dorsal horn laminae I and II are supplied by numerous fibres. The deeper layers of the dorsal horn also receive a serotoninergic innervation, though this is of much lower density than that of the superficial zone. In the ventral horn the cells in the medial and lateral motoneuronal groups are surrounded by networks of serotoninergic fibres. This plexus around the motoneurons is in much closer apposition to the somata in primates than in rodents or carnivores (Kojima and Sano 1983). In the intermediate grey of the spinal cord Kojima and colleagues (1982, 1983) observed two bundles of descending serotoninergic fibres, a small medial one and a much larger lateral one.

The medial bundle, which is situated directly lateral to the central canal, extends throughout the length of the cord. The large lateral bundle is confined to the thoracic and upper lumbar segments and coincides with the nucleus intermediolateralis. The sympathetic preganglionic neurons within that nucleus are densely surrounded by serotoninergic fibre networks.

Experimental neuroanatomical studies which do not distinguish between serotoninergic and non-serotoninergic fibres suggest that most of the raphespinal fibres originate from the nucleus raphes magnus, with smaller contributions from the nuclei raphes pallidus and obscurus (Brodal et al. 1960; Leichnetz et al. 1978). Wessendorf et al. (1981) established experimentally that the nucleus raphes magnus contains numerous serotoninergic cells that project to the spinal cord, and Loewy and McKellar (1981) found that serotoninergic neurons situated in the B1 and B3 groups project to the intermediolateral column and the ventral horn. Finally, it should be mentioned that Bowker and collaborators (1981a, b), using the HRP technique in combination with immunohistochemistry, demonstrated that the nuclei raphes magnus, pallidus and obscurus contain serotoninergic neurons which project to all levels of the spinal cord. These authors established, moreover, that axonal branches of a limited number of serotoninergic neurons situated in the nucleus raphes dorsalis and the adjacent mesencephalic reticular formation attain the cervical part of the spinal cord.

The *supra-ependymal serotoninergic plexus:* Throughout the ventricular system of the brain a supra-ependymal plexus of thin, varicose serotoninergic fibres is found. According to Steinbusch (1981), this plexus is strongly developed on the ventricular surface of the rhombencephalon, in the cerebral aqueduct, and in the lateral ventricles, but only moderately developed in the dorsal part of the third ventricle. In the ventral part of the third ventricle, i.e. the zone adjacent to the hypothalamus, this plexus is entirely lacking. There is evidence suggesting that the supra-ependymal serotoninergic plexus originates mainly from neurons located in the nuclei raphes dorsalis and centralis superior (Aghajanian and Gallager 1975; Cupédo and de Weerd 1980; Parent et al. 1981). Some of its fibres may arise from the supra-ependymal serotoninergic neurons which have been found on the basal surface of the fourth ventricle and in the aqueductus cerebri (Steinbusch and Nieuwenhuys 1983).

As regards the possible functional significance of the various groups of serotoninergic neurons, I will confine myself to a few notes.

1. Perikarya and dendrites of serotoninergic raphe neurons frequently contact blood vessels and tanycyte-like glia cells, i.e. elements with cell bodies forming part of the ependymal ventricular lining and provided with one or a few long, peripherally extending processes. Cummings and Felten (1979; see also Felten and Harragan 1980; Felten et al. 1981; Felten and Sladek 1983), who first described these relationships, considered it likely that the direct neuron-vascular contacts and the tanycyte shafts represent communication channels along which substances carried by the blood and cerebrospinal fluid influence raphe cells. They suggested that raphe neurons might act as both neurons and endocrine-neural transducer cells.
2. Serotoninergic fibres originating from the nuclei raphes dorsalis and centralis superior project to small intraparenchymal blood vessels within the brain (Reinhard et al. 1979), as well as to the pial arteries and arterioles (Edvinsson et al. 1983). These serotoninergic fibres may be involved in the regulation of cerebral blood flow and, furthermore, may well be implicated in the aetiology of serotonin-related cerebrovascular disorders, including migraine, ischaemia and stroke (Welch et al. 1977; Edvinsson et al. 1983).
3. There is evidence indicating that the medullary serotoninergic neurons which project to the sympathetic intermediolateral column in the thoracic spinal cord exert an inhibitory influence on the neurons in that centre, and

that this sympathetico-inhibitory pathway participates in the central control of cardiovascular function (Antonaccio 1977; Cabot et al. 1979; Loewy and Neil 1981).

4. The ascending projections from the mesencephalic and upper rhombencephalic serotoninergic cell groups are involved in the regulation of sleep (see e.g. Puizillout et al. 1981), but the theory of Jouvet (1967, 1969) that the raphe nuclei play a dominant role in the initiation of slow-wave sleep appears to be untenable in its original form.

5. The serotoninergic neuron population located in the nucleus raphes dorsalis projects heavily to the locus coeruleus and the substantia nigra. Because serotonin inhibits catecholaminergic neurons (Dray et al. 1976), it seems likely that the most important serotoninergic centre exerts a strong inhibitory influence on the locus coeruleus and the substantia nigra, i.e. the principal noradrenergic and dopaminergic centres.

6. Axons descending from the medullary raphe nuclei to the spinal cord form part of a system which controls the transmission of nociceptive messages (Fields et al. 1977; Basbaum et al. 1978, 1984). Serotoninergic neurons situated in the nucleus raphes magnus which project to the spinal dorsal horn, particularly to laminae I, II and V, most probably form an important component of this descending pain-control system. It has been shown that the activity of many dorsal horn neurons, including elements situated in laminae I, II and V, is suppressed by direct iontophoresis of serotonin (Engberg and Ryall 1966; Randić and Yu 1976; Weight and Salmoiraghi 1966). La Motte and de Lanerolle (1983b) have presented ultrastructural evidence suggesting that in the dorsal horn serotoninergic fibres terminate on neurons situated in lamina I as well as in lamina II; presumably, the elements in the former layer represent spinothalamic projection neurons and those in the latter arise from local circuit neurons. These authors also observed serotoninergic terminals which were presynaptic to primary afferent fibres possibly containing substance P. Glazer and Basbaum (1984) found that serotonin-radiolabelled axons contact enkephalin-immunoreactive neurons in the superficial dorsal horn.

Two points concerning the raphespinal projection just discussed deserve emphasis: (a) Although serotonin is presumably a very important neurotransmitter in this system, it is by no means the only one. As mentioned before, Bowker et al. (1983) have demonstrated that at least five different putative neurotransmitters are contained within raphe neurons that project to the spinal cord. (b) The fibres passing from the medullary raphe nuclei to the dorsal horn not only inhibit the transmission of noxious stimuli, but presumably also control the transmission of other impulses which enter the cord via the thin-fibred medial portion of the dorsal roots, e.g. those encoding mechanoreceptive and thermal messages.

7. According to Anderson and Proudfit (1981), activation of the serotoninergic neurons that project to the ventral horn facilitates all motoneurons, extensors as well as flexors, and α as well as γ motoneurons. They point out that mass excitation of the bulbospinal serotoninergic projections would simultaneously diminish the influence of the somatic input, and increase motor responsiveness.

8. Concerning the role of the serotoninergic intraventricular neuronal processes, the following suggestions have been made: (a) During neural activity, serotonin is released by these axons into the surrounding cerebrospinal fluid for wide dissemination over the ventricular surfaces of the brain. This serotonin may enter the brain parenchyma by transcellular routes through certain ependymal cells, after which it is free to act upon serotonin receptor sites located within the range of dispersal (Chan-Palay 1977). (b) The serotonin released from these axons may function to regulate the flow of cerebrospinal fluid by modulating the motility of ependymal cell cilia (Aghajanian and Gallager 1975). (c) The supra-ependymal fibres make synapse-like contacts with the ependymal cells, via which they participate in the regulation of the pro-

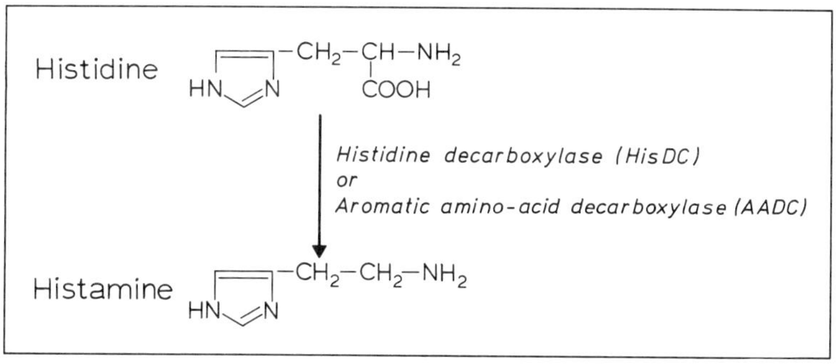

Fig. 10. Synthesis of histamine

duction of cerebrospinal fluid and hormones by these elements (Ribas 1977; Steinbusch and Nieuwenhuys 1983).

Finally, it is worthy of note that in a certain category of patients suffering from endogenous depression the concentration of serotonin and its major metabolite (5-hydroxyindolacetic acid: 5-HIAA) in the cerebrospinal fluid are markedly reduced (Van Praag and Korf 1974; Van Praag 1977). Whether there is in these patients a malfunctioning of the entire population of serotoninergic neurons or only of a certain subset is unknown at present. However, it might be relevant that the supra-ependymal serotoninergic plexus arises mainly from the nucleus raphes dorsalis.

Histamine

The monoamine histamine (HIST) is formed from the amino acid histidine by decarboxylation, a reaction which is catalyzed by the enzyme histidine decarboxylase (HDC) or by aromatic amino acid decarboxylase (AADC; Fig. 10). There is neurochemical, neurophysiological and pharmacological evidence which, taken together, stronlgy suggests that HIST functions as a neurotransmitter in the mammalian central nervous system (Taylor and Snyder 1972; Taylor 1975; Calcutt 1976; Schwartz 1975, 1977; Schwartz et al. 1980a). Moreover, two different types of HIST receptors (H1 and H2) have been shown to be present in the brain (Schwartz 1979; Schwartz et al. 1980b). Neurochemical studies (Taylor and Snyder 1972; Schwartz 1975) have revealed that HIST is distributed widely, though unevenly in the brain. Thus, the HIST content appears to be highest in the hypothalamus, intermediate throughout the telencephalon, and lowest in the brain stem. A similar regional distribution has been demonstrated for HDC (Taylor and Snyder 1972; Schwartz 1975; Barbin et al. 1980). The distribution of histaminergic perikarya and their axonal processes in the central nervous system of the rat has recently been studied by Watanabe and co-workers (1983, 1984), using antibodies against HDC. Their results may be summarized as follows (Fig. 11):

1. HDC-immunoreactive perikarya were concentrated in the caudal part of the hypothalamus, and no HDC-containing cells were observed in other areas. The labelled cells were concentrated in the following four more or less distinct groups: (a) a large group situated mainly within the tuberal magnocellular nucleus but extending rostrally into the dorsomedial nucleus; (b) a cluster of neurons situated in the caudal magnocellular nucleus, extending rostrolaterally along the ventral surface of the hypothalamus; (c) scattered elements situated in the lateral hypothalamic area; (d) some cells located in the posterior hypothalamic nucleus.
2. Widespread but uneven networks of HDC-immunoreactive fibres have been observed in the brain. Most of these fibres were

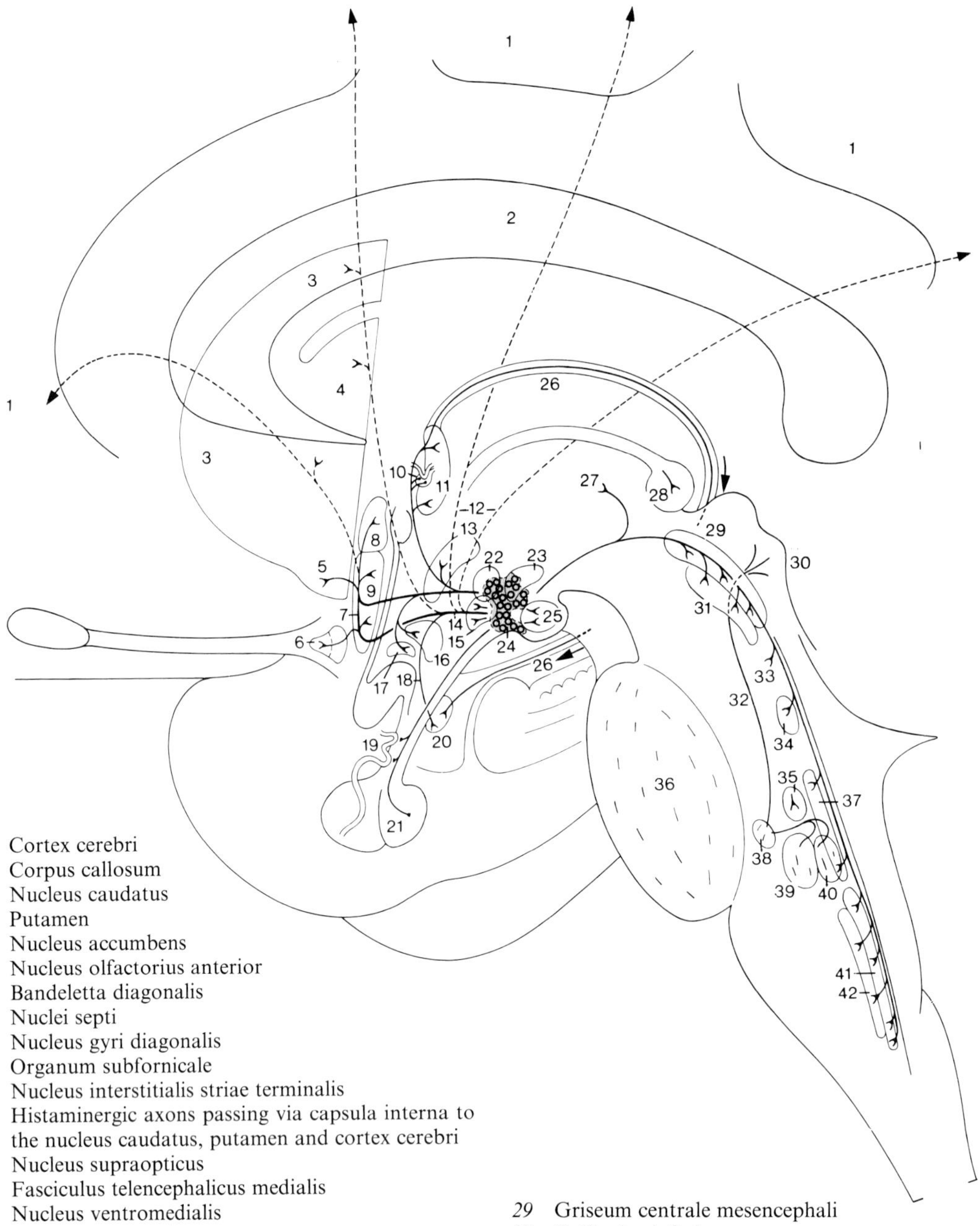

1 Cortex cerebri
2 Corpus callosum
3 Nucleus caudatus
4 Putamen
5 Nucleus accumbens
6 Nucleus olfactorius anterior
7 Bandeletta diagonalis
8 Nuclei septi
9 Nucleus gyri diagonalis
10 Organum subfornicale
11 Nucleus interstitialis striae terminalis
12 Histaminergic axons passing via capsula interna to the nucleus caudatus, putamen and cortex cerebri
13 Nucleus supraopticus
14 Fasciculus telencephalicus medialis
15 Nucleus ventromedialis
16 Nucleus supraopticus
17 Nucleus suprachiasmaticus
18 Fibrae amygdalofugales ventrales
19 Eminentia mediana
20 Nucleus medialis amygdalae
21 Lobus posterior hypophyseos
22 Nucleus dorsomedialis
23 Nucleus posterior hypothalami
24 Group of histaminergic neurons
25 Nuclei mamillares
26 Stria terminalis
27 Nuclei periventriculares thalami
28 Nucleus habenulae medialis
29 Griseum centrale mesencephali
30 Colliculus inferior
31 Nucleus raphes dorsales
32 Lemniscus lateralis
33 Griseum centrale mesencephali
34 Nucleus parabrachialis lateralis
35 Nucleus nervi facialis
36 Nucleus pontis
37 Nucleus vestibularis medialis
38 Nucleus corporis trapezoidei
39 Nucleus cochlearis ventralis
40 Nucleus cochlearis dorsalis
41 Nucleus solitarius
42 Nucleus dorsalis nervi vagi

Fig. 11. Histamine-containing cells and fibres

very thin and had a distinct varicose appearance.
3. Within the diencephalon numerous fibres have been observed in the ventral half of the posterior hypothalamic area, in the ventromedial nucleus and throughout the entire mamillary body. In the thalamus a few fibres appear to be present in the periventricular nucleus and in the lateral geniculate body.
4. From the caudal hypothalamus two groups of HDC-positive fibres have been traced rostrally, one periventricular and the other lateral, the latter ascending through the ventral part of the medial forebrain bundle. Both groups enter the basal telencephalon and unite in the diagonal band of Broca.
5. Within the telencephalon the following areas appear to contain HDC-immunoreactive fibres: the anterior olfactory nucleus (moderate density), the entire cerebral cortex (fibres with moderate density, showing no regional or laminar predominance), the amygdaloid complex – particularly the medial amygdaloid nucleus, the nucleus of the diagonal band (moderate density), the medial and lateral septal nuclei (low to moderate density), the nucleus accumbens (low density) and the caudatus-putamen complex (scattered fibres). From observations of serial sections it was deduced that the periventricular and lateral fibre groups, and the diagonal band of Broca supply the telencephalic centres as follows: The periventricular fibre group issues fibres which pass dorsolaterally and enter the stria terminalis, with which they reach the amygdaloid complex. Fibres dissociating from the lateral group enter the capsula interna to pass to the caudatus-putamen complex and to the cerebral cortex. Other fibres emanating from this group join the ventral amygdalofugal pathway and terminate in the amygdala and prepiriform cortex. The diagonal band of Broca, finally, distributes to its bed nucleus, the septal nuclei, the nucleus accumbens and the cerebral cortex.
6. In the dorsal and periventricular zones of the brain stem, HDC-immunoreactive fibres (mostly in moderate density) have been observed in the mesencephalic and metencephalic central grey, the nucleus raphes dorsalis, the lateral parabrachial nucleus, the nucleus vestibularis medialis, the nucleus solitarius (particularly its caudal, commissural part) and the dorsal nucleus of the vagus nerve. The fibres supplying the periaqueductal grey were observed to emanate from the dense plexus surrounding the groups of HDC-positive perikarya. Fibres proceeding caudally through the pontine and medullary periventricular zone, following the general trajectory of the fasciculus longitudinalis dorsalis of Schütz, presumably project to the remaining brain stem centres mentioned above.
7. Several other structures in the brain stem contain HDC-positive fibres; among these are the pontine nucleus, the motor nucleus of VII, the dorsal and ventral cochlear nucleus, the nucleus of the trapezoid body, the lateral lemniscus and the inferior colliculus. The fibres could not be traced into continuity with the clusters of HDC-containing cells in the hypothalamus. It is noteworthy that several cell masses belonging to the vestibular and auditory systems contain HDC-immunoreactive fibres. However, in the vestibular complex, only one of the four nuclei forming this complex contains such fibres (i.e. the nucleus medialis), and in the auditory system the superior olivary nuclei and the medial geniculate body appeared to be devoid of HDC-containing fibres.

With regard to the telencephalic projections, the observations of Watanabe (1984) summarized above are in general agreement with those of Garbarg et al. (1974) and Ben-Ari et al. (1977). These two groups of investigators demonstrated that, following lesions in the hypothalamus, the HDC activity decreased ipsilaterally in various telencephalic centres. However, the presence of HDC-containing cells in the mesencephalic reticular formation claimed by Pollard et al. (1978) could not be confirmed by Watanabe and co-workers (1984).

Comparison of the distribution of HDC-containing fibres with that of histamine H1 receptors, as determined autoradiographically by Palacios et al. (1981), reveals many similarities, but also some striking differences. Thus, the following centres, which all contained HDC-immunoreactive fibres, also showed high receptor concentrations: the bed nucleus of the stria terminalis, the ventromedial hypothalamic nucleus, the mediocaudal hypothalamic area, the medial amygdaloid nucleus, the pontine nucleus, the motor facial nucleus, the medial vestibular nucleus, the nucleus solitarius and the dorsal vagal nucleus. Similarly, many components of the auditory system had significant densities of receptors. However, some cell masses, as for instance the nucleus preopticus periventricularis and the nucleus raphes magnus, which as far as is known are not supplied by any HDC-containing fibres, nevertheless show high receptor concentrations.

Most recently, Steinbusch and Mulder (1984) carried out an immunohistochemical analysis of the brain of the rat, using antibodies raised against HIST coupled to a carrier protein. These authors confirmed that HIST-containing perikarya are confined to the caudal hypothalamus. As regards the distribution of HIST-containing fibres and terminals, their results closely correspond to those of Watanabe et al. (1984); however, they made the following additional observations: (a) The supraoptic and paraventricular nuclei, the caudal part of the suprachiasmatic nucleus and the medial habenular nucleus all contain HIST-immunoreactive fibres. (b) The central portion of the subfornical organ contains a dense plexus of HIST-positive axons. (c) HIST-containing fibres are also present in the median eminence and in the posterior pituitary; in the median eminence these fibres are found mainly in its lateral parts. (d) Frequently, HIST-positive fibres were observed close to the ventricles, and some of these fibres form a supra-ependymal plexus, suggesting that HIST might be released from these structures into the cerebrospinal fluid. Steinbusch and Mulder (1984) do not mention the presence of histaminergic fibres in the brain stem.

Amino Acids

γ-Aminobutyric Acid

The amino acid γ-aminobutyric acid (GABA) is the most important inhibitory neurotransmitter in the central nervous system (Curtis and Johnson 1974; Fagg and Foster 1983). GABA is produced by the removal of a carboxyl group from glutamate, a reaction which is specifically catalyzed by the enzyme glutamic acid decarboxylase (GAD; Fig. 12). Techniques for localizing GABA in the central nervous system include autoradiography after injection of tritiated GABA, the study of high affinity uptake for GABA, and immunohistochemistry with antibodies directed against GAD (and most recently against GABA itself: Seguela et al. 1984). The following survey will show that, although GABA is primarily synthesized and released by local circuit neurons, a number of long-axoned projection systems also use this inhibitory neurotransmitter (Fig. 13).

There is considerable evidence implicating GABA as an important transmitter in the spinal cord (Curtis and Johnson 1974), used by interneurons to mediate both pre- and postsynaptic inhibition as well as presynaptic facilitation (Barber et al. 1978). This amino acid appears to be present in terminals that form synaptic complexes throughout the cord (McLaughlin et al. 1975) and GABA-containing perikarya have been found in all regions of the spinal grey matter except the motor neuron pools (Barber et al. 1982). GABA is released by endings that are presynaptic to primary afferent terminals (Barber et al. 1978), and it is presumably also involved in Renshaw cell-mediated recurrent inhibition of motoneurons (Cullheim and Kellerth 1981; Polc and Haefely 1982).

It has been established that the Purkinje cells in the cerebellar cortex synthesize GABA and release this inhibitory transmitter in their target areas, i.e. the deep cerebellar nuclei and the lateral vestibular nucleus (Obata et al. 1967, 1970; Fonnum et al. 1970; Kawaguchi and Ono 1973; Saito et al. 1974; McGeer

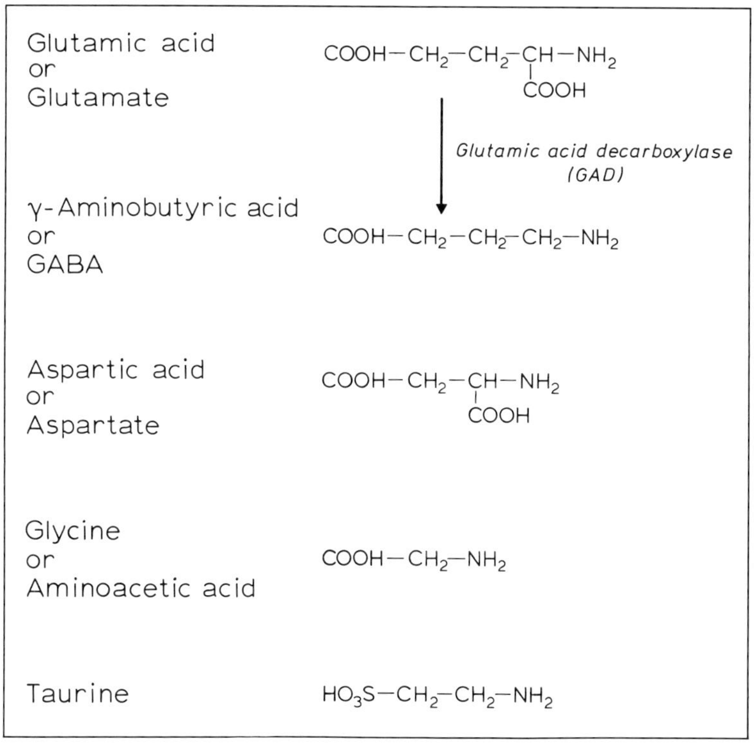

Fig. 12. Amino acid neurotransmitters

et al. 1975). GABA has also been suggested to be the transmitter of the inhibitory local circuit neurons (Golgi, stellate and basket cells) in the cerebellar cortex (Kuriyama et al. 1966; McLaughlin et al. 1980; Saito et al. 1974).

The nucleus raphes dorsalis contains GABA-synthesizing elements which exert a tonic inhibitory control on the serotoninergic neurons present in that centre (Nanopoulos et al. 1982; Scatton et al. 1984). The superficial zone of the superior colliculus contains two different types of small neurons, which accumulate GABA (Mize et al. 1982). Throughout the nucleus reticularis thalami numerous GABA-synthesizing neurons are found. These elements project primarily to other thalamic nuclei on which they probably exert an inhibitory influence (Houser et al. 1980).

GABA is a prominent neurotransmitter in the so-called extrapyramidal motor system. Within this system it occurs in projection neurons as well as in local circuit elements. The nucleus caudatus and the putamen contain numerous GABA-ergic neurons, the axons of which traverse the globus pallidus and terminate in the substantia nigra, pars reticulata (Hattori et al. 1973; Ribak et al. 1976; Brownstein et al. 1977; Fonnum et al. 1978; Jessell et al. 1978), where they are believed to influence, either directly or indirectly, the cells of origin of the dopaminergic nigrostriatal pathway. Several authors (Brownstein et al. 1977; Fonnum et al. 1978; Jessel et al. 1978) have reported that the GABA-ergic strionigral fibres arise preferentially from neurons situated in the posterior parts of the striatum. It has been suggested that these neurons are of the medium-sized, densely spiny type (Ribak 1981; but see Bolam et al. 1983a).

The globus pallidus contains numerous GABA-ergic terminals which belong to striopallidal fibres (Fonnum et al. 1978; Jessell et al. 1978). Whether the cells of origin of these fibres are the same as those of the strionigral projection or rather constitute a

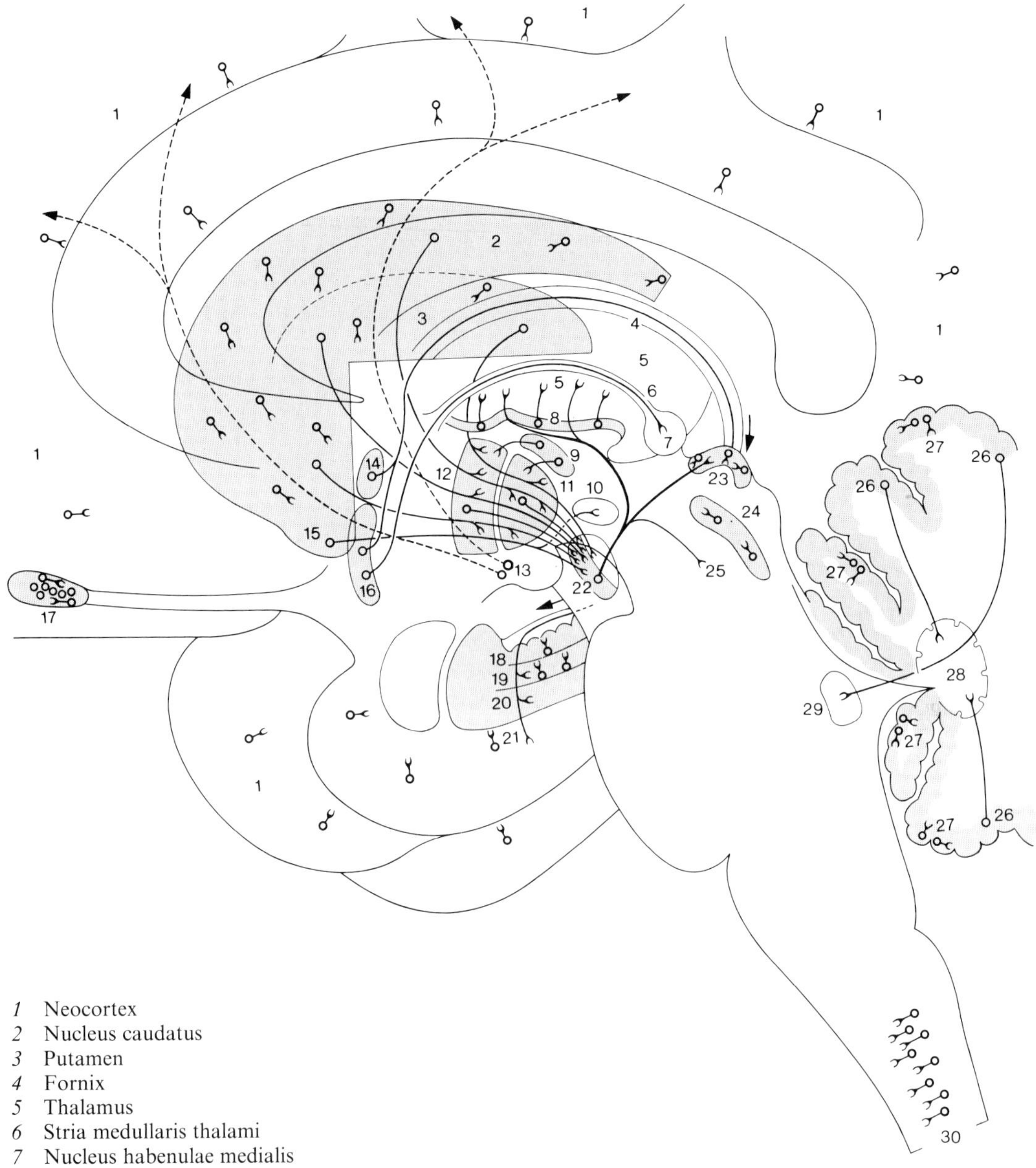

1 Neocortex
2 Nucleus caudatus
3 Putamen
4 Fornix
5 Thalamus
6 Stria medullaris thalami
7 Nucleus habenulae medialis
8 Nucleus reticularis thalami
9 Nucleus subthalamicus
10 Area tegmentalis ventralis
11 Globus pallidus, pars medialis
12 Globus pallidus, pars lateralis
13 GABA-containing cell groups in caudal hypothalamus
14 Nucleus septi medialis
15 Nucleus accumbens
16 Nucleus gyri diagonalis
17 Bulbus olfactorius
18 Fascia dentata
19 Cornu Ammonis
20 Subiculum
21 Cortex entorhinalis
22 Substantia nigra
23 Colliculus superior
24 Nucleus raphes dorsalis
25 Tegmentum mesencephali
26 Purkinje cells
27 Golgi, stellate and basket cells
28 Nuclei centrales cerebelli
29 Nucleus vestibularis lateralis
30 Medulla spinalis

Fig. 13. GABA-containing cell groups and fibres

separate neuron population remains to be determined. In addition to the strionigral and striopallidal projections, GABA has been shown to be the neurotransmitter of a pallidonigral pathway (Fonnum et al. 1978; Jessell et al. 1978).

In the nucleus caudatus and the putamen GABA-accumulating local circuit neurons are present (McGeer and McGeer 1975). These elements, which comprise up to 15% of the total population of striatal neurons, belong to a category characterized as 'medium-sized aspiny' (Bolam et al. 1983a). The nucleus accumbens, which according to Heimer and Wilson (1975) constitutes the ventral part of the striatum, contains GABA-ergic neurons which project to the rostromedial part of the substantia nigra and possibly also to the ventral tegmental area (Walaas and Fonnum 1980a; Strahlendorf and Barnes 1983). The pars reticulata of the substantia nigra contains numerous GABA-synthesizing neurons, which project massively to the ventrolateral and intralaminar thalamic nuclei, to the intermediate and deep layers of the superior colliculus and, diffusely, to the pontomesencephalic tegmental region (Vincent et al. 1978; Di Chiara et al. 1979; Chevalier et al. 1981; Childs and Gale 1983). There is evidence suggesting that GABA is a neurotransmitter in the subthalamo-pallidal pathway (Nauta and Cuenod 1982; Rouzaire-Dubois et al. 1983).

The caudal part of the hypothalamus contains some groups of large neurons (Nagai et al. 1983b; Vincent et al. 1983a) which have been reported to project diffusely to the neocortex (Vincent et al. 1983a). GABA plays a role in regulating the release of several hypophysiotropic peptides and anterior pituitary hormones, including luteinizing hormone and prolactin. However, the neuronal organization of these control mechanisms is not known as yet (Racagni et al. 1982; Lamberts et al. 1983; DeFeudis 1984).

The septum and the nucleus of the diagonal band of Broca are rich in GABA-containing neurons (Köhler and Chan-Palay 1983). A certain proportion of the GABA-ergic cells present in the medial septal nucleus and in the nucleus of the diagonal band project to the hippocampal region and to the entorhinal area (Köhler et al. 1984a). The medial habenular nucleus has also been reported to receive a GABA-ergic input from the nucleus of the diagonal band (Contestabile and Fonnum 1983).

A certain proportion of two types of interneurons present in the olfactory bulb, namely the granule cells and the periglomerular cells, use GABA as a neurotransmitter (Ribak et al. 1977; Halasz et al. 1979; Mugnaini et al. 1984a).

There is strong evidence that GABA is concentrated within the basket cells and other inhibitory intrinsic elements of the hippocampal region (Storm-Mathisen and Fonnum 1971; Storm-Mathisen 1972; Okada and Shimada 1975; Storm-Mathisen et al. 1983), and the same holds true for the neocortex (Ribak 1978; Somogyi et al. 1981; Hendry and Jones 1981).

Addenda:

1. Nagai et al. (1983b) studied the neuronal distribution of GABA-transaminase (GABA-T), i.e. the enzyme which metabolizes GABA, with a histochemical technique in the rat. The stained cells included numerous neuronal groups previously reported to be GABA-ergic on the basis of GAD-immunohistochemistry. In addition, a number of GABA-T-intense cells were found which had not previously been associated with GABA metabolism. These possible GABA-ergic cells include neurons in the following centres: the mesencephalic reticular formation, the rostral aspect of the periaqueductal grey, various thalamic nuclei including the periventricular, intralaminar, rhomboid and subparafascicular, the central, medial and basal nuclei of the amygdala, and the olfactory tubercle. In the thalamus a highly organized pattern of GABA-T staining was found. In this part of the brain GABA-T-positive neurons appeared to form two concentric shells with a dense zone of GABA-T-positive neuropil between.

2. Mogenson and Nielsen (1983) showed that the projection from the nucleus accumbens to the ventral pallidum contains a major GABA-ergic component, and they provided pharmacological evidence indicating that this GABA-ergic projection contributes to locomotor activity.
3. Oertel and Mugnaini (1984) studied the localization of GABA-ergic neurons in the rat basal ganglia by GAD immunohistochemistry. In addition to the various cells already discussed, they noted the presence of GAD-immunoreactive cells in the olfactory tubercle, the ventral pallidum and the central amygdaloid nucleus. The authors stated that the GABA-containing projection neurons in the caudatoputamen belong to the category of medium-sized spiny cells, and they suggested on the basis of parallels in GAD immunohistochemical features and dopaminergic innervation that the central amygdaloid nucleus constitutes a 'striatal' type nucleus of the amygdaloid complex.
4. Aronin et al. (1984) studied the localization of immunoreactive GAD and enkephalin-like immunoreactivity in the caudate nucleus of the rat. Co-localization was found in numerous caudate neurons of medium size. Quantitative analysis revealed that GAD and enkephalin-like immunoreactivity co-exist in about one-half of the caudate cell population containing each of the two substances.
5. Using the GABA-T technique mentioned under point 1 above in combination with retrograde tracing by HRP, Araki et al. (1984) demonstrated that in the rat, GABA-T-positive cells situated in the rostral entopeduncular nucleus and the lateral hypothalamus project to the lateral habenular nucleus.

Glutamate and Aspartate

There is strong electrophysiological and biochemical evidence that the amino acids glutamate (GLU) and aspartate (ASP) are excitatory transmitter substances (Fig. 12) used by a large number of neurons in the central nervous system (for review, see Fagg and Foster 1983; Fonnum 1984). For the mapping of glutamatergic and aspartatergic pathways several techniques have been employed; an autoradiographic technique using retrograde labelling of neurons by tritiated D-aspartate has been worked out (Wiklund et al. 1982), and results of immunohistochemical studies done with the aid of antibodies raised against glutamate coupled to bovine serum albumin (Storm-Mathisen et al. 1983) and against aspartate aminotransferase have been reported (Altschuler et al. 1981). However, the main method employed so far for the localization of these putative neurotransmitters has been to record the high-affinity uptake of radiolabelled GLU and ASP in the synaptosomal fractions of homogenized tissue samples, and to study the effect of lesions of presumed glutamatergic or aspartatergic pathways on this uptake. Lesions of specific pathways are often followed by a reduction of 70%–90% in the uptake of GLU or ASP (Fonnum 1984). These two amino acids probably act as neurotransmitters at separate receptor sites in the central nervous system (Fagg and Foster 1983). However, because the number of studies which discriminate between GLU and ASP in defined pathways is very limited, they are considered together in the following survey (Fig. 14).

Neocortical pyramidal cells contain much glutamate (Storm-Mathisen et al. 1983), and there is evidence suggesting that the following corticofugal projections use this amino acid as a neurotransmitter:

1. The corticostriatal projection, originating from the entire neocortex and terminating throughout the ipsilateral nucleus caudatus and putamen (Divac et al. 1977; McGeer et al. 1977; Fonnum et al. 1981); a limited number of fibres also pass from the frontal cortex to the nucleus accumbens (Walaas 1981)
2. Corticothalamic fibres originating from the whole extent of the ipsilateral neocortex and passing to various parts of the thalamus, including the medial nucleus, the ventrobasal complex, the reticular nucleus and the dorsal

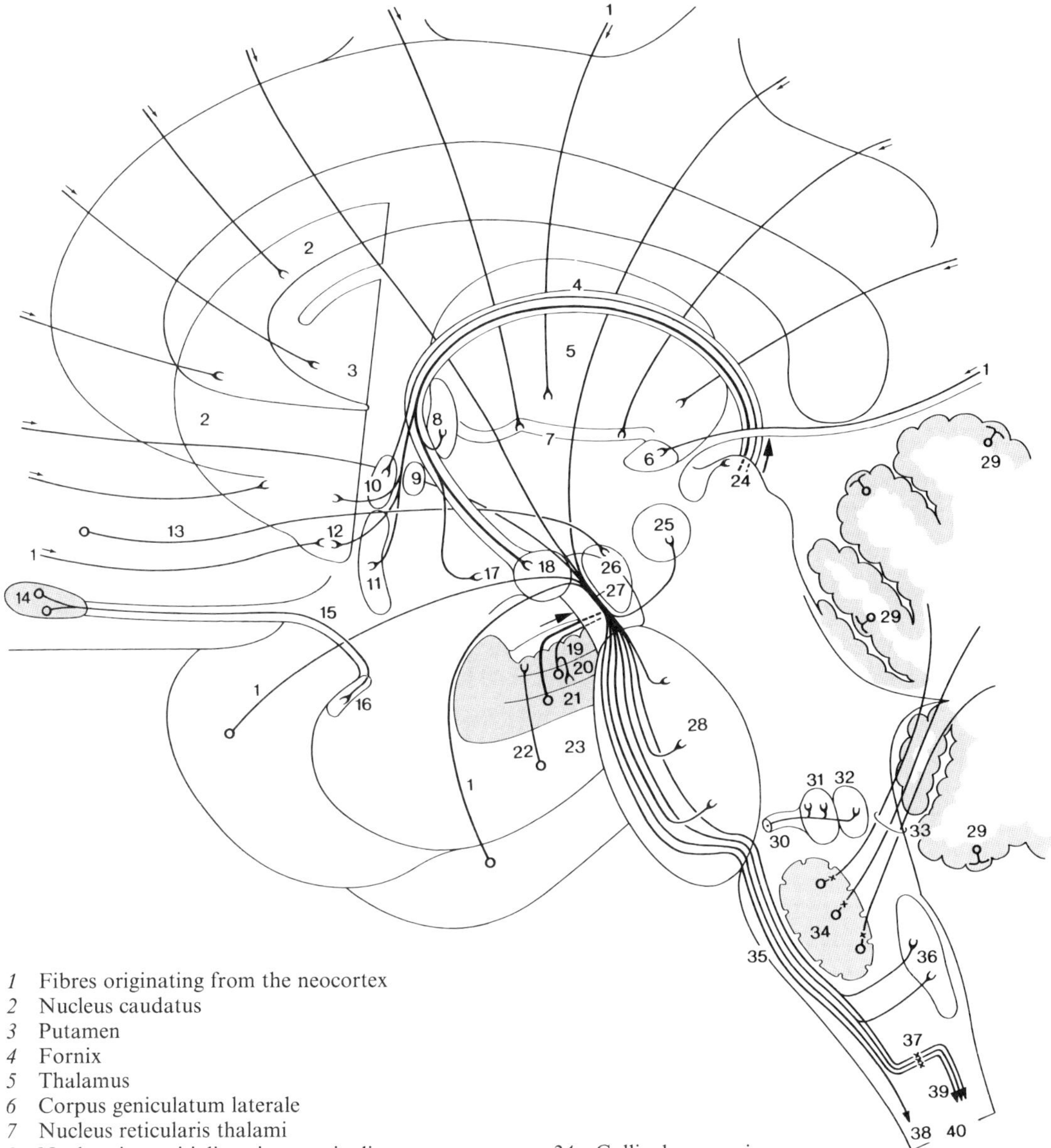

1 Fibres originating from the neocortex
2 Nucleus caudatus
3 Putamen
4 Fornix
5 Thalamus
6 Corpus geniculatum laterale
7 Nucleus reticularis thalami
8 Nucleus interstitialis striae terminalis
9 Commissura anterior
10 Nucleus septi lateralis
11 Area diagonalis
12 Nucleus accumbens
13 Fibres from the rostral and medial parts of the prefrontal cortex, passing to the substantia nigra
14 Bulbus olfactorius
15 Stria olfactoria lateralis
16 Cortex prepiriformis
17 Mediobasal hypothalamus
18 Corpus mamillare
19 Fascia dentata
20 Cornu Ammonis
21 Subiculum
22 'Perforant path'
23 Gyrus parahippocampalis
24 Colliculus superior
25 Nucleus ruber
26 Substantia nigra
27 Pedunculus cerebri
28 Pons
29 Granule cells, giving rise to parallel fibres
30 Nervus cochlearis (VIII)
31 Nucleus cochlearis ventralis
32 Nucleus cochlearis dorsalis
33 Tractus olivocerebellaris
34 Nucleus olivaris inferior
35 Pyramis
36 Nucleus cuneatus medialis (+ nucleus gracilis)
37 Decussatio pyramidum
38 Tractus pyramidalis anterior
39 Tractus pyramidalis lateralis
40 Medulla spinalis

Fig. 14. Glutamate- (and aspartate-) containing cells and pathways

part of the lateral geniculate body (Lund-Karlesen and Fonnum 1978; Fonnum et al. 1981; Young et al. 1981; Rustioni et al. 1982)
3. Corticotectal fibres passing from the visual cortex to the superior colliculus (Lund-Karlesen and Fonnum 1978)
4. Fibres passing from the rostral and medial parts of the prefrontal area to the substantia nigra (Carter 1982)
5. Corticopontine fibres (Thangnipon and Storm-Mathisen 1981; Carter 1982)
6. Fibres arising from the sensorimotor cortex, pursuing the long corticofugal projection and terminating in the red nucleus, the cuneate nucleus and the cervical and lumbar regions of the spinal cord (Young et al. 1981)

The following pathways related to the hippocampus have been reported to contain glutamatergic and/or aspartatergic fibres:

1. The so-called perforant path which, originating from the lateral part of the entorhinal cortex, traverses the subiculum and terminates in the cornu ammonis and the fascia dentata (Nadler et al. 1976, 1978; Fonnum et al. 1979)
2. Fibres originating from the cornu ammonis, which project, via the fornix, bilaterally to the lateral septal nucleus (Fonnum and Walaas 1978; Storm-Mathisen and Woxen-Opsahl 1978; Zaczek et al. 1979; Walaas and Fonnum 1980b)
3. Fibres presumably arising mainly from the subiculum which, after having pursued the fornix, terminate in the bed nucleus of the stria terminalis, the nucleus of the diagonal band of Broca, the nucleus accumbens, the ventral part of the nucleus caudatus and the putamen, the mediobasal hypothalamus and the mamillary body (Walaas 1981; Storm-Mathisen and Woxen-Opsahl 1978; Walaas and Fonnum 1980b)

The terminals of the lateral olfactory tract, which passes from the olfactory bulb to the prepiriform cortex, have been suggested to utilize GLU and/or ASP as their neurotransmitters (Harvey et al. 1975; Collins 1979; Collins et al. 1981).

In the cerebellar cortex two major excitatory inputs converge upon the dendritic trees of the Purkinje cells, i.e. the parallel fibres arising from the cerebellar granule cells and the climbing fibres derived from the inferior olive. It is well established that GLU is the neurotransmitter released by the parallel fibre boutons (Young et al. 1974; McBride et al. 1976a, b; Sandoval et al. 1978), and there is also evidence suggesting that ASP is the transmitter of the climbing fibres (Wiklund et al. 1982).

The auditory nerve consists of fibres which peripherally contact the hair cells in the organ of Corti and centrally terminate in the ventral and dorsal cochlear nuclei. Available evidence suggests that at least a certain proportion of the auditory nerve fibres utilize GLU and/or ASP as their transmitter(s) (Wenthold and Gulley 1977; Wenthold 1978; Altschuler et al. 1981).

It has been suggested that GLU and/or ASP may be the transmitter(s) of large, myelinated, non-nociceptive primary afferent fibres entering the central nervous system via the trigeminal nerve and the spinal dorsal roots (for review, see Salt and Hill 1983). However, the role of these amino acids in primary afferent transmission has been qualified by others (Fagg and Foster (1983) as inadequately substantiated.

Addenda:
1. Either GLU or ASP has tentatively been identified as the neurotransmitter of two local projections from hippocampal pyramidal cells, namely from fields CA3 to CA1 and from CA4 to the fascia dentata (Aamodt et al. 1984).
2. Davies et al. (1984) provided evidence indicating that in the rat the nucleus basalis/substantia innominata region possesses a substantial glutamatergic innervation. They demonstrated that this innervation does not arise within the ipsilateral frontoparietal cortex. This finding does not preclude the possibility of a glutamatergic cortical projection

to the nucleus basalis/substantia innominata. Mesulam and Mufson (1984) demonstrated that in the rhesus monkey the nucleus basalis does receive a cortical input, which, however, arises mainly from limbic and paralimbic areas.

3. Some cerebellorubral and cerebellothalamic fibres probably use GLU as their neurotransmitter (Nieoullon et al. 1984).

Glycine

Glycine, the simplest of all amino acids (Fig. 12), is well established as a major inhibitory transmitter in the spinal cord and lower brain stem (for review, see Pycock and Kerwin 1981; Fagg and Foster 1983), where it is released mainly by local circuit neurons (Curtis et al. 1968; Ljungdahl and Hökfelt 1973). It has been suggested that glycinergic interneurons also occur in two 'extrapyramidal' centres: the substantia nigra and the neostriatum (nucleus caudatus + putamen). In the substantia nigra these neurons would exert a tonic inhibitory influence on the cells of origin of the nigrostriatal dopaminergic pathway, and the glycinergic elements would in turn be inhibited by GABA-containing axons descending from the striatum (Pycock and Kerwin 1981). In the striatum the glycinergic local circuit neurons would be interposed in an analogous way between GABA- and dopamine-containing elements (Pycock and Kerwin 1981). Finally, it should be mentioned that several authors (Ohta and Oomura 1979; McGeer and Singh 1980; Kita and Oomura 1982) have suggested the existence of a glycinergic cortico-hypothalamic tract. This inhibitory pathway has been reported to originate from the prefrontal cortex, in particular from area 10, and to terminate in the lateral hypothalamic area (Kita and Oomura 1982) and in the ventromedial hypothalamic nucleus (Ohta and Oomura 1979).

Taurine

The sulphur-containing amino acid taurine (Fig. 12) is widely but heterogeneously distributed in the mammalian central nervous system. It is abundant in the cerebral cortex and the cerebellum, while its concentration is rather low in the pons, the medulla oblongata and the spinal cord (Lombardini 1976).

Electrophysiological experiments (e.g. Curtis and Watkins 1960; Haas and Hösli 1973; Chan-Palay et al. 1982c; Okamoto et al. 1983) have shown that taurine is able to exert a prominent inhibitory influence on neurons in several parts of the central nervous system. However, it is unclear at present whether this amino acid serves as a true neurotransmitter or instead modulates the excitability of neuronal membranes and/or affects the release of other neuroactive substances, and thus may be more appropriately designated as a neuromodulator (Kuriyama et al. 1983).

Chan-Palay et al. (1982a, b) studied the distribution of taurine in the cerebellum of several mammals by autoradiography after in vivo injections of tritiated taurine directly in the cerebellum, and by immunohistochemistry with antibodies against cysteine-sulphinic acid decarboxylase (CSADCase), the enzyme responsible for taurine synthesis in the brain. Uptake and sequestration of tritiated taurine labelled numerous Purkinje cells, many granule cells and, moreover, basket, stellate and Golgi cells, as well as large projection neurons in the deep cerebellar nuclei. Immunocytochemical labelling with antibodies against CSADCase showed a similar distribution. It was also found that the labelled neurons in the cerebellar cortex are arranged in longitudinally oriented microzones. The findings of Chan-Palay et al. (1982a, b; see also Chan-Palay 1984), summarized above, call for four brief comments: (a) The indication that cerebellar stellate cells are taurinergic is in accord with the results of biochemical (Nadi et al. 1977) and neurophysiological studies (Okamoto et al.1983). (b) The indication that Purkinje, Golgi and basket cells contain taurine is surprising, because these elements are generally considered to be GABA-ergic (Schulman 1983; Ito 1984; Ottersen and Storm-Mathisen 1984). However, Purkinje, Golgi and basket cells are known to have inhibitory roles (Eccles et al. 1967; Ito 1984), and GABA and taurine are both inhibitory neuromediators. (c) The finding that taurine may be present in granule cells and in deep cerebellar projection neurons – elements which are generally considered to be excitatory – is puzzling. (d) In general, it may be stated that the remarkable results of Palay et al. (1982a, b) on the cellular localization of taurine in the cerebellum require further substantiation. Confirmation is required that taurine is present within the cells under consideration in normal physiological conditions, and that it acts at these sites as a true neuromediator. When these two points have been unequivocally established, but not before, it would be reasonable to conduct a full re-examination of the roles of the various neuronal elements in the cerebellar circuits, as suggested by Chan-Palay (1984).

Gut-Brain Peptides

Substance P

Substance P (SP) was discovered more than 50 years ago in extracts of brain and intestine. In the early 1970s it was identified as an undecapeptide and its structure was determined (Fig. 15). Shortly thereafter, synthetic SP was prepared, permitting the study of its physiological and pharmacological properties and the development of histochemical techniques for its localization. In the central nervous system SP has a slowly developing but long-lasting potent excitatory effect on neurons (Otsuka and Konishi 1983). Our knowledge concerning its central distribution may be summarized as follows (Fig. 16):

About 20% of the cell bodies in the spinal dorsal root ganglia contain SP. These cells have small somas and small, unmyelinated and finely myelinated axons (Hökfelt et al. 1975c, 1977a; De Lanerolle and LaMotte 1983). Their peripheral processes have been found in the epidermis and in the walls of blood vessels and glands; their central processes project primarily to the superficial layers of the dorsal horn (laminae I and II, outer zone) (Ljungdahl et al. 1978; DiFiglia et al. 1982b; Charnay et al. 1983). Similarly, the trigeminal ganglion contains a considerable number of SP-immunoreactive cells which are connected to unmyelinated fibres (Lehtosalo et al. 1984). These elements project to the caudal part of the spinal nucleus of the trigeminal nerve (Cuello and Kanazawa 1978; Cuello et al. 1978; DelFiacco and Cuello 1980), and primary afferents containing the same peptide enter the central nervous system by way of the VIIth, IXth and Xth nerves to terminate in the nucleus of the solitary tract (Gillis et al. 1980). The SP-containing spinal and trigeminal primary afferent neurons are probably involved in pain perception (for review, see Nicoll et al. 1980), whereas those terminating in the nucleus of the solitary tract have been suggested to convey baroreceptive and chemoreceptive information (Gillis et al. 1980).

Three streams of SP-containing fibres descend from the brain stem to the spinal cord, one originating from the nucleus of Edinger-Westphal, a second arising from the mesencephalic periaqueductal grey and a third emanating from the nucleus raphes magnus. The cells of origin of the first of these projections are numerous and located throughout the length of the Edinger-Westphal nucleus. The efferents of these elements extend caudally as far as the lumbar level of the cord, but their site of termination remains to be determined (Phipps et al. 1983). The fibres descending from the periaqueductal grey contain, in addition to SP, a second neuroactive compound, CCK (Skirboll et al. 1982). Two different neuroactive principles are also found in the raphespinal fibres originating from the nucleus raphes magnus. In these fibres SP appears to co-exist with serotonin (Chan-Palay et al. 1978; Hökfelt et al. 1978c). They project to the ventral and probably also to the dorsal horn of the spinal cord (Gilbert et al. 1982).

SP-containing fibres have been shown to project to the nucleus raphes magnus from a

Gut-brain peptides

Substance P

H-Arg^{1}-Pro-Lys-Pro-Gln-Gln-Phe-Phe-Gly-Leu-Met^{11}-NH_2

Vasoactive intestinal peptide [VIP] (human, bovine, porcine)

H-His^{1}-Ser-Asp-Ala-Val-Phe-Thr-Asp-Asn-Tyr^{10}-Thr-Arg-Leu-Arg-Lys-Gln-Met-Ala-Val-Lys^{20}-Lys-Tyr-Leu-Asn-Ser-Ile-Leu-Asn^{28}-NH_2

Cholecystokinin [CCK]

CCK-8

H-Asp^{1}-Tyr (SO_3H)-Met-Gly-Trp-Met-Asp-Phe^{8}-NH_2

CCK-33 (porcine)

H-Lys^{1}-Ala-Pro-Ser-Gly-Arg-Val-Ser-Met-Ile^{10}-Lys-Asn-Leu-Gln-Ser-Leu-Asp-Pro-Ser-His^{20}-Arg-Ile-Ser-Asp-Arg-Asp-Tyr-(SO_3H)-Met-Gly-Trp^{30}-Met-Asp-Phe^{33}-NH_2

CCK-39 (porcine)

H-Tyr-Ile-Gln-Gln-Ala-Arg- etc.

Neurotensin

pGlu^{1}-Leu-Tyr-Glu-Asn-Lys-Pro-Arg-Arg-Pro-Tyr-Ile-Leu^{13}-OH

Avian pancreatic polypeptide [APP] (= human PP)

H-Ala^{1}-Pro-Leu-Glu-Pro-Val-Tyr-Pro-Gly-Asp^{10}-Asn-Ala-Thr-Pro-Glu-Gln-Met-Ala-Gln-Tyr^{20}-Ala-Ala-Asp-Leu-Arg-Arg-Tyr-Ile-Asn-Met^{30}-Leu-Thr-Arg-Pro-Arg-Tyr^{36}-NH_2

Fig. 15. Primary structure of some gut-brain peptides. For significance of the amino acid code, see Table 7 on p. 198

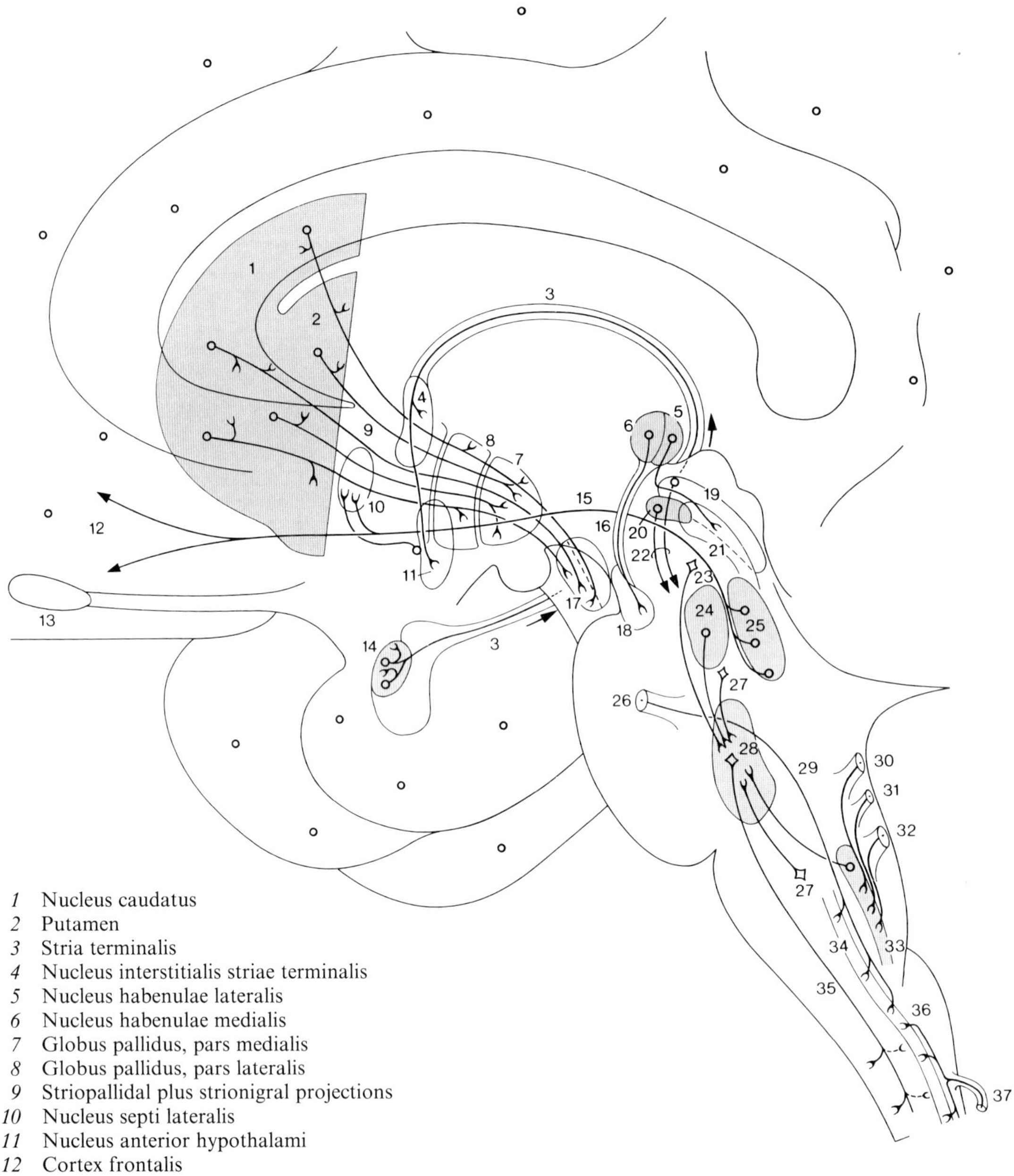

1 Nucleus caudatus
2 Putamen
3 Stria terminalis
4 Nucleus interstitialis striae terminalis
5 Nucleus habenulae lateralis
6 Nucleus habenulae medialis
7 Globus pallidus, pars medialis
8 Globus pallidus, pars lateralis
9 Striopallidal plus strionigral projections
10 Nucleus septi lateralis
11 Nucleus anterior hypothalami
12 Cortex frontalis
13 Bulbus olfactorius
14 Corpus amygdaloideum, pars medialis
15 Fasciculus telencephalicus medialis
16 Tractus habenulointerpeduncularis
17 Substantia nigra, pars reticulata
18 Nucleus interpeduncularis
19 Griseum centrale mesencephali
20 Nucleus accessorius nervi oculomotorii (Edinger-Westphal)
21 Nucleus raphes dorsalis
22 Spinal projections originating from 19 and 20
23 Nucleus cuneiformis
24 Nucleus centralis superior
25 Area tegmentalis dorsolateralis
26 Radix sensoria nervi trigemini
27 Formatio reticularis rhombencephali
28 Nucleus raphes magnus
29 Tractus spinalis nervi trigemini
30 Nervus facialis
31 Nervus glossopharyngeus
32 Nervus vagus
33 Nucleus solitarius
34 Nucleus spinalis nervi trigemini, pars caudalis
35 Raphespinal projection
36 Substantia gelatinosa
37 Radix dorsalis nervi spinalis

Fig. 16. Substance P-containing cell groups and fibres

considerable number of brain stem nuclei, including the mesencephalic and rhombencephalic reticular formation, the central superior nucleus and the nucleus solitarius (Beitz 1982b).

SP-containing neurons are found in both the medial and lateral habenular nuclei. The fibres originating from the medial nucleus descend with the tractus habenulointerpeduncularis to the interpeduncular nucleus (Hökfelt et al. 1975c; Hong et al. 1976; Mroz et al. 1976), while those from the lateral nucleus project to the nucleus raphes dorsalis (Neckers et al. 1979; Vincent et al. 1980). (For the pattern of distribution of the SP-containing fibres within the various subnuclei of the interpeduncular nucleus, see Hamill et al. 1984.)

The periventricular grey of the rostral rhombencephalon in rodents contains a fairly distinct cell group, known as the nucleus laterodorsalis tegmenti of Castaldi. In this cell mass, SP-containing neurons are found which give rise to a long, ascending pathway. Via the medial forebrain bundle, the fibres of this pathway project to the lateral septal nucleus and to the medial parts of the frontal cortex (Paxinos et al. 1978; Sakanaka et al. 1981b, 1982b, 1983). The projection to the lateral septal nucleus is augmented by a number of SP-containing fibres originating from the area situated between the anterior hypothalamic nucleus and the lateral hypothalamic region (Sakanaka 1982).

In the caudate nucleus and the putamen a population of medium-sized SP-containing cells has been found (Bolam et al. 1983b). These elements contribute substantially to the large strionigral projection (Kanazawa et al. 1977; Hong et al. 1977; Gale et al. 1977; Brownstein et al. 1977; Staines et al. 1980; Pioro et al. 1984). The SP-containing strionigral fibres issue collaterals within the neostriatum itself and terminate in the pars reticulata of the substantia nigra, which shows the highest SP concentration of the entire brain (Mroz et al. 1977). The concentration of this peptide in the substantia nigra is markedly reduced in patients suffering from Huntington's chorea (Buck et al. 1981) and in parkinsonian patients (Mauborgne et al. 1983; Tenovuo et al. 1984). However, the mechanism of this reduction might be quite different in these two diseases. In Huntington's chorea the reduction is presumably associated with neuronal degeneration in the nucleus caudatus and putamen, but in Parkinson's disease the striatum is considered to be intact. Here, the reduction in SP content in the substantia nigra might well be a metabolic consequence of the loss of nigrostriatal dopaminergic neurons (Mauborgne et al. 1983).

It has been reported that in the rat the SP-containing fibres arise predominantly from the rostral part of the caudatoputamen complex (Gale et al. 1977; Jessell et al. 1978). Recently, Kohno et al. (1984) provided experimental immunohistochemical evidence suggesting that in the rat the SP-containing cells present in various parts of the caudatoputamen (CP) project differently to the substantia nigra (SN). According to their observations, SP-positive neurons in the posterior portion of the CP project to the SN pars lateralis, and SP-positive cells in the lateroventral part of the anterior portion of the CP extend to the SN pars compacta and pars reticulata, but SP-positive cells in the dorsal part of the anterior portion of the CP do not innervate the SN.

A plexus of SP-containing fibres is present in both parts of the globus pallidus; however, the density of the plexus is much higher in the medial part than in the lateral part (Paxinos et al. 1978; Staines et al. 1980; Haber and Elde 1981; Rønnekleiv et al. 1984). There is evidence suggesting that the plexus in the lateral pallidal part is constituted by collaterals of the strionigral fibres, but that the plexus in the medial part consists of the terminal segments of a separate striopallidal projection, originating mainly from the rostromedial striatum (Kanazawa et al. 1980; Mauborgne et al. 1983).

The medial part of the amygdaloid complex contains a number of SP-positive cells (Sakanaka et al. 1981a). These elements give rise

to a dense intrinsic plexus in the amygdala (Emson et al. 1978b), but they also project via the stria terminalis to the bed nucleus of that bundle and to the ventrolateral part of the anterior hypothalamic nucleus (Sakanaka et al. 1981a).
The olfactory bulb contains SP-positive neurons; these elements are situated in the glomerular layers and have been identified as external tufted cells (Burd et al. 1982a).
Substance P-immunoreactive neurons have been observed in several neocortical areas in the baboon. These elements are mostly small in size and vary in shape. They occur in laminae III–VI but are most common in laminae V and VI (Beach and McGeer 1983).

Addenda:
1. Several important findings of Cuello and Kanazawa (1978) have been unjustly omitted from the preceding text and from Fig. 16: (a) The nucleus interpeduncularis contains a large number of SP-immunoreactive cell bodies. (b) Certain cell masses and fibre systems contain SP-positive fibres and/or terminals – the rhombencephalon: periventricular zone, locus coeruleus, parabrachial nuclei, nucleus sensorius principalis of the trigeminal nerve, nucleus mesencephalicus of the same nerve (densely innervated), reticular formation, nucleus olivaris superior, pontine nuclei (sparsely innervated), nucleus facialis and nucleus ambiguus; the mesencephalon: superior colliculus (particularly the stratum griseum profundum), periaqueductal grey, medial lemniscus (limited number of fibres), and area tegmentalis ventralis; the diencephalon: lateral habenular nucleus, nucleus anterior dorsalis thalami, nucleus periventricularis thalami, nucleus ventralis thalami, corpus geniculatum laterale, subthalamic nucleus, periventricular hypothalamic zone, nucleus supraopticus, nucleus paraventricularis, nucleus dorsomedialis hypothalami, nucleus arcuatus, area hypothalamica lateralis and medial forebrain bundle; the telencephalon: nucleus preopticus periventricularis, nucleus preopticus medialis, tuberculum olfactorium and, in the amygdaloid complex, the nuclei medialis and corticalis (both densely innervated) and the nucleus centralis.
2. DelFiacco et al. (1984) studied the localization of substance P-immunoreactive structures in the human brain stem. Perikarya, fibres and terminals containing this peptide showed a location similar to that described in other mammals. However, some differences were detectable – for instance, the presence of a dense plexus of SP-immunoreactive fibres, the cuneate fasciculus. The trajectory of the fasciculus longitudinalis dorsalis of Schütz and the mesencephalic central grey appeared to contain numerous SP-positive fibres and/or terminals.
3. Substance P-immunoreactive fibres, originating from an as yet undetermined population of cells situated in the ventral part of the rhombencephalic reticular formation, join the lateral lemniscus. A certain proportion of these fibres terminate in the lateral parabrachial nucleus (Takatsuki et al. 1983).
4. The lateral habenular nucleus contains a dense plexus of SP-immunoreactive fibres which can be divided into two parts: medial and lateral. The fibres in the lateral part originate from cells situated in the rostral entopeduncular nucleus and adjacent areas, while those in the medial part arise from the medial habenular nucleus (Shinoda et al. 1984).
5. Vincent et al. (1983c) demonstrated that a certain proportion of the fibres present in the cholinergic dorsal tegmental pathway described by Shute and Lewis (1967; cf. Fig. 2) project directly to the frontal cortex; they also provided evidence indicating that in a subpopulation of the cells of origin of these fibres (which are situated mainly within the nucleus laterodorsalis tegmenti; cf. Fig. 16) acetylcholine co-exists with substance P.
6. Rønnekleiv et al. (1984) studied the distribution of SP-containing cell bodies and fibres in the hypothalamus and pituitary gland of the rhesus monkey, using an immunohistochemical technique. Their results may be summarized as follows: SP cell bodies are present in the infundibular nucleus, in the area lateral to that centre, and in the median

eminence. SP fibres are concentrated in the infundibular region and in the external zone of the median eminence. Furthermore, numerous SP fibres are present in the neural lobe of the pituitary gland.

7. An immunohistochemical study of the distribution of SP-containing neurons in the preoptic-hypothalamic region of the rat by Panula et al. (1984) revealed the presence of numerous SP-positive cells in the nucleus preopticus medialis and lateralis, nucleus anterior, nucleus ventromedialis and nucleus dorsomedialis.

8. Gall and Moore (1984) studied the distribution of substance P, enkephalin, tyrosine hydroxylase and serotonin immunoreactivity in the septal region of the rat. Numerous SP-positive cell bodies were found in the dorsal part of the lateral septal nucleus, whereas enkephalinergic cell bodies appeared to be concentrated in a central septal field. The immunoreactive structures studied showed a more or less distinct laminar arrangement. The dorsolateral SP-immunoreactive perikaryal field appeared to be overlapped by a density of serotonin-immunoreactive axons, while the group of enkephalin-immunoreactive perikarya was found to be co-extensive with a field of tyrosine hydroxylase-positive (i.e. catecholaminergic) fibres.

Vasoactive Intestinal Polypeptide

Vasoactive intestinal polypeptide (VIP) was first isolated from extracts of the porcine duodenum (Said and Mutt 1970a, b). It appears to be present in large quantities throughout the gastrointestinal tract. Apart from having a powerful vasodilatatory effect, it stimulates the conversion of glycogen into glucose, enhances lipolysis and insulin secretion, inhibits the production of gastric acid, and stimulates exocrine secretion by the pancreas and small intestine (Mutt and Said 1974). Four years after its isolation VIP was sequenced as a 28-amino acid polypeptide (Mutt and Said 1974; Fig. 15), and antibodies were raised against it to be used in radioimmunology and immunohistochemistry. With the help of these techniques it was established that this 'gut hormone' is present in the central nervous system, where it has been detected in highest concentrations in the cerebral cortex (Bryant et al. 1976; Larsson et al. 1976; Said and Rosenberg 1976; Fuxe et al. 1977). Although the exact role of VIP in the central nervous system is still unknown, physiological and biochemical, as well as pharmacological evidence indicates that this polypeptide may be an excitatory neurotransmitter and/or neuromodulator (Giachetti et al. 1977; Quik et al. 1978; Philips et al. 1978; Emson et al. 1978a; Dodd et al. 1979). Our probably as yet very fragmentary knowledge of the VIP-containing cell groups and their efferents is surveyed below (Fig. 17). Because VIP-containing cell bodies and fibres show a parallel distribution, the suggestion has been made that most of the VIP-immunoreactive elements represent local circuit neurons (Lorén et al. 1979b). However, there is experimental evidence indicating the presence of at least a few long VIP-containing projections (Roberts et al. 1980a, b; Marley et al. 1981).

VIP-immunoreactive cell bodies are found throughout the neocortex. Most of these elements are bipolar, with their long axis oriented perpendicular to the cortical surface (Fuxe et al. 1977; Lorén et al. 1979b; Sims et al. 1980; Morrison et al. 1984). VIP-containing fibres and terminals are also present in the neocortex. They are abundant in layers II–IV, but scattered in the remaining layers (Fuxe et al. 1977; Roberts et al. 1980a). This distribution pattern of fibres and terminals, the absence of VIP in the major cortical efferent projection systems, and the finding that cortical undercutting does not modify the VIP content of the cortex (Emson 1979a, b) all indicate that most of the VIP-positive neocortical neurons are intrinsic to the cortex. However, the presence of VIP-containing fibres in the corpus callosum suggests that some could be engaged in transcallosal cortico-cortical connections as well (Lorén et al. 1979b).

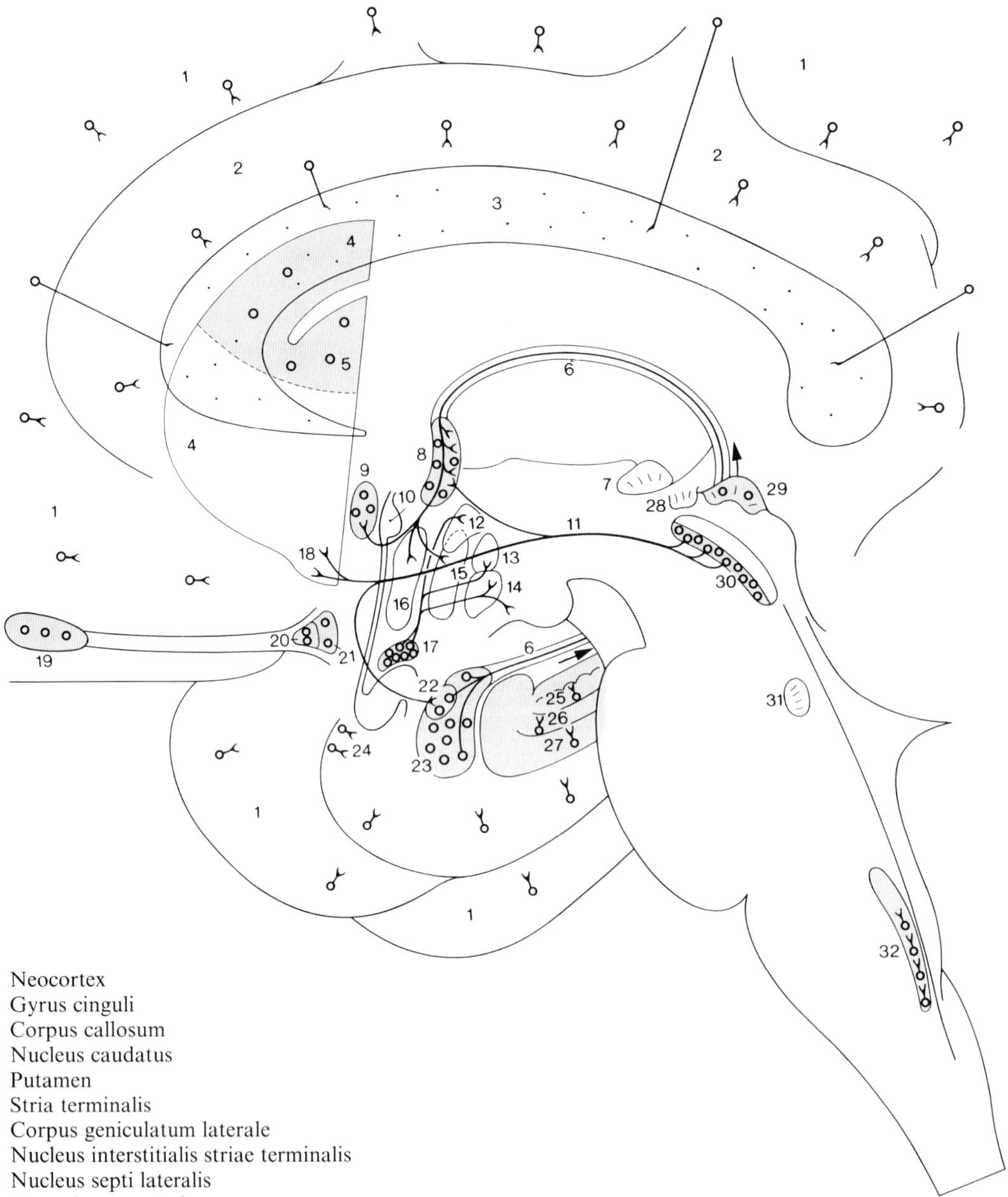

1 Neocortex
2 Gyrus cinguli
3 Corpus callosum
4 Nucleus caudatus
5 Putamen
6 Stria terminalis
7 Corpus geniculatum laterale
8 Nucleus interstitialis striae terminalis
9 Nucleus septi lateralis
10 Commissura anterior
11 Fasciculus telencephalicus medialis
12 Nucleus paraventricularis
13 Nucleus dorsomedialis
14 Nucleus ventromedialis
15 Nucleus anterior hypothalami
16 Regio preoptica
17 Nucleus suprachiasmaticus
18 Nucleus accumbens
19 Bulbus olfactorius
20 Nucleus olfactorius anterior
21 Tuberculum olfactorium
22 Nucleus centralis amygdalae
23 Corpus amygdaloideum
24 Cortex praepiriformis
25 Gyrus dentatus
26 Cornu Ammonis
27 Subiculum
28 Area pretectalis
29 Colliculus superior
30 Griseum centrale mesencephali
31 Nucleus parabrachialis lateralis
32 Nucleus solitarius

Fig. 17. Vasoactive intestinal polypeptide-containing cells and fibres

Morrison and colleagues (1984) thorougly studied the intracortical VIP-positive cells and their distribution pattern. Their analysis revealed that: (a) approximately 1% of the cortical neurons are VIP-positive, (b) their distribution is fairly uniform and statistically random, and (c) each VIP-containing cortical cell occupies a similarly sized radial unit, which is generally between 15 μm and 60 μm in diameter, and is contiguous with that of other VIP-containing cells in the neocortex. It is known that VIP stimulates the formation of cyclic AMP through specific membrane receptors (Quik et al. 1978), and that the resultant increase in cyclic AMP via some intermediate steps leads to the breakdown of glycogen, and in turn to an increase in the glucose available for the generation of energy. There is evidence indicating that VIP also elicits such a cyclic AMP-mediated glycogenolysis in the cortex (Magistretti et al. 1981; Magistretti and Schorderet 1984). Given the morphological characteristics and distribution of the VIP-containing cells, the suggestion has been made that VIP could locally regulate cortical energy metabolism within radially oriented cortical units (Morrison et al. 1984; Magistretti and Schorderet 1984).

VIP-containing cell bodies and terminal plexuses are also found in the prepiriform, cingulate and entorhinal cortical areas. Within these cortical areas most of the VIP-positive elements are concentrated in layer II (Lorén et al. 1979b; Sims et al. 1980).

In the various parts of the hippocampal formation, VIP-immunoreactive cell bodies and fibres have been observed. Most of these elements presumably represent local circuit neurons (Lorén et al. 1979b; Roberts et al. 1980; Léranth et al. 1984).

Several parts of the corpus striatum contain VIP-positive structures. Roberts and collaborators (1980a) reported the presence of a small number of VIP-containing perikarya in the dorsomedial part of the caudato-putamen and numerous VIP-positive fibres have been observed in the nucleus accumbens as well as in the so-called ventral pallidum (Lorén et al. 1979b).

Scattered VIP-containing perikarya have been reported to be present in most nuclei of the amygdaloid complex (Lorén et al. 1979b; Roberts et al. 1980a; Sims et al. 1980). There is experimental evidence indicating that VIP-immunoreactive cells situated in the central, lateral and basolateral amygdaloid nuclei project massively, via the stria terminalis, to the dorsocaudal portion of the bed nucleus of the stria terminalis and, less intensely, to the preoptic and anterior hypothalamic areas (Roberts et al. 1980a). It has been suggested that VIP-containing fibres arising from the stria terminalis also terminate in the lateral septal nucleus (Sims et al. 1980; Woodhams et al. 1983).

Several basal telencephalic structures, including the olfactory bulb, the anterior olfactory nucleus, the olfactory tubercle, the bed nucleus of the stria terminalis and the lateral septal nucleus, have been observed to contain VIP-positive cell bodies (Lorén et al. 1979b; Roberts et al. 1980a; Sims et al. 1980).

The suprachiasmatic nucleus contains a very dense concentration of VIP-immunoreactive perikarya (Sims et al. 1980; Stopa et al. 1984). Efferents from these VIP-containing elements course dorsally for some distance and then split into a moderately dense rostrodorsal component and a more diffuse caudal component. The rostrodorsally directed fibres presumably terminate in the paraventricular nucleus, whereas the caudal fibres distribute to the dorsomedial, ventromedial and premamillary nuclei.

The ventral portion of the mesencephalic central grey contains an extensive network of VIP-immunoreactive cell bodies and processes lying immediately adjacent to the ependymal cells lining the cerebral aqueduct (Marley et al. 1981). The superior colliculus has also been reported to contain some VIP-positive cell bodies (Lorén et al. 1979b; Roberts et al. 1980a; Sims et al. 1980). There is experimental evidence suggesting that at least a certain proportion of the VIP-positive cells in the periaqueductal grey project rostrally and attain, via the medial forebrain bundle, a number of forebrain areas, includ-

ing the central amygdaloid nucleus, the bed nucleus of the stria terminalis, the nucleus accumbens and the rostral portions of the hypothalamus (Marley et al. 1981).
The lateral geniculate body, the pretectum, the superior colliculus and the lateral parabrachial nucleus contain plexuses of VIP-positive fibres (Lorén et al. 1979b).
Numerous VIP-immunoreactive cells and fibres are found in the nucleus solitarius. These elements are concentrated in the medial and commissural portions of the nucleus, i.e. the area corresponding to the primary baroreceptive centre (Palkovits et al. 1982b). Small VIP-immunoreactive cells are present in the nodose ganglion (Lundberg et al. 1978), but it has been found that the concentration of VIP in the nucleus solitarius is lowered only insignificantly by uni- or bilateral transections of the solitary tract (Palkovits et al. 1982b). There is evidence suggesting that the majority of the VIP-positive elements in the nucleus solitarius represent local circuit neurons (Palkovits et al. 1982b).
VIP is contained in spinal primary afferent fibres which arise from small spinal ganglion cells (Lundberg et al. 1978; Kawatani et al. 1982, 1983; Anand et al. 1983). The great majority of these fibres enter the cord at the sacral level (Kawatani et al. 1982, 1983; Anand et al. 1983; Basbaum and Glazer 1983). They concentrate in the fasciculus dorsolateralis of Lissauer and distribute numerous collaterals to the superficial zone of the substantia gelatinosa. In longitudinal sections the VIP-containing collaterals and terminals exhibit a periodic organization, in which clusters occur at regular intervals along the length of the cord (Kawatani et al. 1982). Limited numbers of collaterals arising from the VIP-containing primary afferents penetrate into deeper parts of the spinal grey, as for example lamina V and the region around the central canal. Because the distribution pattern of the dorsal root fibres just discussed closely parallels that of the visceral primary afferent projection from the pelvic organs (Morgan et al. 1981), it is reasonable to assume that VIP is contained in part in (possibly nociceptive) visceral afferents (Basbaum and Glazer 1983).

Addendum: For a review of the neurobiological and neuroendocrine functions in which VIP is implicated, see Rostène (1984). Some interesting aspects are put forward in this review. (a) In contrast to what is observed in the rat, the human median eminence has been reported to contain high amounts of VIP. (b) High concentrations of VIP have been found in the hypophyseal portal blood, suggesting that the peptide, released into the portal blood, could reach the pituitary and affect hormonal secretion. (c) VIP is able to induce a dose-dependent inhibition of SST release, and this effect is obtained only in the mediobasal hypothalamus. (d) VIP is able to stimulate prolactin release by direct action on the pituitary. (e) The neural pathways along which VIP reaches the mediobasal hypothalamus and the median eminence remain to be determined.

Cholecystokinin

Certain cells in the glands of the stomach produce gastrin, a polypeptide hormone which stimulates the formation of gastric juice. Cholecystokinin is a related hormone which is synthesized in the duodenum and brings about secretion of pancreatic juice and the ejection of bile. In 1975 Vanderhaeghen and colleagues reported the occurrence of gastrin-like immunoreactivity in the central nervous system. Subsequent studies (Dockray 1976; Dockray et al. 1978; Muller et al. 1977; Rehfeld 1978a, b; Rehfeld et al. 1979) revealed that this immunoreactivity corresponds principally to the carboxyl-terminal octapeptide of cholecystokinin, which shares the same COOH-terminal pentapeptide with gastrin (Fig. 15). Because of this structural similarity between cholecystokinin and gastrin, most immunohistochemical techniques are unable to distinguish between these two substances. However, detailed immunohistochemical studies suggest that, whereas the terminal octapeptide of cholecystokinin oc-

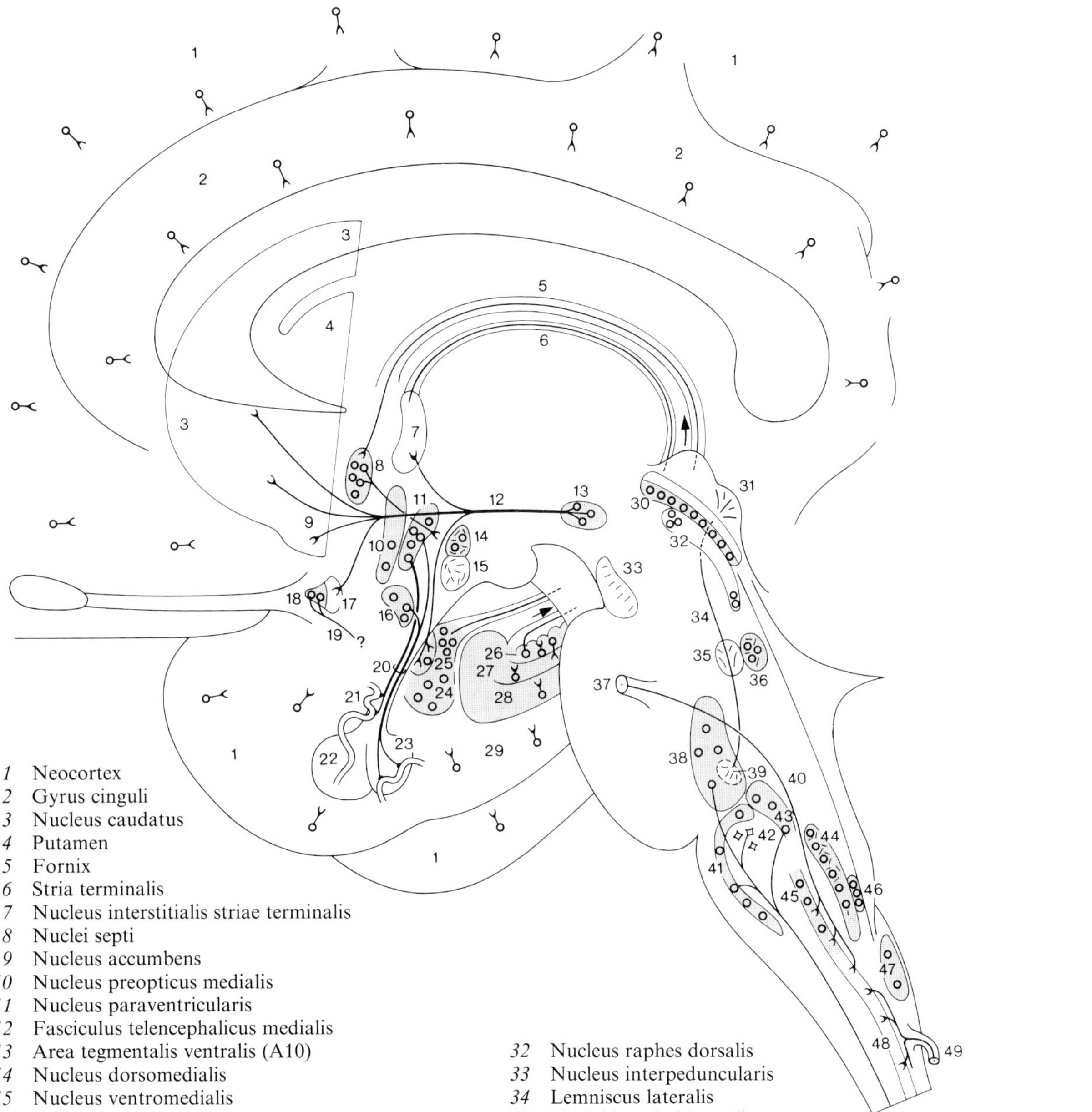

1 Neocortex
2 Gyrus cinguli
3 Nucleus caudatus
4 Putamen
5 Fornix
6 Stria terminalis
7 Nucleus interstitialis striae terminalis
8 Nuclei septi
9 Nucleus accumbens
10 Nucleus preopticus medialis
11 Nucleus paraventricularis
12 Fasciculus telencephalicus medialis
13 Area tegmentalis ventralis (A10)
14 Nucleus dorsomedialis
15 Nucleus ventromedialis
16 Nucleus supraopticus
17 Tuberculum olfactorium
18 Nucleus olfactorius anterior
19 Stria olfactoria lateralis
20 Tractus paraventriculo-supraoptico-hypophyseos
21 Eminentia mediana
22 Lobus anterior hypophyseos
23 Lobus posterior hypophyseos
24 Corpus amygdaloideum
25 Nucleus corticalis amygdalae
26 Fascia dentata
27 Cornu Ammonis
28 Subiculum
29 Gyrus parahippocampalis
30 Griseum centrale mesencephali
31 Colliculus inferior
32 Nucleus raphes dorsalis
33 Nucleus interpeduncularis
34 Lemniscus lateralis
35 Nuclei lemnisci lateralis
36 Nucleus parabrachialis lateralis
37 Nervus trigeminus
38 Nucleus raphes magnus
39 Oliva superior
40 Tractus spinalis nervi trigemini
41 Nucleus raphes pallidus
42 Nucleus reticularis gigantocellularis
43 Nucleus raphes obscurus
44 Nucleus solitarius
45 Substantia gelatinosa nuclei spinalis nervi trigemini, pars caudalis
46 Area postrema
47 Nucleus cuneatus medialis
48 Substantia gelatinosa spinalis
49 Radix dorsalis nervi spinalis

Fig. 18. Cholecystokinin-containing cell groups and fibres

curs throughout the central nervous system, gastrin-like peptides are found only in the pituitary and hypothalamus (Rehfeld 1978b; Rehfeld et al. 1979; Larsson and Rehfeld 1981). In this brief survey these substances will be taken together as cholecystokinin-like peptides (CCK). CCK-containing neurons and CCK-containing fibre networks have been demonstrated in many parts of the neuraxis, but the number of known CCK projections is very limited indeed (Fig. 18). Although several studies based upon physiological, pharmacological and biochemical approaches suggest that CCK may act as a neurotransmitter or neuromodulator (Pinget et al. 1979; Dodd and Kelly 1979; Emson et al. 1980b; Golterman et al. 1980; Rehfeld 1980; Phillis and Kirkpatrick 1980; Saito et al. 1980; Innis and Snyder 1980), its exact role in the central nervous system is still obscure.

The cerebral cortex contains relatively very high concentrations of cholecystokinin, yet the CCK-positive neurons detected form only a small percentage of the total population of cortical neurons (Larsson and Rehfeld 1979; Hendry et al. 1982). In the neocortex these elements occur in all layers, but they are concentrated superficially, in layers I–III (Larsson and Rehfeld 1979; Lorén et al. 1979a; Innis et al. 1979; Hendry et al. 1982; Morrison and Magistretti 1983). Most of these elements are non-pyramidal, small neurons, but in the human pericentral cortex some CCK-positive, medium-sized and small pyramidal cells have been observed (Sakamoto et al. 1984). There is some evidence suggesting that the axons of the CCK-containing neocortical elements pass downward, away from the brain surface, and terminate predominantly in layer VI (Larsson and Rehfeld 1979; Hendry et al. 1982). Lorén et al. (1979a) observed single CCK-positive perikarya in the piriform and the entorhinal cortex only occasionally, but according to Vanderhaeghen et al. (1980) such elements are more numerous in limbic structures such as the cingulate, prepiriform, periamygdaloid and entorhinal cortex than in the neocortex.

CCK-positive cells have been observed in all parts of the hippocampal formation, i.e. the fascia dentata, the cornu ammonis and the subiculum (Lorén et al. 1979a; Vanderhaeghen et al. 1980; Greenwood et al. 1981). There is evidence indicating that CCK is present in both intrinsic and extrinsic hippocampal fibres (Greenwood et al. 1981; Handelmann et al. 1981). The intrinsic fibres include mossy fibre axons of the dentate gyrus granule cells (Gall 1984). The extrinsic fibres enter the fornix and terminate in the lateral septal nucleus (Greenwood et al. 1981).

In the striatum, including the nucleus accumbens, the concentration of CCK is higher than in any other part of the brain (Beinfeld et al. 1981). The ventromedial part of the striatum and the nucleus accumbens contain a plexus of thin CCK-positive fibres which, as is discussed below, originates from the area tegmentalis ventralis (Vanderhaeghen et al. 1980; Hökfelt et al. 1980a, b).

The anterior olfactory nucleus contains numerous CCK-positive cells, and it has been suggested that fibres originating from these cells enter the lateral olfactory tract (Cho et al. 1983).

Both the lateral and medial septal nuclei contain CCK-immunoreactive cells, and a substantial projection possibly arising from these cell masses passes around the anterior commissure and down into the dorsomedial hypothalamus (Woodhams et al. 1983).

The medial forebrain bundle, the stria terminalis, the anterior commissure and the diagonal band of Broca all contain CCK-positive fibres (Vanderhaeghen et al. 1980; Cho et al. 1984). The fibres in the diagonal band belong to an ascending system, because they disappear after hemitransection at the level of the caudal hypothalamus (Hökfelt et al. 1980c).

The amygdaloid complex, particularly the nucleus corticalis, contains CCK-positive perikarya (Lorén et al. 1979a; Innis et al. 1979; Vanderhaeghen et al. 1980). Dense CCK-positive fibres were also seen in the amygdala, with the highest concentration in the central nucleus (Innis et al. 1979).

In the magnocellular portion of the paraventricular nucleus and in the supraoptic nucleus numerous CCK-containing perikarya have been observed, and a dense collection of CCK-positive fibres descends from these cells to the median eminence and the posterior pituitary (Lorén et al. 1979a; Innis et al. 1979; Vanderhaeghen et al. 1980). Beinfeld et al. (1980) have shown that lesioning of the paraventricular nuclei reduces the CCK content of the posterior pituitary by 60% and demonstrated, moreover, that the CCK content of the posterior pituitary is considerably decreased by physiological perturbations which stimulate the release of vasopressin and oxytocin. They proposed that CCK may be either co-secreted with vasopressin and oxytocin to act on peripheral targets or involved in the regulation of vasopressin or oxytocin neurosecretion. Recently, Palkovits et al. (1984b) concluded from radioimmunoassay of CCK after paraventricular lesions that most, if not all, of the CCK in the posterior pituitary and in the median eminence originates from the paraventricular nucleus.

Apart from the supraoptic and paraventricular nuclei several other cell masses in the preoptico-hypothalamic region contain CCK-positive cells. These cell masses include the medial preoptic, dorsomedial and supramamillary nuclei (Vanderhaeghen et al. 1980; Greenwood et al. 1981). The dorsomedial and ventromedial hypothalamic nuclei have been reported to contain a dense network of CCK-immunoreactive fibres (Vanderhaeghen et al. 1980).

It is known that many cells located in the ventral tegmental area, which corresponds roughly to the A10 area of Dahlström and Fuxe, are dopaminergic and project via the medial forebrain bundle to a large number of limbic forebrain structures, including the nucleus accumbens, the olfactory tubercle, the bed nucleus of the stria terminalis and the central amygdaloid nucleus. These cells and their processes constitute together the so-called mesolimbic dopaminergic system. The area tegmentalis ventralis also appears to contain numerous CCK-positive cells (Vanderhaeghen et al. 1980; Hökfelt et al. 1980a, b; Cho et al. 1983; Kubuta et al. 1983) which project to the same areas as the neurons of the dopaminergic mesolimbic system (Hökfelt et al. 1980a, b), and the experimental work of Hökfelt and colleagues (1980a, b) has shown that CCK is located in cells which also contain dopamine. Because not all mesolimbic dopaminergic cells contain this peptide, the CCK-dopamine neurons form a subpopulation within the area tegmentalis ventralis. It is noteworthy that malfunction of the area tegmentalis ventralis and its limbic outflow may well be implicated in schizophrenia, and that there is evidence suggesting that CCK could be useful as an antipsychotic agent (Nair et al. 1983; Voigt and Wang 1984).

The ventral part of the mesencephalic central grey contains a densely packed collection of CCK-positive cell bodies (Lorén et al. 1979a; Innis et al. 1979; Kubuta et al. 1983). Neurons in a well-defined cell group in the rostroventral part of this area show both substance P- and CCK-immunoreactivity (Skirboll et al. 1982). The mesencephalic central grey is a very effective stimulation site for producing analgesia, and it is worthy of note that CCK has been shown to be a potent analgesic (Zetler 1980; Jurna and Zetler 1981).

The dorsal raphe nucleus contains CCK-immunoreactive cells. These elements are situated in two groups, one rostral and the other caudal to the serotoninergic dorsal raphe neurons. There is evidence suggesting that these dorsal raphe CCK-containing cells are local circuit neurons (Kooy et al. 1981).

Several mesencephalic areas, including the inferior colliculus (Kubuta et al. 1983), the dorsal and lateral parts of the interpeduncular nucleus (Lorén et al. 1979a; Hamill et al. 1984) and the parabigeminal nucleus (Kiyama et al. 1983) contain a dense plexus of CCK-positive fibres.

In the rhombencephalon the following cell masses have been shown to contain CCK-positive cells: the nucleus of the solitary tract and the adjacent area postrema, the lateral parabrachial nucleus, the cuneate nucleus, the caudal part of the nucleus spinalis of the trigeminal nerve, the nucleus raphes magnus, the nucleus raphes pallidus, the nucleus raphes obscurus, and a certain portion of the nucleus reticularis magnocellularis. Moreover, rich networks of CCK-immunoreactive fibres have been observed in the lateral lemniscus and its nuclei, the nucleus of the trapezoid body (which belongs to the superior olivary complex), the nucleus of the solitary tract (particularly its rostral portion), the lateral parabrachial nucleus and the caudal part of the nucleus descendens of the trigeminal nerve (Vanderhaeghen et al. 1980; Kubuta et al. 1983; Kiyama et al. 1983; Mantyh and Hunt 1984a). It is notable that several centres and pathways belonging to the auditory system of the brain stem contain numerous CCK-immunoreactive fibres. However, the source of these fibres is unknown at present. There is experimental evidence indicating that the CCK-immunoreactive axons found in the superficial division of the caudal part of the nucleus descendens of the trigeminal nerve represent primary afferent fibres (Maderdrut et al. 1982), and it has also been shown that a certain proportion of the CCK-containing elements in the reticular formation and in the three raphe nuclei mentioned project to the spinal cord (Mantyh and Hunt 1984a). Intererstingly, double-labelling experiments have established that some of the CCK-immunoreactive neurons in the nucleus raphe pallidus and nucleus raphes obscurus also contain serotonin (Mantyh and Hunt 1984a).

A dense network of CCK-immunoreactive fibres is present in laminae I and II of the spinal cord (Larsson and Rehfeld 1979; Vanderhaeghen et al. 1980; Lundberg et al. 1978; Gibson et al. 1981). There is evidence suggesting that this network is formed by primary afferent fibres arising from a population of small dorsal root ganglion cells, and that all of these CCK-positive ganglion cells also contain substance P, and vice versa (Jancsó et al. 1981; Dalsgaard et al. 1982a). So far as the presence of CCK in primary afferent neurons is concerned, the findings just summarized are in harmony with those of Maderdrut et al. (1982); however, they are at variance with those of Marley and colleagues (1982), who remained unable to demonstrate the presence of CCK in sensory roots, and found that dorsal rhizotomy does not affect the CCK level in the spinal cord.

Apart from the superficial dorsal horn, CCK-immunoreactive fibres are present in other parts of the spinal grey as well. Loose plexuses of positive fibres have been found around motor neurons of the ventral horn, particularly in the sacral spinal cord, around neurons of the thoracic intermediolateral column, and around the central canal (Larsson and Rehfeld 1979; Gibson et al. 1981; Mantyh and Hunt 1984a).

Addenda:

1. Kiss et al. (1984) studied the distribution and projections of CCK-immunoreactive neurons in the nucleus paraventricularis hypothalami of the rat. They found that CCK immunoreactivity is present in two distinguishable neuronal populations in the paraventricular nucleus. More than 60% of these cells were found to be typical parvocellular neurons. From experiments in which immunohistochemistry was combined with microsurgical intervention they inferred that the magnocellular cells send their axons to the pituitary, whereas axons of CCK-immunoreactive parvocellular neurons terminate in the median eminence.
2. Using radioimmunoassay, immunocytochemistry, retrograde tracing of HRP and anterograde degeneration techniques, Záborski et al. (1984) demonstrated that axons originating from CCK-positive cells situated in the lateral parabrachial nucleus ascend in the brain stem, enter the dorsolateral part of the medial forebrain bundle, and terminate in the nucleus ventromedialis hypothalami.
3. By combining immunofluorescence and fluorescence retrograde labelling techniques, Fallon and Seroogy (1984) demonstrated that in the rat, CCK-positive cells in the dorsal lateral geniculate nucleus project to area

17 of the visual cortex, and that the axons of elements containing the same peptide and situated in the superior olive pass to the inferior colliculus.

4. Using a combination of techniques similar to that mentioned under point 3, Maciewicz et al. (1984) have shown that the nucleus of Edinger-Westphal in the cat contains numerous CCK-immunoreactive cells throughout its length, and that some of these elements project to the spinal cord where they presumably terminate principally in the dorsal horn. The distribution and frequency of these cells appeared to be similar to the pattern of the substance P-immunoreactive neurons in the nucleus of Edinger-Westphal.

Neurotensin

Neurotensin (NT) is a tridecapeptide which was originally isolated from bovine hypothalamic extracts (Fig. 15). It was discovered, sequenced and synthesized by Carraway and Leeman (1973, 1975a, b). The same authors also developed a specific radioimmunoassay by which they confirmed the presence of this peptide in the hypothalamus and the gastrointestinal tract (Carraway and Leeman 1976). Subsequent radioimmunological (Uhl and Snyder 1976; Kobayashi et al. 1977; Kataoka 1979; Cooper et al. 1981; Langevin and Emson 1982; Manberg et al. 1982) and immunohistological studies (Uhl et al. 1977, 1979b; Uhl 1982; Kahn et al. 1980; Jennes et al. 1982; Hara et al. 1982; Minagawa et al. 1983) have shown that NT is distributed widely but unevenly in the central nervous system.

There is evidence suggesting that NT plays a role in the control of the pituitary gland. Thus, intraventricular and/or intracisternal administration has been shown to influence the release of several anterior pituitary hormones, such as growth hormone, follicle-stimulating hormone, luteinizing hormone, prolactin and thyrotropin (Rivier et al. 1977; Vale et al. 1977; Makino et al. 1978; Maeda and Frohman 1978; Vijayan and McCann 1979, 1980). Whether NT exerts its neuroendocrine influences at the level of the median eminence or, via the portal vessels, on the anterior pituitary itself is unknown. NT also affects the release of SST from the hypothalamus (Sheppard et al. 1979; Shimatsu et al. 1982). Moreover, biochemical and pharmacological studies have shown that NT displays properties consistent with a neurotransmitter role in the central nervous system (Uhl and Snyder 1976, 1977; Lazarus et al. 1977; Iversen et al. 1978). Injection of NT into the ventricular system produces, apart from the neuroendocrine effects already mentioned, hypothermia (Nemeroff et al. 1977), changes in blood pressure, inhibition in gastric secretion (see Higgins et al. 1984), and a marked decrease in sensitivity to pain, the last of these being unresponsive to naloxone (Clineschmidt and McGuffin 1977).

NT-containing perikarya and fibres have been observed in many different parts of the brain (Fig. 19). There is good evidence for an amygdalofugal NT-containing tract (Uhl and Snyder 1979), and it is also known that the nucleus raphes magnus receives NT-containing fibres from several centres in the brain stem (Beitz 1982a). On the whole, however, our knowledge of the NT-ergic projections is still very limited. Before presenting a brief survey of the neurotensin-containing cell groups and fibres it is worthy of note that the structure of the NT precursor is unknown and that this peptide may well be only one member of a larger peptide family (Goedert 1984).

A band of NT-ergic fibres is observed in the deeper layers of the neocortex, with the highest density in layer VI (Uhl et al. 1977; Jennes et al. 1982).

The deeper part of the cingulate cortex contains a longitudinal bundle of NT-positive fibres. However, there is no unanimity with regard to the origin of this bundle. Roberts and colleagues (1981) and Polak and Bloom (1982) believe that it originates from the subiculum and terminates in the anterior cingulate cortex, but Jennes et al. (1982) hold that the hippocampal region is free of NT-immunoreactive perikarya. The latter au-

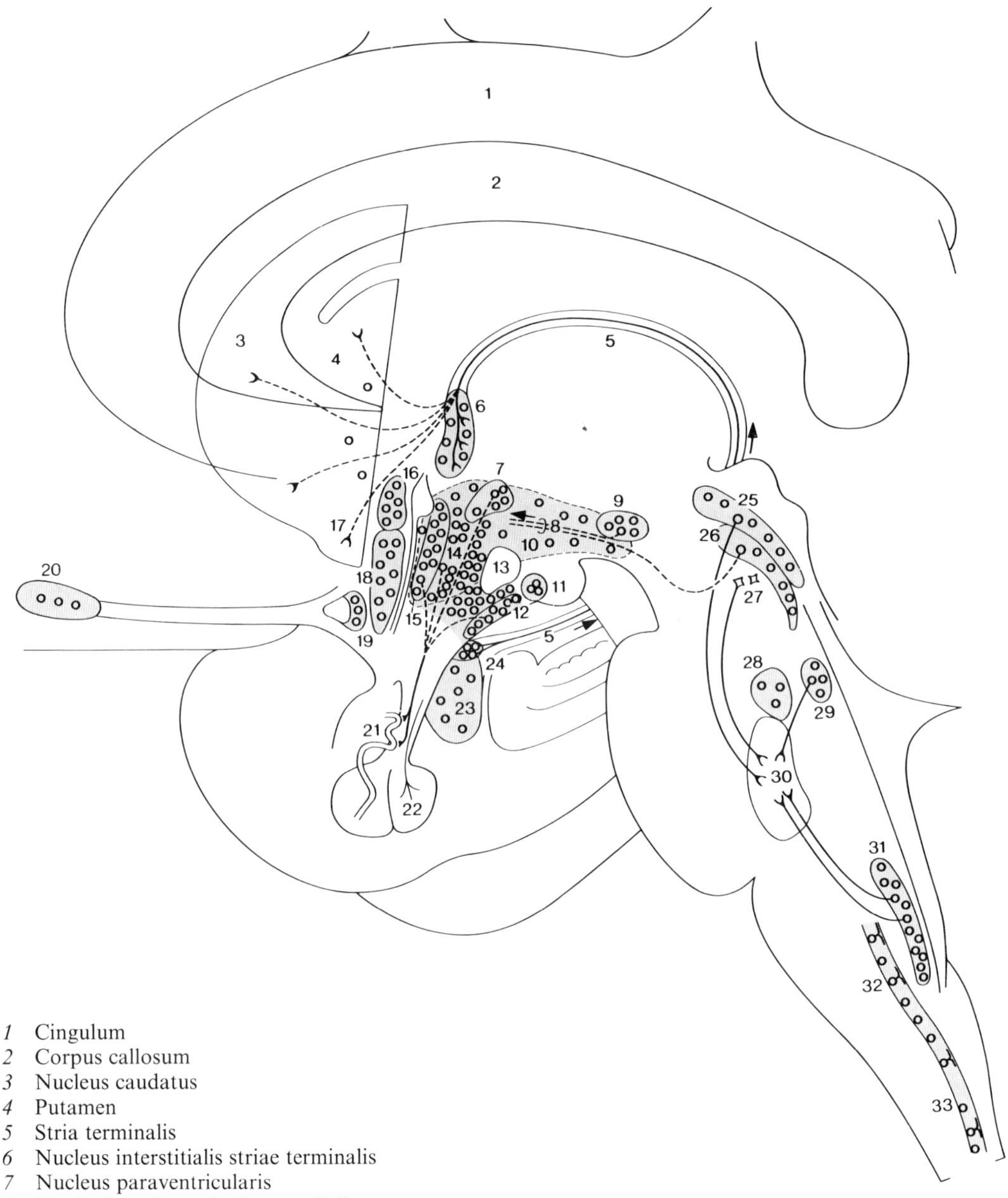

1 Cingulum
2 Corpus callosum
3 Nucleus caudatus
4 Putamen
5 Stria terminalis
6 Nucleus interstitialis striae terminalis
7 Nucleus paraventricularis
8 Fasciculus telencephalicus medialis
9 Area tegmentalis ventralis
10 Area lateralis hypothalami
11 Nucleus premamillaris ventralis
12 Nucleus infundibularis
13 Nucleus ventromedialis
14 Periventricular neurotensin-containing cells in the preoptic and rostral hypothalamic areas
15 Nucleus preopticus medialis
16 Nucleus septi lateralis
17 Nucleus accumbens
18 Nucleus gyri diagonalis
19 Tuberculum olfactorium
20 Bulbus olfactorius
21 Eminentia mediana
22 Lobus posterior hypophyseos
23 Corpus amygdaloideum
24 Nucleus centralis amygdalae
25 Griseum centrale mesencephali
26 Nucleus raphes dorsalis
27 Nucleus cuneiformis
28 Nucleus raphes pontis
29 Nucleus parabrachialis lateralis
30 Nucleus raphes magnus
31 Nucleus solitarius
32 Nucleus spinalis nervi trigemini, pars caudalis
33 Substantia gelatinosa

Fig. 19. Neurotensin-containing cells and fibres

thors consider it likely that the supracallosal fibres in the cingulate cortex originate from the parabrachial nuclei. NT-ergic fibres occur throughout the prepiriform cortex, especially in its medial portion (Jennes et al. 1982).

Some NT-positive cell bodies have been observed in the rostromedial area of the caudate-putamen complex, i.e. the area directly adjacent to the bed nucleus of the stria terminalis (Jennes et al. 1982). In the remaining portions of the striatum only the presence of NT-immunoreactive fibres has been reported. In the rat these fibres show a low-to-moderate density (Uhl et al. 1977; Uhl 1982), except for the most medial, dorsal and ventral parts of the striatum, where a higher density is observed (Jennes et al. 1982). The concentration of NT-immunoreactivity is much higher in the striatum of the cat than in the rat. In the former species this peptide shows a patch-like, mosaic distribution, which corresponds to that of enkephalin (Goedert et al. 1983).

The nucleus accumbens has also been reported to contain a dense plexus of NT-ergic fibres (Uhl et al. 1977; Uhl 1982).

With regard to the presence of NT in the globus pallidus there is unanimity in the literature. Uhl and collaborators (1977; Uhl 1982) observed evenly distributed NT-containing fibres of moderate density in the globus pallidus of the rat, but according to Jennes et al. (1982), this centre in the rat contains hardly any NT-ergic fibres. Goedert et al. (1983) found a particularly rich NT-ergic innervation of the globus pallidus in the cat.

The site of origin of the NT-containing fibres projecting to the basal ganglia is unknown at present. However, NT-positive cell bodies have been found in several regions which project to the striatum, including the amygdala, the ventral tegmental area and the dorsal raphe nucleus (Jennes et al. 1982). Fibres originating from these last two centres might reach the basal ganglia via the medial forebrain bundle, a pathway which contains large numbers of NT-ergic fibres (Jennes et al. 1982).

Interestingly, NT has several actions in common with neuroleptic drugs. This property, in combination with the finding that NT levels in the cerebrospinal fluid appear grossly diminished in about half of untreated schizophrenic patients, has led to the hypothesis that NT acts in the so-called mesolimbic system as an endogenous neuroleptic substance (Prange and Nemeroff 1982; Nemeroff et al. 1983).

In all parts of the amygdaloid complex, NT-positive cell bodies and fibres are found, and both show their highest concentration in the nucleus centralis amygdalae (Uhl 1982; Jennes et al. 1982; Hara et al. 1982). Experimental evidence (Uhl and Snyder 1979) indicates that NT-containing cell bodies in this nucleus project via the stria terminalis and terminate, at least in part, in the bed nucleus of that pathway. Jennes and colleagues (1982) considered it likely that the stria terminalis issues NT-positive fibres to the dorsal, medial and ventral portions of the caudate-putamen complex, as well as to the nucleus accumbens.

As mentioned before, NT may strongly influence the sensitivity of pain. In this context it is worthy of note that an antinociceptive response can be consistently produced by local injections of NT into the nucleus centralis amygdalae (Kalivas et al. 1982).

NT-containing cell bodies and fibres have been observed in the olfactory bulb, the olfactory tubercle, the lateral and medial septal nuclei, the bed nucleus of the stria terminalis and the nucleus of the diagonal band of Broca (Uhl 1982; Jennes et al. 1982; Hara et al. 1982; Köhler and Eriksson 1984). The NT-positive cell bodies in the nucleus last mentioned form a continuum with those in the periventricular preoptic area.

The medial zone of the thalamus, including the midline nuclei, contains a moderately dense plexus of NT-immunoreactive fibres (Uhl et al. 1977; Uhl 1982; Jennes et al. 1982).

Radioimmunological as well as immunohistochemical studies have shown that, of all parts of the central nervous system, the pre-

optico-hypothalamic region has by far the highest concentration of NT-immunoreactivity (Kobayashi et al. 1977; Kataoka 1979; Uhl 1982; Jennes et al. 1982; Hara et al. 1982).
Numerous NT-positive perikarya are found in a continuum which includes the periventricular zone of the preoptic and anterior hypothalamic areas, the medial preoptic nucleus, the magnocellular and parvocellular parts of the paraventricular nucleus, the infundibular nucleus and strands of cells surrounding the dorsomedial nucleus. In addition, NT-immunoreactive somata occur in the lateral preoptic and hypothalamic areas, and in the ventral premamillary nucleus (Kahn et al. 1980, 1982; Uhl 1982; Hara et al. 1982; Jennes et al. 1982; Ibata et al. 1984). Interestingly, numerous NT-positive neurons in the infundibular nucleus also contain dopamine (Hökfelt et al. 1984a). In all of the nuclei and areas mentioned, more or less dense plexuses of NT-containing fibres are also present. The data on the occurrence of NT-immunoreactive cell bodies and fibres in the hypothalamus are all derived from the rat. A radioimmunological study (Langevin and Emson 1982) suggested that in the human hypothalamus there is a focus of high NT concentration in the anterior hypothalamic area, which clearly does not extend into the preoptic region. Moderate concentrations of NT were found in paraventricular, ventromedial and infundibular nuclei.
The lateral part of the external zone of the median eminence contains an accumulation of NT-immunoreactive fibres (Kahn et al. 1980; Uhl 1982; Jennes et al. 1982) and in the posterior pituitary some scattered fibres containing the same peptide have been observed (Uhl et al. 1977; Kahn et al. 1980; Kahn et al. 1982). Some authors (Uhl et al. 1977; Goedert et al. 1982) have mentioned the presence of a distinct population of NT-containing cells in the anterior pituitary, but others (Kahn et al. 1980, 1982) remained unable to confirm this observation.
Several of the areas where NT was found to be concentrated are known to be important in anterior pituitary regulation. These include the perikarya in the periventricular zone, the medial preoptic nucleus, the parvocellular portion of the paraventricular nucleus and the infundibular nucleus. There is experimental evidence indicating that all of these areas project to the median eminence. However, whether NT-containing cells in one, several, or all of these areas are the source of NT-immunoreactive fibres in the median eminence is not yet known. It also remains to be elucidated whether the NT-ergic fibres in the median eminence contact blood vessels or other nerve fibres. These uncertainties, and the fact that NT may well be produced and released in the anterior pituitary itself, render it difficult to design a picture as to how the NT-containing neurons in the various hypothalamic centres may exert their control of the secretion of anterior pituitary hormones. Although there is no unanimity, most authors agree that NT inhibits the release of growth hormone, prolactin and thyrotropin (Maeda and Frohman 1978; McCann et al. 1982; Frohman et al. 1982). It has been suggested that NT may exert this inhibitory influence indirectly, i.e. by stimulating hypothalamic dopaminergic neurons. The dopamine released would then inhibit the secretion of the pituitary hormones mentioned (McCann et al. 1982; Frohman et al. 1982). The recently discovered co-existence of NT and dopamine in many infundibular neurons (Hökfelt et al. 1984a) further complicates the issue. The function of the NT-containing axons in the posterior pituitary is entirely obscure. The paraventricular nucleus seems to be the most likely source of these fibres (Kahn et al. 1982). Finally, it should be mentioned that the preoptico-hypothalamic region appears to be involved in thermoregulation and in the control of nociception. Hypothermia can be induced by NT injections into the area preoptica medialis, contiguous with the anterior hypothalamic area, whereas an antinociceptive response can be produced by NT in the rostral area preoptica medialis (Kalivas et al. 1982).
The following mesencephalic centres have been shown to contain moderately dense col-

lections of NT-positive perikarya and fibres: the periaqueductal grey, the nucleus raphes dorsalis, the mesencephalic reticular formation and the area tegmentalis of Tsai (Uhl et al. 1979b; Uhl 1982; Jennes et al. 1982; Beitz 1982a; Beitz et al. 1983). In the area last mentioned most NT-immunoreactive neurons also contain dopamine (Hökfelt and Everitt 1984). The plexus of NT-ergic fibres present in the area tegmentalis ventralis merges laterally with a lower density of fibres and terminals in the substantia nigra. The nigra, however, has been demonstrated to contain large amounts of NT receptors, particularly in the pars compacta. There is evidence indicating that these receptors are located on the somata and dendrites of nigral dopaminergic neurons (Uhl 1982; Uhl et al. 1984). Manberg and colleagues (1982) have found that in man the pars compacta of the substantia nigra contains a relatively very large amount of NT.

In the rhombencephalon the inner part of the substantia gelatinosa of the caudal nucleus of the trigeminal nerve and the nucleus solitarius contain large numbers of NT-positive perikarya and a dense plexus of NT-immunoreactive axons. NT-containing cells and fibres have also been observed in the lateral parabrachial nucleus, the nucleus raphes pontis and the locus coeruleus (Uhl et al. 1977; Uhl 1982; Jennes et al. 1982). It should be noted, however, that one group of investigators (Minagawa et al. 1983) denied the presence of NT-ergic perikarya in the locus coeruleus. A caudal continuation of the NT-positive cell population in the nucleus raphes dorsalis extends into the rostral rhombencephalon, surrounding the nucleus tegmenti dorsalis dorsolaterally and ventrolaterally (Jennes et al. 1982).

A certain number of NT-immunoreactive cells in the dorsolateral part of the nucleus solitarius also contain adrenaline, whereas in some neurons situated in the medial part of the same nucleus NT co-exists with noradrenaline (Hökfelt et al. 1984a).

Experimental studies (Beitz 1982a; Beitz et al. 1983) have revealed that NT-containing cell bodies in the periaqueductal grey, the mesencephalic reticular formation, the lateral parabrachial nucleus and the nucleus solitarius project to the nucleus raphes magnus. Higgins et al. (1984) provided experimental evidence indicating that NT-containing neurons in the nucleus solitarius project to the dorsal motor vagal nucleus and form part of the path along which the aortic baroreceptor reflexes are effected. Other NT-ergic solitarius cells may be involved in the regulation of gastric acid secretion.

In the spinal cord the substantia gelatinosa contains large numbers of NT-positive neurons and a dense feltwork of NT-immunoreactive fibres (Uhl 1982; Polak and Bloom 1982; Jennes et al. 1982; Difiglia et al. 1984). The neuronal perikarya are concentrated in the outer one-third of lamina II, the inner one-third of lamina II, and the outer lamina III (Seybold and Elde 1982). The plexus of fibres present in the substantia gelatinosa extends with decreasing density ventrally over laminae III and IV. Probably, this entire plexus is of intrinsic spinal origin. Evidence for the presence of NT-ergic primary afferent fibres is lacking (Polak and Bloom 1982).

A moderately dense network of NT-immunoreactive fibres and sparsely scattered perikarya containing the same peptide are found in the area surrounding the spinal central canal (Uhl 1982; Jennes et al. 1982).

Addenda:

1. Kalivas and Miller (1984) have provided experimental evidence indicating that the NT-containing fibres present in the nucleus accumbens and in the diagonal band of Broca originate from the area tegmentalis ventralis.
2. In a recent electron-microscopic immunocytochemical study Ibata et al. (1984) mention the presence of numerous NT-immunoreactive terminals in the vicinity of pericapillary spaces of the external layer of the median eminence. In their opinion, this strongly suggests that an NT-like immunoreactive substance is released into the portal capillaries. Ibata et al. (1984) also observed that small

axons containing NT-immunoreactive granules and clear vesicles terminate on small dendrites or dendritic spines in the internal layer of the median eminence.

Hypophysiotropic Peptides

Corticotropin-Releasing Factor

A 41-amino acid peptide that probably corresponds to the corticotropin-releasing factor (CRF) was recently isolated from ovine hypothalamic tissue (Vale et al. 1981; Fig. 20). This peptide has been synthesized and antisera against it have been raised. With the aid of these sera Swanson et al. (1983) have studied the distribution of CRF-immunoreactive cells and fibres in the brain of the rat. The following synopsis is based entirely on the work of these authors (Fig. 21).

CRF-immunoreactive cells are distributed throughout the nucleus paraventricularis hypothalami. Some of these elements are located in the magnocellular part of this nucleus, but most of them are concentrated in its parvocellular division. The CRF-containing cells in the paraventricular nucleus give rise to a massive projection to the neurohaemal (external) zone of the median eminence (Makara 1981). This pathway is probably involved in the control of ACTH and β-endorphin release from the anterior lobe of the pituitary by way of the hypothalamo-hypophyseal portal system.

A series of cell groups in the basal telencephalon, hypothalamus and brain stem appear to contain CRF-stained neurons. In the basal telencephalon these elements form a massive, more or less continuous group which extends over the bed nucleus of the stria terminalis, the substantia innominata and the central nucleus of the amygdala. In the telencephalon impar the nucleus preopticus medialis contains numerous CRF-immunoreactive cells. Moreover, CRF-stained cell bodies are found scattered over the lateral preoptico–lateral hypothalamic continuum. In the midbrain two CRF-stained cell groups have been found; one consists of scattered cells in the dorsal part of the reticular formation, the other lies in the periaqueductal grey and occupies a position rostral to the dorsal raphe nucleus and medial to the nucleus of Edinger-Westphal. In the rhombencephalon the following cell masses have been found to contain CRF-immunoreactive perikarya: the lateral tegmental area of the rostral rhombencephalon, the lateral parabrachial nucleus, the medial vestibular nucleus and the nucleus of the solitary tract. Moreover, some rhombencephalic areas known to contain considerable numbers of noradrenergic elements have been found to harbour CRF-immunoreactive cells. These areas include the ventral half of the locus coeruleus, an area situated just lateral to the superior olivary complex corresponding to the A5 catecholaminergic cell group, and a number of cells lying in the ventrolateral part of the medullary reticular formation coinciding with the A1 catecholaminergic cell group.

The brain contains numerous thin, varicose CRF-immunoreactive axons, which in several places are concentrated in distinct fibre streams. The preliminary findings with regard to the distribution of these fibres of Swanson et al. (1983) may be summarized as follows.

1. Fibres possibly originating from the bed nucleus of the stria terminalis ascend to the lateral septal nucleus, and other fibres appear to enter the medial septal–diagonal band continuum from the region of the substantia innominata and the preoptic region.
2. Many fibres interconnect the rostral hypothalamus and adjacent parts of the basal telencephalon with the amygdala. The two large conduction channels of the latter structure, i.e. the stria terminalis and the ventral amygdalofugal pathway, contain numerous CRF-immunoreactive fibres.
3. Large numbers of fibres, apparently emanating from the large cluster of telodiencephalic CRF-positive cells, course caudally through the medial forebrain bundle. In the caudal hypothalamus these fibres split up

Hypophysiotropic peptides

Corticotropin-releasing factor [CRF]

CRF (ovine)

H-Ser^{1}-Gln-Glu-Pro-Pro-Ile-Ser-Leu-Asp-Leu^{10}-Thr-Phe-His-Leu-Leu-Arg-Glu-Val-Leu-Glu^{20}-Met-Thr-Lys-Ala-Asp-Gln-Leu-Ala-Gln-Gln^{30}-Ala-His-Ser-Asn-Arg-Lys-Leu-Leu-Asp-Ile^{40}-Ala-NH_2

CRF (rat)

Glu^{2} , Ala^{22} , Arg^{23} , Glu^{25} , Met^{38} , Glu^{39} , Ile^{41}

Luteinizing hormone-releasing hormone [LH-RH]

pGlu^{1}-His-Trp-Ser-Tyr-Gly-Leu-Arg-Pro-Gly^{10}-NH_2

Somatostatin

H-Ala^{1}-Gly-Cys-Lys-Asn-Phe-Phe-Trp-Lys-Thr-Phe-Thr-Ser-Cys^{14}-OH

Somatostatin - 28 (ovine)

H-Ser^{1}-Ala-Asn-Ser-Asn-Pro-Ala-Met-Ala-Pro^{10}-Arg-Glu-Arg-Lys-Ala-Gly-Cys-Lys-Asn-Phe^{20}-Phe-Trp-Lys-Thr-Phe-Thr-Ser-Cys^{28}-OH

Thyrotropin-releasing hormone [TRH]

pGlu-His-Pro-NH_2

Fig. 20. Primary structure of some hypophysiotropic peptides. For significance of the amino acid code, see Table 7 on p. 198

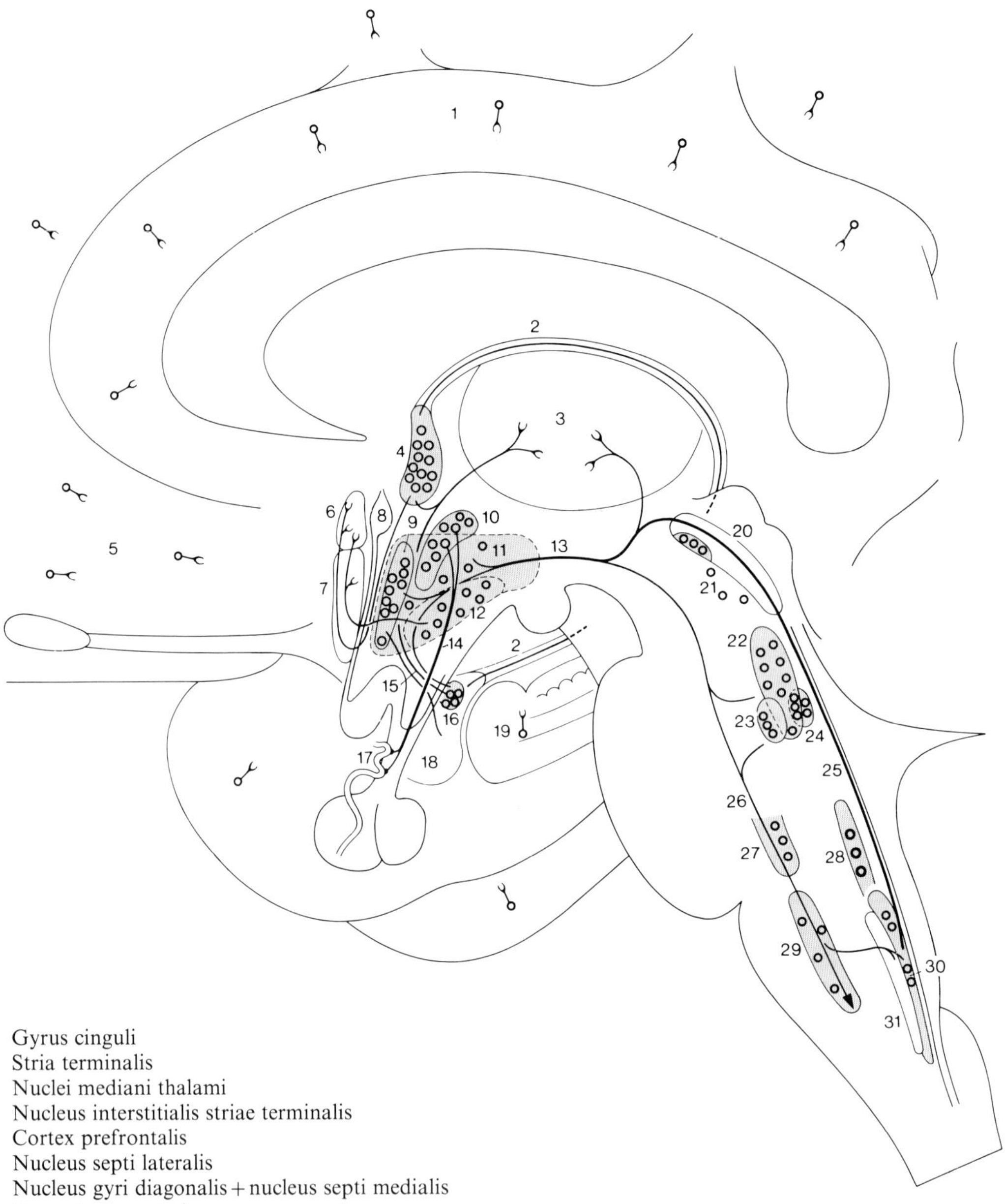

1 Gyrus cinguli
2 Stria terminalis
3 Nuclei mediani thalami
4 Nucleus interstitialis striae terminalis
5 Cortex prefrontalis
6 Nucleus septi lateralis
7 Nucleus gyri diagonalis + nucleus septi medialis
8 Commissura anterior
9 Nucleus preopticus medialis
10 Nucleus paraventricularis
11 Area lateralis hypothalami
12 Substantia innominata
13 Fasciculus telencephalicus medialis
14 Tractus paraventriculoinfundibularis
15 Fibrae amygdalofugales ventrales
16 Nucleus centralis amygdalae
17 Eminentia mediana
18 Corpus amygdaloideum
19 Hippocampus
20 Griseum centrale mesencephali
21 Formatio reticularis mesencephali
22 Area laterodorsalis tegmenti
23 Locus coeruleus
24 Nuclei parabrachiales
25 Dorsal (periventricular) stream of CRF-containing fibres
26 Ventral stream of CRF-containing fibres
27 Cell group A5
28 Nucleus vestibularis medialis
29 Cell group A1
30 Nucleus solitarius
31 Nucleus dorsalis nervi vagi

Fig. 21. Corticotropin-releasing factor-containing cell groups and fibres

into two contingents, a larger dorsal one and a smaller ventral one. The dorsal contingent distributes to the midline region of the thalamus and then takes a caudal course, forming a periventricular bundle throughout the brain stem. The ventral contingent enters the ventral tegmental area and passes caudally through the lateral part of the reticular formation. The two fibre contingents connecting the prosencephalon with the brain stem have been described here as descending. However, the direction of fibres in these two bundles is unclear, because they appear to interconnect areas that contain CRF-immunoreactive cell bodies. Thus, the periaqueductal grey, the locus coeruleus, the parabrachial nuclei, the medial vestibular nucleus and the dorsal vagal complex lie in or near the dorsal periventricular fibre stream, whereas the CRF-positive perikarya lying in the areas of the A1 and A5 catecholaminergic cell groups are located close to the ventral fibre stream. It is uncertain which of these cell groups contributes to each of these two pathways, and which of them receives input from them, particularly as the dorsal and ventral streams are interconnected at several levels.

Preliminary experiments with intraventricular injections suggest that CRF is able to influence neurons within the central nervous system, and that it plays an important role in central mechanisms that regulate the autonomic nervous system (Brown et al. 1982; Fisher et al. 1982). The CRF-stained cells and fibres in the basal forebrain, diencephalon and brain stem, discovered by Swanson et al. (1983) and briefly described above, most probably constitute the morphological substrate of these mechanisms.

The superficial layers of the cerebral cortex contain CRF-labelled cells which, judging from their shape, represent local circuit neurons. These elements are most common in limbic regions, such as the prefrontal areas and the cingulate gyrus. Similar cells are found in the hippocampal formation. Whether CRF-containing cells and/or fibres occur in the spinal cord is unknown at present.

Addenda:

1. The distribution of CRF-immunoreactive neurons in the central nervous system of the rat has also been studied by Cummings et al. (1983) and Pilcher and Joseph (1984). Both groups confirmed most of the findings of Swanson et al. (1983) summarized above, but noted the presence of a number of additional groups of CRF-positive cells. The additional cell groups mentioned by Cummings et al. (1983) include the nucleus supraopticus, the nucleus arcuatus, the nucleus raphes dorsalis, the medullary reticular formation and the nucleus cuneatus externus, whereas Pilcher and Joseph (1984) observed additional neuronal populations in the following centres: nucleus accumbens, hippocampus (not specified), organum vasculosum laminae terminalis, nucleus reuniens thalami, scattered neurons throughout the premamillary, mamillary and posterior hypothalamic nuclei, lateral tegmental nucleus (situated directly ventral to the medial geniculate body), Kölliker-Fuse nucleus, A7 noradrenergic group, rhombencephalic periventricular grey, nucleus prepositus hypoglossi and the area around the central canal of the cervical spinal cord. Pilcher and Joseph (1984), who also examined the distribution of ACTH-immunoreactive fibres in the central nervous system, noted throughout the brain a remarkable concordance of localization of CRF-immunoreactive perikarya and ACTH-immunoreactive fibres. In view of the well known interactions of CRF with the ACTH-secreting cells in the pituitary, they suggested the possibility of a similar relationship of CRF and ACTH neuronal populations in the central nervous system.
2. Vincent and Satoh (1984) demonstrated that a cluster of CRF-immunoreactive cells situated in the metencephalic periventricular grey projects to the sacral spinal cord. Also, CRF-positive varicose fibres were detected in the intermediolateral column of the sacral cord. They conjectured that the cluster of CRF-immunoreactive cells corresponds to the micturition reflex centre of Barrington.

Luteinizing Hormone-Releasing Hormone

Luteinizing hormone-releasing hormone (LHRH) is a decapeptide (Fig. 20), the ability of which to stimulate the secretion of luteinizing hormone and follicle-stimulating hormone from the gonadotropic cells in the anterior lobe of the pituitary is well known. Equally well documented is the fact that LHRH is produced in the mediobasal hypothalamus and delivered to the gonadotrophs via the hypothalamo-hypophyseal portal system after its release from nerve terminals in the median eminence. Immunohistochemical studies have shown, however, that LHRH-synthesizing cells are by no means confined to the mediobasal hypothalamus, and that their axons project not only to the median eminence, but also to many other intra- as well as extra-hypothalamic sites. LHRH has been iontophorized onto neurons in a variety of brain areas; in most of them its application appeared to stimulate neuronal firing (Moss 1977). It is also worthy of note that in the central nervous system LHRH-containing axons have been shown to form typical synaptic terminals (Silverman 1984). In autonomic ganglia a peptide closely resembling LHRH has been reported to elicit prolonged EPSPs with long latencies, and it was found to operate on a much larger range than 'classical' neurotransmitters (Jan and Jan 1983).

In the brain the LHRH-positive perikarya are not grouped into discrete nuclei; rather they are scattered over a continuum extending from the septal region anteriorly to the premammillary region posteriorly (Fig. 22). LHRH-containing elements have been observed in the medial septal nucleus, the nucleus of the diagonal band of Broca, the bed nucleus of the stria terminalis (many cells), the lamina terminalis, the medial preoptic nucleus (many cells), the anterior hypothalamic area, the supraoptic nucleus (small cells), the infundibular nucleus and the areas lateral, dorsal and caudal to this cell group (many cells in most mammals, but not in the rat) (Silverman et al. 1977, 1982; Barry 1977, 1978; Silverman and Krey 1978). The projections described below emanate from these elements (Fig. 22).

Perikarya in the nucleus infundibularis and adjacent mediobasal hypothalamic areas give rise to LHRH-containing axons, which join the tubero-infundibular tract. These axons pass to the infundibular stalk, the fibres of which form a dense plexus around the capillary loops of the pituitary portal circulation in the median eminence (Silverman et al. 1977; Barry 1977; Silverman and Krey 1978; Anthony et al. 1984). In several mammalian species (man, monkey, ferret, bat) LHRH-containing tubero-infundibular fibres have been observed to traverse the pituitary stalk and enter the posterior pituitary. Most of these fibres terminate in the border zone with the adenohypophysis, but others penetrate the deeper portion of the neural lobe (Silverman et al. 1977; Bary 1977; Anthony et al. 1984; King et al. 1984). Because in the adenohypophysis the gonadotrophs are concentrated in the zone adjacent to the neural lobe, it is conceivable that the release of hormone from these elements is – at least in part – regulated by these tubero-hypophyseal fibres (Anthony et al. 1984). The description just given does not hold for the rat. In this species the tubero-infundibular tract contains few if any LHRH-positive fibres (Merchenthaler et al. 1980; Bennett-Clarke and Joseph 1982).

LHRH-containing cells in the medial preoptic nucleus, particularly its basal part, project to the suprachiasmatic nucleus and the median eminence (Silverman and Krey 1978). The preoptico-infundibular projection of LHRH-positive fibres is particularly large in the rat (Ibata et al. 1979; Liposits and Sétáló 1980; Merchenthaler et al. 1980). In this species it has been shown experimentally that LHRH-containing neurons situated in the medial septal nucleus and in the nucleus of the diagonal band of Broca contribute to this projection (Kawano and Daikoku 1981). It is important to note that in the rat the LHRH-positive fibres, passing from the preoptic region to the infundibulum, constitute

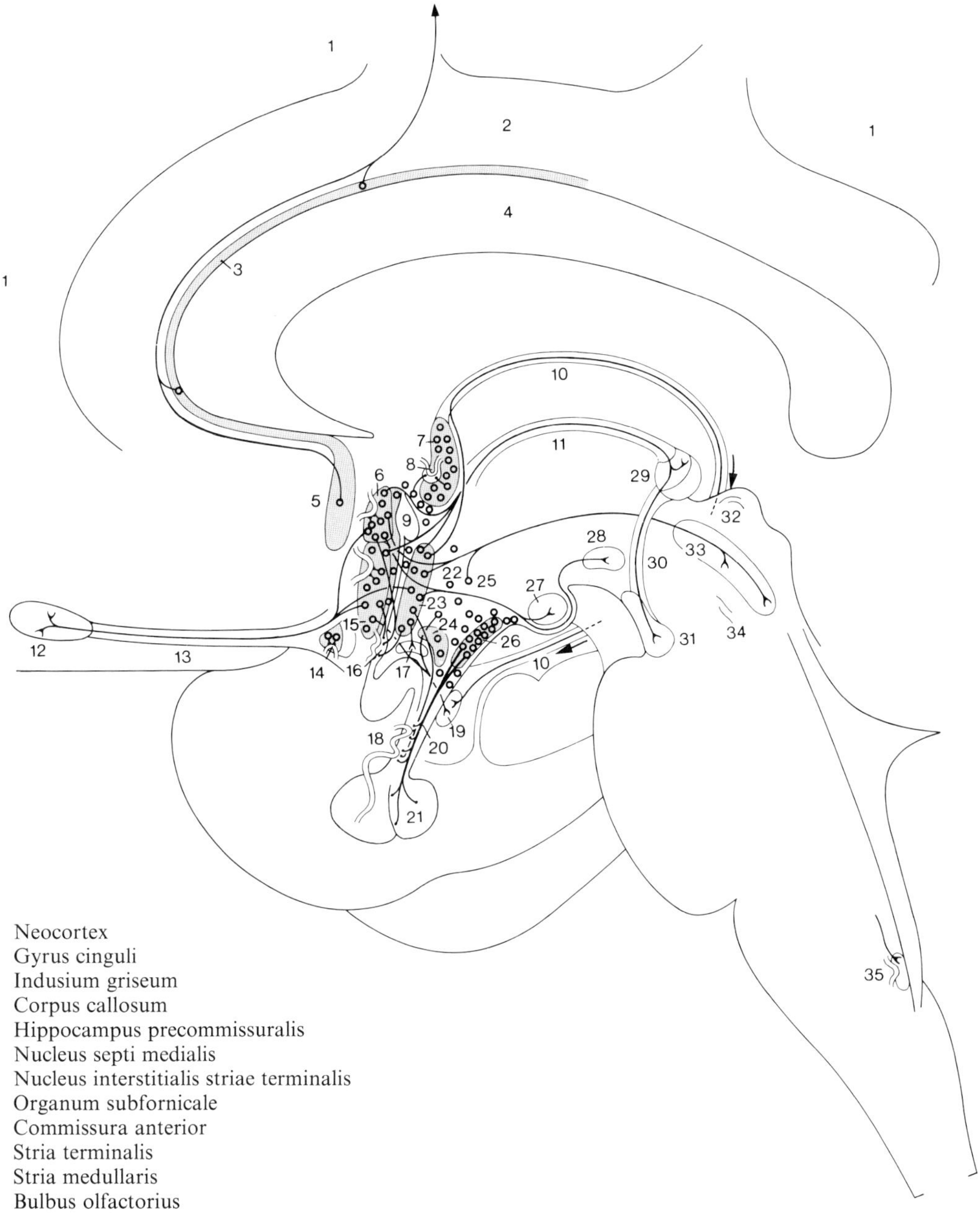

1 Neocortex
2 Gyrus cinguli
3 Indusium griseum
4 Corpus callosum
5 Hippocampus precommissuralis
6 Nucleus septi medialis
7 Nucleus interstitialis striae terminalis
8 Organum subfornicale
9 Commissura anterior
10 Stria terminalis
11 Stria medullaris
12 Bulbus olfactorius
13 Tractus olfactorius
14 Substantia perforata anterior
15 Nucleus gyri diagonalis
16 Organum vasculosum laminae terminalis
17 Nucleus suprachiasmaticus
18 Eminentia mediana
19 Nucleus medialis amygdalae
20 Tractus tuberoinfundibularis
21 Lobus posterior hypophyseos
22 Area hypothalamica anterior
23 Nucleus preopticus medialis
24 Nucleus supraopticus
25 Area hypothalamica lateralis
26 Nucleus infundibularis
27 Corpus mamillare, nucleus medialis
28 Area tegmentalis ventralis
29 Nucleus habenulae medialis
30 Tractus habenulointerpeduncularis
31 Nucleus interpeduncularis
32 Colliculus superior
33 Griseum centrale mesencephali
34 Nucleus cuneiformis
35 Area postrema

Fig. 22. Luteinizing hormone-releasing hormone-containing cells and fibres

a medial contingent which courses caudally in a periventricular position and a lateral contingent which joins the medial forebrain bundle and then swings medially to reach the median eminence (Kawano and Daikoku 1981; King et al. 1982; Palkovits et al. 1984a).

LHRH-positive cells situated in the medial preoptic nucleus, the nucleus of the diagonal band of Broca and the medial septal nucleus have been observed to pass to the organum vasculosum laminae terminalis, where their terminal branches form a dense plexus around the capillary network (Barry 1977; Burchanowski and Sternberger 1980; Kawano and Daikoku 1981; Witkin et al. 1982). A comparable projection to the subfornical organ arises from LHRH-containing perikarya situated in the bed nucleus of the stria terminalis and the area surrounding the anterior commissure (Burchanowski and Sternberger 1980; Witkin et al. 1982).

Numerous LHRH-containing fibres and terminals have been observed in the ependymal lining of the third and lateral ventricles. The presence of these fibres, which originate from cells situated in the septum, the bed nucleus of the stria terminalis and the medial preoptic nucleus, suggests that the cerebrospinal fluid may participate in the transport of LHRH (Barry 1977; Burchanowski and Sternberger 1980; Witkin et al. 1982).

In the medial septal nucleus, the nucleus of the diagonal band of Broca and the anterior perforated substance, or olfactory tubercle, small groups of LHRH-positive cells are clustered around large blood vessels which penetrate into these areas. The axons of these cells have been observed to terminate around the blood vessels rather than contact other neurons (Silverman and Krey 1978).

LHRH-positive elements situated in the medial preoptic nucleus, medial septal nucleus and the nucleus of the diagonal band have been reported to project to the olfactory bulb, where they terminate in the external plexiform and glomerular layers (Burchanowski and Sternberger 1980; Bennett-Clarke and Joseph 1982; Witkin et al. 1982). Some LHRH-immunoreactive elements have been observed in the nervus terminalis (Silverman et al. 1982).

Witkin et al. (1982) reported that in the rat, LHRH-positive cells situated in the anterior hippocampus and in the indusium griseum (i.e. the hippocampal rudiment along the dorsum of the corpus callosum), send their axons through the cingulate cortex and the neocortex toward the subarachnoid space on the dorsal surface of the brain.

The medial amygdaloid nucleus contains a plexus of LHRH-positive fibres. Silverman and Krey (1978) observed that these fibres originate from the medial preoptic nucleus and reach the amygdala by way of the stria terminalis, but according to Witkin et al. (1982), the amygdala is supplied by LHRH-positive fibres which arise from the nucleus of the diagonal band and the lateral hypothalamic area and follow a direct lateral course.

LHRH-immunoreactive neurons situated in the medial septal nucleus, the nucleus of the diagonal band and the medial preoptic nucleus project, via the stria medullaris and the fasciculus retroflexus, to the medial habenular nucleus and to the medial and caudal portions of the interpeduncular nucleus (Barry 1978; Silverman and Krey 1978; Silverman et al. 1982; Witkin et al. 1982).

LHRH-positive cells located in the medial septal and medial preoptic nuclei give rise to two projections which pass caudally through the hypothalamus, one dorsal and the other ventral. The dorsal projection, which also includes axons of LHRH-containing cells situated in the lateral hypothalamic area, passes to the periaqueductal grey and reaches the pontomesencephalic junction (Liposits and Sétáló 1980; Witkin et al. 1982). The ventral projection supplies the medial mammillary nucleus and continues, via the capsule of the mammillary body, to the area tegmentalis ventralis (Silverman and Krey 1978; Silverman et al. 1982; Witkin et al. 1982).

Scattered LHRH-immunoreactive fibres of unknown origin have been observed in the superior colliculus, the mesencephalic reticular formation and the area postrema (Barry 1978; Silverman and Krey 1978).

In primates the LHRH-containing cells in the infundibular nucleus and their projections to the median eminence (and posterior pituitary?) are essential for both the tonic and cyclic components of gonadotropin secretion to occur (Silverman et al. 1982), but the function of the remaining LHRH-containing projections is less clear. It has been repeatedly suggested that these projections may play a role in the regulation of sexual behaviour and reproduction (Silverman et al. 1977; Silverman and Krey 1978; Witkin et al. 1982; King et al. 1984), and it has been pointed out that several of the structures innervated, such as the olfactory system, the habenula, the medial amygdaloid nucleus and the subfornical organ, are implicated in the control of reproduction (Witkin et al. 1982; King et al. 1984). In the female rat, the mesencephalic central grey, which is also supplied by LHRH-containing fibres, is presumably involved in lordosis behaviour (Shivers et al. 1983). However, the possibility remains that certain groups of LHRH-containing neurons subserve functions that are not related to reproduction or sexual behaviour.

Somatostatin

Somatotropin release-inhibiting factor, or somatostatin (SST), was first isolated and characterized by Brazeau et al. (1973) from the ovine hypothalamus. It is a tetradecapeptide (Fig. 20) which, as its names indicate, inhibits the secretion of somatotropin or growth hormone from the anterior pituitary (for review, see Vale et al. 1977). In addition, it blocks the release of pituitary thyrotropin and prolactin (Snyder 1980). Apart from its role as a neuroendocrine release inhibiting factor, SST has appeared to be widely distributed in the central nervous system, and there is now strong evidence suggesting that it may be a central neurotransmitter or neuromodulator (for review, see Elde 1979; McCann 1982). Antibodies against SST have been raised, and these have rendered it possible to study the localization of neurons containing this peptide in the central nervous system using immunohistochemical techniques (Hoffman and Hayes 1979; Bennett-Clarke et al. 1980; Krisch 1981; Finley et al. 1981 b; Shiosaka et al. 1982). Vincent and colleagues (Vincent and Johansson 1983; Vincent et al. 1983 b) recently reported that neurons containing both SST and avian pancreatic polypeptide are selectively stained by the histochemical technique for nicotinamide adenine dinucleotide phosphate (NADPH)-diaphorase activity. SST has appeared to be present in both interneurons and projection neurons. Some SST-containing pathways have been studied by combining microsurgical intervention with immunohistochemical (Sakanaka et al. 1981 a) or radioimmunological (Palkovits et al. 1982 a) techniques. However, studies of this type are very limited in number and the following survey shows that the exact sites of origin and/or termination of many somatostatinergic projections are still unknown (Fig. 23).

All parts of the neocortex as well as the cingulate cortex contain SST-positive perikarya (Bennett-Clarke et al. 1980; Finley et al. 1981 b; Shiosaka et al. 1982). Although these perikarya are scattered throughout layers II–VI, the majority are restricted to layers V and VI (Finley et al. 1981 b; Shiosaka et al. 1982). Because the corona radiata contains very few somatostatinergic fibres, Finley et al. (1981 b) considered it likely that most of the SST-containing cortical elements represent local circuit neurons. The piriform cortex and the various parts of the hippocampal formation (fascia dentata, cornu Ammonis, subiculum) have also been reported to contain SST-positive perikarya (Finley et al. 1981 b; Shiosaka et al. 1982).

Throughout its extent the neostriatum (nucleus caudatus plus putamen) contains a population of evenly distributed SST-positive perikarya (Graybiel et al. 1981; Finley et al. 1981 b; Takagi et al. 1983). These elements appear to belong to a set of medium-sized aspiny interneurons reported previously in

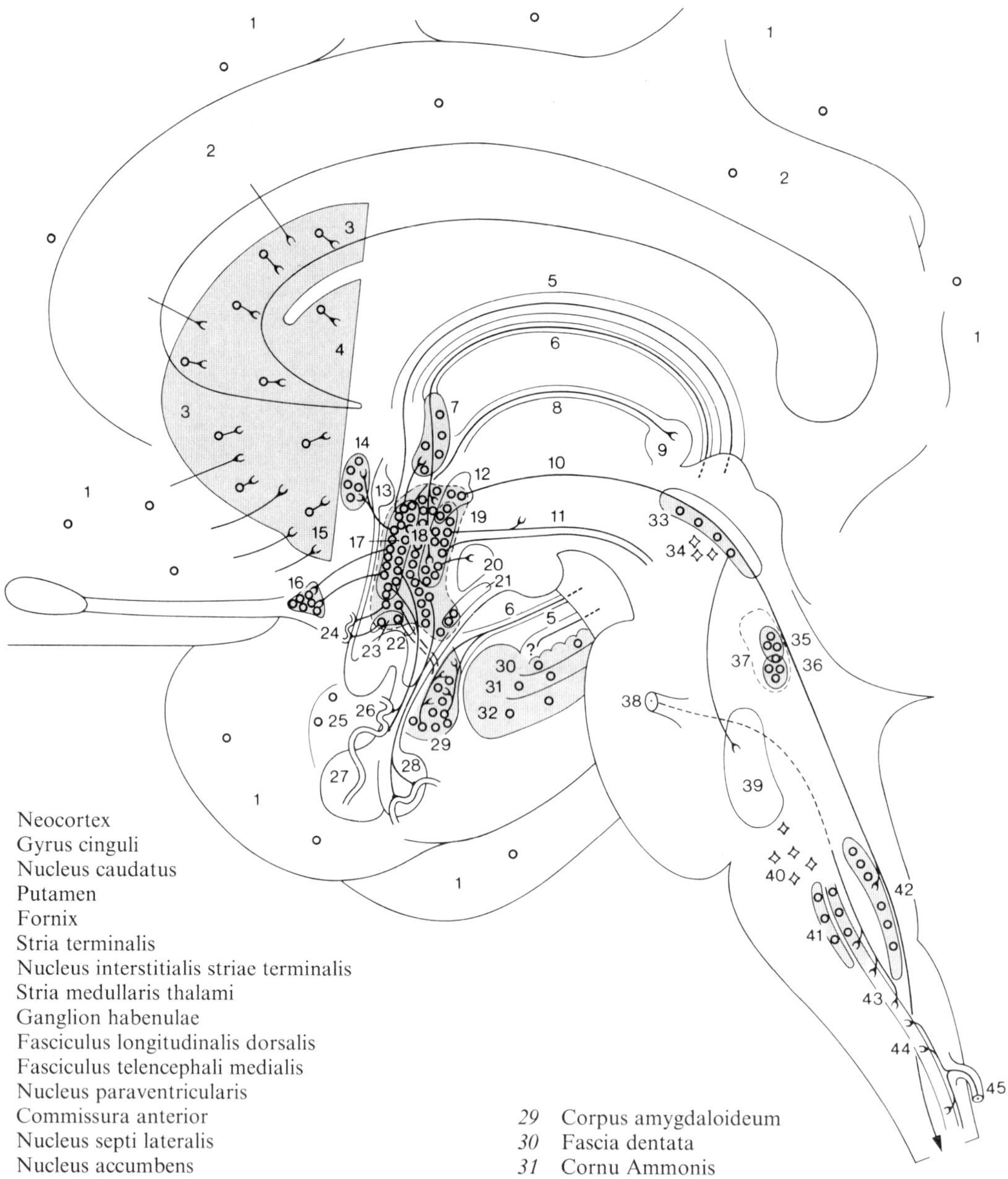

1 Neocortex
2 Gyrus cinguli
3 Nucleus caudatus
4 Putamen
5 Fornix
6 Stria terminalis
7 Nucleus interstitialis striae terminalis
8 Stria medullaris thalami
9 Ganglion habenulae
10 Fasciculus longitudinalis dorsalis
11 Fasciculus telencephali medialis
12 Nucleus paraventricularis
13 Commissura anterior
14 Nucleus septi lateralis
15 Nucleus accumbens
16 Nucleus olfactorius anterior + tuberculum olfactorium
17 Nucleus preopticus medialis
18 Preoptico-hypothalamic group of SST-containing neurons
19 Nucleus anterior hypothalami
20 Nucleus ventromedialis
21 Nucleus infundibularis
22 Fibrae amygdalofugales ventrales
23 Nucleus suprachiasmaticus
24 Organum vasculosum laminae terminalis
25 Cortex prepiriformis
26 Eminentia mediana
27 Lobus anterior hypophyseos
28 Lobus posterior hypophyseos
29 Corpus amygdaloideum
30 Fascia dentata
31 Cornu Ammonis
32 Subiculum
33 Griseum centrale mesencephali
34 Nucleus cuneiformis
35 Nucleus lemnisci lateralis dorsalis
36 Nucleus parabrachialis medialis
37 Area dorsolateralis tegmenti
38 Nervus trigeminus
39 Nucleus raphes magnus
40 Formatio reticularis rhombencephali
41 Nucleus ambiguus
42 Nucleus solitarius
43 Substantia gelatinosa nuclei spinalis nervi trigemini, pars caudalis
44 Substantia gelatinosa spinalis
45 Radix dorsalis nervi spinalis

Fig. 23. Somatostatin-containing cells and pathways

Golgi and electron-microscopical studies of the neostriatum (Takagi et al. 1983; Vincent and Johansson 1983; see, however, Pickel et al. 1980). Vincent et al. (1983b) have shown that a certain proportion of the somatostatinergic striatal neurons contain, in addition, avian pancreatic polypeptide. The elements containing these two neuropeptides appear to be specifically and intensely stainable by a histochemical technique which demonstrates NADPH-diaphorase activity.

The nucleus accumbens and the adjacent ventromedial parts of the nucleus caudatus and the putamen, which have been designated together as the limbic striatum (Kelley et al. 1982), show a much higher level of SST-immunoreactivity than do the remaining parts of the striatum (Davies et al. 1981; Beal et al. 1983). Burd et al. (1982b) have presented evidence suggesting that 50% of the total SST may be attributed to the neurons with perikarya within the striatum and that the remaining 50% may be contained in afferent terminals. Since the density of SST-positive perikarya is not greater in the limbic striatum than in the remainder of the striatum, Beal et al. (1983) suggested that the SST-containing afferent terminals may be preferentially localized in the former region. They pointed out that there is some support for this notion in previous immunohistochemical studies (Elde 1979; Bennett-Clarke et al. 1980), which reported a preponderance of SST-staining terminals in the ventromedial striatum and nucleus accumbens. The source of these terminals is unknown. However, it has been experimentally established that the frontal cortex, the cingulate cortex, the hippocampal formation and the amygdala project preferentially to the limbic striatum, and all of these structures are known to contain SST-immunoreactive cells (Bennett-Clarke et al. 1980; Finley et al. 1981b; Shiosaka et al. 1982; Sakanaka et al. 1981a).

In the rat the nucleus entopeduncularis, i.e. the homologue of the primate medial pallidal segment, contains SST-immunoreactive perikarya.

The amygdaloid complex contains numerous SST-positive cells and a dense plexus of somatostatinergic fibres (Bennett-Clarke et al. 1980; Finley et al. 1981b; Shiosaka et al. 1982; Sakanaka et al. 1981a). The densest aggregation of SST-containing perikarya is found in the central amygdaloid nucleus (Finley et al. 1981b). There is experimental evidence suggesting that SST-immunoreactive fibres originating from the amygdala project via the stria terminalis to the lateroventral part of the anterior hypothalamic nucleus and, less intensively, to the lateral hypothalamus (Sakanaka et al. 1981a). SST-positive cell bodies have also been observed in the anterior olfactory nucleus, the olfactory tubercle, the lateral septal nucleus and the bed nucleus of the stria terminalis (Finley et al. 1981b; Shiosaka et al. 1982).

The periventricular zones of the preoptic region and the anterior hypothalamus have been shown to contain a large population of SST-immunoreactive perikarya. This field of neurons is dorsocaudally continuous with a group of SST-positive perikarya situated in the parvocellular portion of the paraventricular nucleus. Ventrally, it extends into the suprachiasmatic and infundibular nuclei (Hökfelt et al. 1978b; Hoffman and Hayes 1979; Krisch 1979; Bennett-Clarke et al. 1980; Finley et al. 1981b; Shiosaka et al. 1982). In addition to these periventricular cells, the hypothalamus contains numerous scattered SST-containing perikarya located distant to the third ventricle (Finley et al. 1981b; Shiosaka et al. 1982).

The large group of periventricular, SST-positive cells gives rise to the following projections:

1. Numerous fibres pass to the neurohaemal contact zone of the median eminence. The lesion experiments of Critchlow et al. (1978) have shown that the preoptic region is an important source of these fibres. Other fibres project to the posterior lobe of the pituitary and to the organum vasculosum of the lamina terminalis, apparently terminating on capillaries (Hoffman and Hayes 1979; Finley et al. 1981b).

2. Several short-distance connections to other hypothalamic centres, including the ventromedial nucleus and the ventral premammillary nucleus, have been described (Hoffman and Hayes 1979; Krisch 1979).
3. Fibres have been reported to pass dorsally and to enter the bed nucleus of the stria terminalis (Hoffman and Hayes 1979; Woodhams et al. 1983).
4. A large number of fibres take an initially lateral course, to join the medial forebrain bundle and follow this bundle either rostrally or caudally (Palkovits et al. 1982a). The rostrally coursing fibres terminate in the olfactory tubercle (Krisch 1981; Palkovits et al. 1982a) and in the lateral septal nucleus (Palkovits et al. 1982a). The caudally coursing fibres supply, according to Krisch (1981), the various parvocellular hypothalamic nuclei, the supramamillary nuclei and the lateral mamillary nucleus. Finley et al. (1981b) were able to trace SST-containing fibres in the medial forebrain bundle as far caudal as the mesencephalic tegmentum. They considered it possible that the medial forebrain bundle conveys ascending fibres from SST-immunoreactive perikarya in the mesencephalic tegmentum to the septal region.
5. Fibres arising mainly from the preoptic portion of the periventricular complex of SST-containing cells pass via the stria medullaris to the epithalamus (Palkovits et al. 1982a), to terminate in the lateral habenular nucleus (Finley et al. 1981b). Krisch (1981), on the other hand, reported that these fibres terminate in the medial habenular nucleus; she also observed that some of the SST-containing stria medullaris fibres join the fasciculus retroflexus and reach the peripheral parts of the interpeduncular nucleus.
6. Palkovits et al. (1982a) have presented experimental evidence suggesting that fibres originating from the preoptico-hypothalamic complex of SST-immunoreactive cells enter the fornix and pass to the hippocampal region. Because this region contains SST-positive neurons, it is possible that the fornix also contains hippocampofugal somatostatinergic fibres.
7. Krisch (1978) believed that SST-containing fibres arising from the hypothalamus pass by way of the stria terminalis and the ventral amygdalofugal pathway to the amygdaloid complex. Both of these pathways have been observed to contain SST-immunoreactive fibres (Finley et al. 1981b). The experimental work of Palkovits et al. (1982a), however, suggests that, contrary to the views of Krisch (1978), most of these fibres are amygdalofugal.
8. According to Krisch (1981), the group of periventricular SST-immunoreactive hypothalamic cells gives rise to a large descending pathway which attains the most caudal parts of the neuraxis. Her observations may be summarized as follows: The fibres of this pathway follow a periventricular course and enter the fasciculus longitudinalis dorsalis of Schütz. Throughout the length of the brain stem this bundle issues somatostatinergic fibres which supply a variety of centres, including the medial geniculate body, the nucleus of Darkschewitsch, the inferior colliculus, the dorsal tegmental nucleus, the nucleus of the lateral lemniscus, the lateral parabrachial nucleus, the nucleus of the trapezoid body, the nucleus praepositus hypoglossi, the nucleus spinalis of the trigeminal nerve, the nucleus solitarius and the reticular formation. A certain proportion of these fibres enter the spinal cord to form a bundle which passes caudally along the surface of the dorsal horn. In all parts of the spinal cord this superficial bundle gives rise to a dense fibre plexus in the adjacent substantia gelatinosa. In the lower lumbar and upper sacral segments numerous fibres from the superficial bundle bend inward and converge on the central region of the cord. In the sacral segments these fibres form a caudally directed central bundle, situated immediately dorsal to the central canal. Some of the fibres of this central bundle form a terminal plexus in the central grey; others radiate laterally into the intermediolateral area.

The following comments on these findings should be made: (a) Krisch herself (1981) has emphasized that her observations require experimental verification. (b) Contrary to other authors (see below), Krisch remained unable to find any SST-positive cells in the brain stem, spinal cord or sensory ganglia; hence, she assumed that all somatostatinergic fibres present in the caudal part of the neuraxis are derived from the hypothalamic groups of SST-immunoreactive neurons. (c) With regard to the sites of termination of SST-containing fibres, several of Krisch's findings have been confirmed by other authors; however, because the brain stem, the cord and the sensory ganglia do contain SST-immunoreactive neurons, many of the terminal plexuses observed by Krisch may well have a local source. (d) Nevertheless, there is experimental evidence indicating that a limited number of SST-positive cells located in the parvocellular portion of the paraventricular nucleus project directly to the region of the dorsal vagal complex and to the spinal cord (Sawchenko and Swanson 1982b). These fibres may well descend through the brain stem via the fasciculus longitudinalis dorsalis of Schütz.

According to Finley et al. (1981b), several areas in the brain stem contain SST-positive perikarya. These areas include: the ventral and lateral regions of the periaqueductal grey and the adjacent parts of the mesencephalic reticular formation; a large, cytoarchitectonically ill-defined area in the lateral part of the rostral rhombencephalic tegmentum, extending into the dorsal nucleus of the lateral lemniscus and the medial parabrachial nucleus; the medial, large-celled part of the rhombencephalic reticular formation; the nucleus of the solitary tract; the nucleus ambiguus; the spinal trigeminal nucleus. All of these areas have been observed to contain, in addition, more or less dense plexuses of SST-immunoreactive fibres. SST-containing neuronal processes appear to be present as well in the following centres and fibre tracts: the medial pretectal area, the superior and inferior colliculi, the oculomotor, trochlear and abducens nuclei, the ventral tegmental area, all parts of the substantia nigra, the interpeduncular nucleus, the ventral and dorsal tegmental nuclei, the superior central nucleus, the pontine raphe nucleus, the nucleus raphes magnus, the pontine nuclei, certain parts of the medial and lateral rhombencephalic reticular formation, the nucleus of the trapezoid body, the nucleus fastigii, the nucleus praepositus hypoglossi, the vestibular nuclear complex, the cochlear nuclei, the caudal part of the spinal trigeminal nucleus, the nucleus gracilis, the nucleus cuneatus (lateralis and medialis), the facial nucleus, the hypoglossal nucleus, the fasciculus longitudinalis dorsalis, the brachium of the inferior colliculus, the lateral lemniscus and certain portions of the cerebellar white matter. (It appeared to be impossible to include all of these structures in Fig. 23.) So far as the distribution of SST-immunoreactive fibres in the myelencephalon is concerned, the results of Finley et al. (1981b) agree largely with those of Forssmann et al. (1979).

Our knowledge concerning the exact origin, course and termination of SST-containing fibres in the brain stem is very limited indeed. All we know at present is that, as already mentioned, the dorsal vagal complex receives some somatostatinergic afferents from the paraventricular nucleus (Sawchenko and Swanson 1982b), and that a number of SST-positive cells in the periaqueductal grey project to the nucleus raphes magnus (Beitz et al. 1983). Finley et al. (1981b) have pointed out that the presence of plexuses of SST-immunoreactive fibres in the external eye muscle nuclei, in conjunction with the localization of such plexuses in the pretectal area and the superior colliculus, suggests that somatostatinergic neurons may play a role in the coordination of eye movements. These authors also emphasized that the brain stem auditory system contains numerous SST-positive perikarya and fibres.

There is evidence suggesting that the spinal ganglia contain small, SST-immunoreactive perikarya and that the central processes of these elements enter the fasciculus dorsolateralis of Lissauer to give rise to a dense terminal plexus in the substantia gelatinosa (Hökfelt et al. 1976b; Forssmann 1978; Burnweit and Forssmann 1979; Ho and Berelowitz 1984).

SST-positive perikarya and networks of somatostatinergic fibres have been observed in the nucleus intermediolateralis of the thoracic cord (Forssmann et al. 1979) and in the area surrounding the central canal (Burnweit and Forssmann 1979). Plexuses of SST-containing fibres have also been reported to be present in the entire zona intermedia and in the columna ventralis of the lower lumbar and sacral segments (Forssmann et al. 1979).

Addenda:

1. Helke (1984) studied the SST content of the nucleus of the solitary tract in normal rats and in animals which had sustained various denervation procedures related to the nucleus. Because neither midbrain hemisection nor nodose ganglionectomy reduced the SST content of the nucleus studied, Helke concluded that the SST innervation of the nucleus of the solitary tract originates mainly from rhombencephalic centres.
2. Kalia et al. (1984a) provided physiological evidence suggesting that SST-containing neurons located in the nucleus of the solitary tract are involved in reflex circuits along which an inhibitory influence is exerted on respiratory efferent centres.
3. Schrøder (1984) made an immunohistochemical study of the distribution of SST in the spinal cord of the rat with particular reference to the localization in the caudal centres that innervate the pelvic organs. His main results may be summarized as follows: (a) Deafferentiation experiments showed that the bulk of the spinal SST has an intrinsic spinal origin. (b) The marginal layer and particularly the substantia gelatinosa contain a dense immunoreactivity in terminal-like structures. Such structures were also found along the medial border of the dorsal horn and in the nucleus of the dorsolateral funiculus. In all of these regions SST-positive perikarya were also observed. (c) Many terminals were observed in the sacral parasympathetic intermediolateral nucleus, but in the sympathetic nuclei only a few of these structures appeared to be present. (d) Immunoreactive somata are present in the area surrounding the central canal at all levels. (e) The intermediolateral nucleus of the sacral cord contains SST-positive parasympathetic preganglionic neurons. (f) The ventral horn generally contains few SST terminals; however, a dense network of SST-containing fibres was found in the sixth lumbar segment in relation to the neurons in Onuf's nucleus X complex, i.e. the nucleus that innervates the small pelvic muscles including the striated sphincters.

Thyrotropin-Releasing Hormone

Thyrotropin-releasing hormone (TRH) was the first hypothalamic releasing hormone to be isolated and chemically identified (Burgus et al. 1970; Nair et al. 1970). It is a tripeptide (Fig. 20) which regulates the release of thyrotropin, or thyroid-stimulating hormone (TSH), and prolactin by the anterior pituitary. Antibodies against TRH have been produced and used to study the distribution of this peptide in the central nervous system by radioimmunoassay, and its localization in neurons and their processes by immunohistochemistry.

Radioimmunological studies have shown that TRH is widespread throughout the central nervous system of mammals and man (Brownstein et al. 1974; Jackson and Reichlin 1974; Oliver et al. 1974; Winokar and Utiger 1974; Okon and Koch 1976; Parker and Porter 1983). In fact, the hypothalamus contains only about one-third of the total amount of this peptide that is present in the brain. These findings suggest that TRH exerts influences in the brain which may well be independent of its neuroendocrine functions. It is worthy of note in this context that TRH inhibits neurons when applied iontophoretically (Dyer and Dyball 1974; Renaud et al. 1975; Winokur and Beckman 1978).

Regional analyses have shown that the greatest concentration of TRH is in the median eminence (Brownstein et al. 1974). High concentrations have also been found in several hypothalamic nuclei, including the ventromedial, dorsomedial and infundibular (Brownstein et al. 1974). Outside the hypothalamus relatively large amounts of TRH have been found in the preoptic and septal areas, in the nucleus solitarius and in the motor nuclei of cranial nerves III, V, VII, X and XII (Brownstein 1974; Kubek et al. 1983; Eskay et al. 1983).

Immunohistochemical studies have revealed the presence of TRH-positive perikarya in the hypothalamus, i.e. in the periventricular area, the periparaventricular area, the dorsomedial and ventromedial nuclei, a basal hypothalamic-suprachiasmatic complex, the perifornical area and the lateral hypothalamus; cell bodies are also present in the preoptic suprachiasmatic nucleus and in the rhombencephalic raphe magnus and raphe pallidal nuclei (Hökfelt et al. 1975a, b; Johansson and Hökfelt 1980). At least some of the TRH-positive raphe cells also contain both substance P and serotonin (Johansson and Hökfelt 1980).

The highest concentration of TRH-immunoreactive terminals has been found in the medial part of the external layer of the median eminence. Dense networks of fibres have also been seen in the parvocellular part of the paraventricular nucleus, the dorsomedial nucleus, the perifornical region, the dorsal part of the nucleus accumbens and the bed nucleus of the stria terminalis, particularly its ventral part. Less dense networks of TRH-immunoreactive fibres have been observed in many parts of the brain, including the periventricular hypothalamic area, the medial part of the ventromedial nucleus, the zona incerta, the organum vasculosum laminae terminalis, the ventral part of the lateral septal nucleus, the ventral part of the lateral parabrachial nucleus, the peripheral part of the nucleus solitarius, the rhombencephalic reticular formation and the motor nuclei of III, V, VII, X and XII (Hökfelt et al. 1975a, b; Johansson and Hökfelt 1980).

TRH-immunoreactive fibres also occur in the spinal cord. Such fibres are found in the area surrounding the central canal, but the densest region of TRH-immunoreactivity in the cord is observed in laminae VII, VIII and IX. Many fibres are seen in close association with α-motoneurons (Hökfelt et al. 1975b; Gibson et al. 1981).

Some experimental studies on the origin, course and termination of TRH-containing pathways have been done. After a total deafferentiation of the medial basal hypothalamus in the rat, the concentration of TRH within the isolated island of hypothalamic tissue appeared to be 76% lower than in tissue from sham-operated control animals (Brownstein et al. 1975). Thus, much of the TRH that is normally present in the medial basal hypothalamus is presumably synthesized by cells outside of this region. Hökfelt et al. (1978a, b) reported that after a lesion in the hypothalamic periventricular anterior area all TRH-containing nerve terminals in the median eminence disappear. Aizawa and Greer (1981) provided experimental evidence indicating that only the immediate area of the paraventricular nucleus is important for TSH secretion and showed, furthermore, that the medial preoptic area tonically inhibits TSH secretion. The course of the TRH-containing fibres that pass to the median eminence was investigated by Palkovits et al. (1982c), by means of various surgical transections. Their results suggest that these fibres reach the medial basal hypothalamus from an anterolateral direction, just as several other classes of peptide-containing fibres do. Hökfelt et al. (1978a) demonstrated that TRH-positive terminals in the ventral horn disappear below the lesion after a total transection of the spinal cord, indicating the existence of supraspinal descending TRH-containing projections. The paraventricular nucleus does not contribute significantly to the TRH-innervation of the spinal cord (Lechan et al. 1983), but double-labelling experiments have shown that TRH-positive neurons present in the nucleus raphes magnus and in the ventrolateral part of the reticular formation project to the spinal cord (Bowker et al. 1983).

Neurohypophyseal Peptides

Vasopressin and Oxytocin

It has been long known that the nonapeptide hormones arginine-vasopressin (AVP) and oxytocin (OXT; Fig. 24) are produced in the nucleus supraopticus and the nucleus paraventricularis, two magnocellular nuclei situated in the anterior hypothalamus (Bargmann and Scharrer 1951). Both hormones are formed in the perikarya and conjugated with a larger carrier protein called a neurophysin. Individual supraoptic and paraventricular neurons synthesize either AVP or OXT, but not both (Dierickx and Vandesande 1979; Vandesande and Dierickx 1979; Sofroniew et al. 1979). Their axons descend through the infundibular stalk to the posterior pituitary, thus forming the paraventriculo- and supraoptico-hypophyseal pathways. The neurohormones AVP and OXT and their respective neurophysins are transported along the axons of these pathways and released into the blood vessels of the neurohypophysis. AVP is responsible for antidiuresis, promoting reabsorption of water by the kidney; OXT stimulates smooth muscle in the uterus and mammary glands to contract.

The structure and functions of the classical hypothalamo-hypophyseal neurosecretory system just outlined are well established (Sofroniew et al. 1979; Swanson and Sawchenko 1983; Silverman and Zimmerman 1983). However, recent investigations using antibodies raised against AVP, OXT and their associated neurophysins, and in particular studies combining immunohistochemistry with fibre-tracing techniques, have revealed that (a) neurons synthesizing AVP and OXT are not confined to the supraoptic and paraventricular nuclei, and (b) axons transporting these neurohormones are not directed only to the neurohypophysis; rather they form an extensive network in the central nervous system, innervating a large number of functionally diverse centres. The most important results of these recent investigations are summarized below (Fig. 25; for reviews, see Buijs et al. 1983; Sofroniew 1983).

The hypothalamo-hypophyseal pathway is reinforced by fibres originating from small groups of large AVP and OXT cells which are scattered over the rostral hypothalamus. These so-called accessory nuclei (which are not depicted in Fig. 25) contain, as do the supraoptic and paraventricular nuclei, intermingled populations of AVP and OXT neurons (Peterson 1966; Sofroniew 1983).

The paraventricular nucleus sends AVP-containing fibres to the median eminence, where

Neurohypophyseal peptides

Arginine vasopressin

H-Cys-Tyr-Phe-Gln-Asn-Cys-Pro-Arg-Gly-NH_2

Oxytocin

H-Cys-Tyr-Ile-Gln-Asn-Cys-Pro-Leu-Gly-NH_2

Fig. 24. Primary structure of arginine vasopressin and oxytocin. For significance of the amino acid code, see Table 7 on p. 198

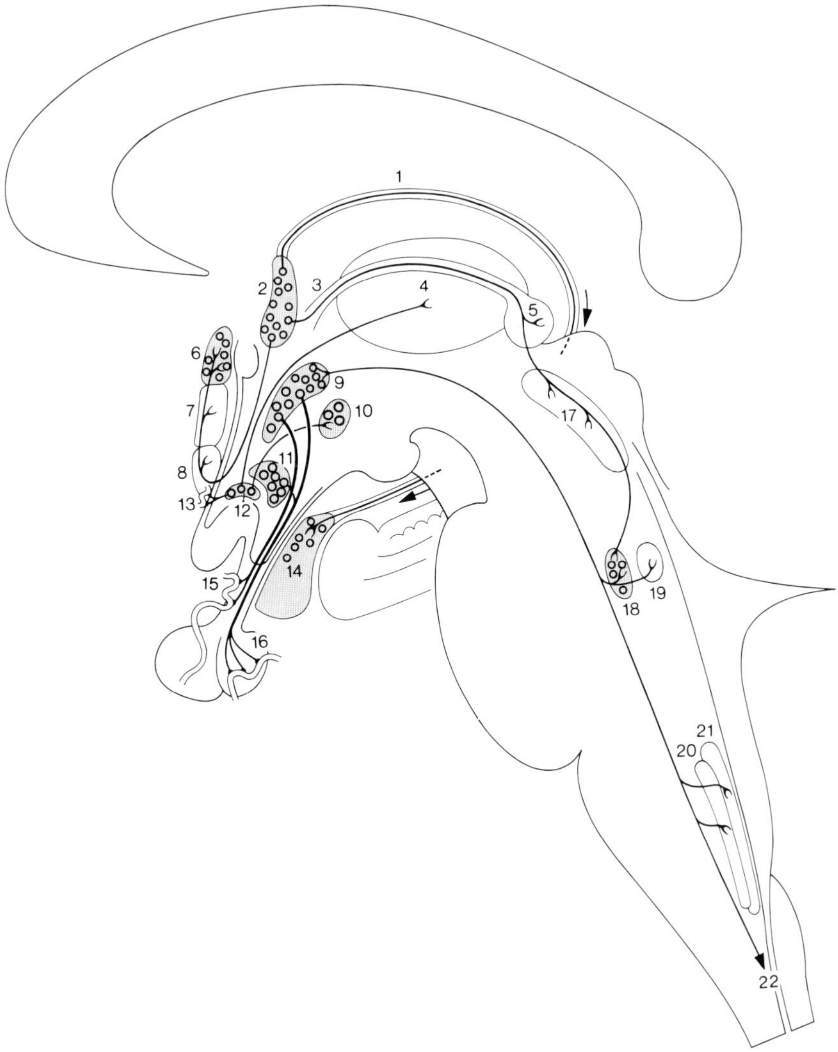

1 Stria terminalis
2 Nucleus interstitialis striae terminalis
3 Stria medullaris thalami
4 Thalamus
5 Nucleus habenulae lateralis
6 Nucleus septi lateralis
7 Nucleus gyri diagonalis, pars dorsalis
8 Nucleus gyri diagonalis, pars ventralis
9 Nucleus paraventricularis
10 Nucleus dorsomedialis
11 Nucleus supraopticus
12 Nucleus suprachiasmaticus
13 Organum vasculosum laminae terminalis
14 Corpus amygdaloideum
15 Eminentia mediana
16 Hypophysis, lobus posterior
17 Griseum centrale mesencephali
18 Locus coeruleus
19 Nuclei parabrachiales
20 Nucleus dorsalis nervi vagi
21 Nucleus solitarius
22 Nucleus intermediolateralis; substantia gelatinosa

Fig. 25. Vasopressin- and oxytocin-containing cell groups and pathways

their terminals contact the hypophyseal portal capillaries (Antunes et al. 1977; Sofroniew et al. 1979). It has been suggested that the AVP released from these terminals is involved in the regulation of adenohypophyseal ACTH secretion (Bugnon et al. 1982), but the validity of this proposal has been questioned by Silverman and Zimmerman (1983).

The caudal portion of the paraventricular nucleus contains numerous peptidergic cells which are clearly smaller than the conspicuous, large neurosecretory elements concentrated in the rostral part of this centre. Most of these small elements produce OXT, but some synthesize AVP. This parvocellular part of the paraventricular nucleus gives rise to a descending projection which distributes fibres to the locus coeruleus, the parabrachial nuclei, the dorsal motor vagal nucleus and the nucleus of the solitary tract (Swanson 1977; Buijs 1978; Buijs et al. 1978; Sofroniew and Weindl 1978; Hosoya and Matsushita 1979; Swanson and Hartman 1980; Swanson and Kuypers 1980; Sawchenko and Swanson 1982b; De Vries and Buijs 1983). A certain proportion of these descending fibres enter the dorsal parts of the lateral funiculus of the spinal cord. Within the spinal grey matter these fibres terminate predominantly in the intermediolateral column, the central grey and the marginal zone of the dorsal horn (Swanson and McKellar 1979). In the vagal complex the amount of OXT exceeds that of AVP to an even greater extent (Jenkins et al. 1984). The role of AVP and OXT in the connection between the paraventricular nucleus and the autonomic centres in the lower brain stem and spinal cord may well lie in the regulation of processes in which these peptides are also involved in the periphery, such as the control of blood pressure and lactation (Buijs et al. 1983). This coupling of central and peripheral actions of AVP and OXT has been suggested as a general feature of hypothalamic peptides (see Swaab 1982).

A certain proportion of the small neurons contained within the suprachiasmatic nucleus contain AVP and its related neurophysin (Buijs et al. 1978; van Leeuwen et al. 1978; Sofroniew and Weindl 1980; Stopa et al. 1984). In the rat these fibres have been traced to the organum vasculosum of the lamina terminalis, the dorsomedial hypothalamic nucleus and the periventricular thalamic nucleus (Hoorneman and Buijs 1982).

A considerable number of small AVP cells has been found in the bed nucleus of the stria terminalis (van Leeuwen and Caffé 1983). De Vries and Buijs (1983) have shown that fibres originating from these cells pass via the diagonal band of Broca to the nucleus of the same name and to the lateral part of the septum. Moreover, these authors presented experimental evidence suggesting that AVP-containing fibres arising from the bed nucleus of the stria terminalis also project to the anterior area of the amygdala, the lateral habenular nucleus, the mesencephalic central grey and the locus coeruleus.

Several other cell masses, including the lateral septal nucleus, the medial part of the amygdaloid complex, the dorsal portion of the nucleus dorsomedialis hypothalami and the locus coeruleus, have been reported to contain numerous AVP-immunoreactive perikarya (Caffé and van Leeuwen 1983). However, the efferents and areas of termination of these elements are unknown.

It has already been mentioned that the peptidergic projections to the parabrachial nuclei, the vagal complex and the spinal nucleus intermediolateralis are probably involved in central autonomic regulation. The fibres passing to the superficial part of the dorsal horn may well modulate the processing of nociceptive information. Much remains to be learned concerning the functional significance of the various AVP and OXT pathways in the higher parts of the brain. However, it has been suggested that in the septum, AVP may be involved in thermoregulation (Kasting et al. 1981, 1982), and the extensive behavioural studies of De Wied and collaborators (De Wied 1976, 1983; De Wied et al. 1976, 1984) have shown that AVP and related hormones affect memory and learning processes, at least in rats. In this context it is interesting that AVP and OXT fibres are present in several areas thought to be involved in memory processes, such as the mediodorsal thalamic nucleus, the hippocampus and the neocortex. (For a recent review on the possible role of AVP in central integrative processes, see Doris 1984.)

Addendum: Mason et al. (1984) have presented experimental neuroanatomical and electrophysiological evidence indicating that in addition to their functional role in neurosecretion, the magnocellular neurons of the supraoptic nucleus communicate via axon collaterals with other neurons in the lateral hypothalamus.

Pro-Opiomelanocortin Derivatives

Corticotropin, or adrenocorticotropic hormone (ACTH), is a member of the opiomelanocortin family of peptides (Fig. 26). The members of this family share a common large-molecular-weight precursor, pro-opiomelanocortin (POMC). This common precursor is cleaved to yield equimolecular amounts of ACTH, β-lipotropin (β-LPH), and a third moiety known as the N-terminus region, or 16-K fragment. By enzymatic processing all three of these peptides can be cleaved further into smaller biologically active peptides. Thus, ACTH is the source of α-melanocyte-stimulating hormone (α-MSH) and corticotropin-like intermediate lobe peptide (CLIP), while β-LPH can be cleaved to yield β-endorphin (β-END) and γ-lipotropin (γ-LPH). These last two peptides are, in turn, the precursors of two still smaller peptides, the former giving rise to β-END 1–27, the latter to β-melanocyte-stimulating hormone (β-MSH). A third melanocyte-stimulating hormone, known as γ-MSH, stems from the N-terminus region of the POMC molecule (Fig. 26; Mains et al. 1977; Eipper and Mains 1980; Akil et al. 1984).

Peptides derived from POMC are found in the pituitary as well as in the brain. It has been established that in the corticotrophs of the anterior pituitary ACTH and β-LPH are the main products of the precursor, while in the cells of the intermediate lobe of the pituitary the smaller fragments of the POMC molecule, α-MSH, CLIP, γ-LPH and β-END, predominate. Thus, it appears that the POMC-containing cells in the anterior lobe and those in the intermediate lobe of the pituitary process the precursor quite differently (Moon et al. 1973; Pelletier et al. 1977; Mains et al. 1977; Bloom et al. 1977; Guillemin et al. 1977; Mains and Eipper 1979; Akil et al. 1984).

In the brain, neurons containing peptides derived from POMC are found in the mediobasal hypothalamus, in a region largely coinciding with the nucleus infundibularis. Indeed, the presence in this region of compounds identical to or immunologically related to all of the known members of the POMC family has been established by radioimmunological and/or immunohistochemical studies: POMC (Liotta et al. 1979), 16-K fragment (Guy et al. 1980), ACTH (Krieger et al. 1977a; Watson et al. 1978; Pelletier and Leclerc 1979; Léranth et al. 1980; Joseph 1980; Knigge et al. 1981), β-LPH (Watson et al. 1977; Zimmerman et al. 1978; Pelletier et al. 1980), γ-MSH (Kawai et al. 1984), α-MSH (Dubé et al. 1978; Jacobowitz and O'Donohue 1978; Oliver and Porter 1978; Watson and Akil 1980; Gramsch et al. 1980), CLIP (Emson et al. 1984), γ-LPH (Pique et al. 1981; Emson et al. 1984), β-END (Gramsch et al. 1980; Pique et al. 1981; Finley et al. 1981a), β-MSH (Bloch et al. 1979; Bugnon et al. 1979), α-END (Bloch et al. 1979; Bugnon et al. 1979). Moreover, a large number of immunohistochemical studies have shown that cell bodies in the infundibular nucleus may simultaneously contain two or more of these peptides, e.g. ACTH + β-LPH (Sofroniew 1979; Nilaver et al. 1979), ACTH + α-MSH + 16-K fragment (Guy et al. 1980) and β-LPH + CLIP + α-MSH + β-END (Bloch et al. 1978, 1979; Bugnon et al. 1979). According to a recent radioimmunological study (Emson et al. 1984), in the human

Pro-opiomelanocortin and its derivatives (human)

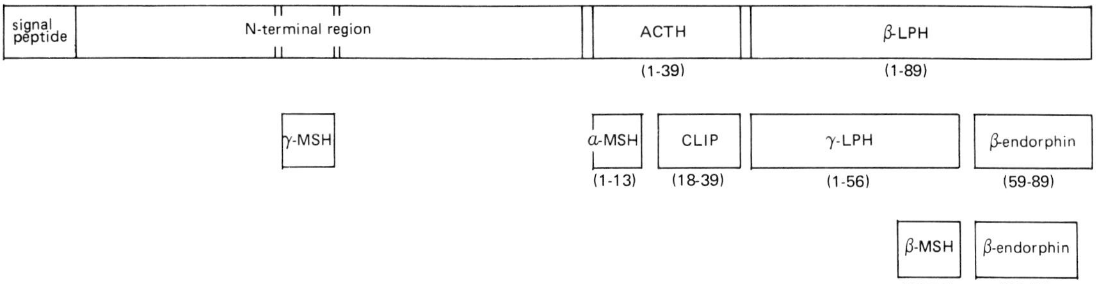

Adrenocorticotropic hormone [ACTH (1 - 39)] = Corticotropin

H-Ser^{1}-Tyr-Ser-Met-Glu-His-Phe-Arg-Trp-Gly^{10}-Lys-Pro-Val-Gly-Lys-Lys-Arg-Arg-Pro-Val^{20}-Lys-Val-Tyr-Pro-Asn-Gly-Ala-Glu-Asp-Glu^{30}-Ser-Ala-Glu-Ala-Phe-Pro-Leu-Glu-Phe^{39}-OH

α-Melanocyte-stimulating hormone [α-MSH] = Ac - ACTH (1 - 13) - NH_2

Corticotropin-like intermediate lobe peptide [CLIP] = ACTH (18 - 39)

β-Lipotropin [β-LPH (1 - 89)]

H-Glu^{1}-Leu-Thr-Gly-Gln-Arg-Leu-Arg-Glu-Gly^{10}-Asp-Gly-Pro-Asp-Gly-Pro-Ala-Asp- Asp-Gly^{20}-Ala-Gly-Ala-Gln-Ala-Asp-Leu-Glu-His-Ser^{30}-Leu-Leu-Val-Ala-Ala-Glu-Lys-Lys-Asp-Glu^{40}-Gly-Pro-Tyr-Arg-Met-Glu-His-Phe-Arg-Trp^{50}-Gly-Ser-Pro-Pro-Lys-Asp-Lys-Arg-Tyr-Gly^{60}-Gly-Phe-Met-Thr-Ser-Glu-Lys-Ser-Gln-Thr^{70}-Pro-Leu-Val-Thr-Leu-Phe-Lys-Asn-Ala-Ile^{80}-Ile-Lys-Asn-Ala-Tyr-Lys-Lys-Gly-Glu^{89}-OH

γ-LPH = β-LPH (1 - 56)

β-Endorphin = β-LPH (59 - 89)

H-Tyr^{1}-Gly-Gly-Phe-Met-Thr-Ser-Glu-Lys-Ser^{10}-Gln-Thr-Pro-Leu-Val-Thr-Leu-Phe-Lys-Asn^{20}-Ala-Ile-Ile-Lys-Asn-Ala-Tyr-Lys-Lys-Gly^{30}-Glu-OH

β-MSH = β-LPH (39 - 56)

γ-MSH

H-Tyr^{1}-Val-Met-Gly-His-Phe-Arg-Trp-Asp-Arg-Phe-Gly^{12}-OH

Fig. 26. Schematic representation of the human pro-opiomelanocortin (POMC) precursor molecule and some of its cleavage products (*above*); simplified from Udenfriend and Kilpatrick (1983). Primary structure of some POMC derivatives (*below*). For significance of the amino acid code, see Table 7 on p. 198

brain the smaller fragments of POMC, β-END, γ-LPH, α-MSH, CLIP and ACTH predominate. How the neurons in the infundibular nucleus process POMC is at present a matter of controversy (Akil et al. 1984). It cannot be excluded that different sets of infundibular neurons follow different patterns of enzymatic cleavage, leading to different mixtures of peptides. Ultrastructural studies have shown that some of the peptides mentioned, such as α-MSH (Pelletier and Dube 1977) and ACTH (Pelletier and Leclerc 1979), are stored in dense core vesicles, suggesting that they may play a role as central neurotransmitters or neuromodulators. Others, however, may only be present as intermediate products, i.e. as precursors of smaller peptides.

Following these introductory notes is a discussion of the ACTH-containing neurons and their projections. Subsequently, consideration is given to the distribution of some other members of the opiomelanocortin family, i.e. β-LPH, α-, β- and γ-MSH, and β-END, and finally a brief comment on the functional significance of these substances is made.

Corticotropin, or Adrenocorticotropic Hormone (ACTH)

Immunohistochemical studies have shown that numerous cells in the mediobasal hypothalamus contain ACTH (Sofroniew 1979; Pelletier and Leclerc 1979; Léranth et al. 1980; Joseph 1980; Knigge et al. 1981). These elements form a group of evenly distributed cells which largely coincides with the infundibular nucleus (Fig. 27). However, it extends rostrally into the retrochiasmatic area, caudally into the submamillary region, and dorsally into the zone between the ventricular surface and the ventromedial hypothalamic nucleus (Knigge et al. 1981). That ACTH is actually formed in the brain, rather than transported there from the pituitary, appears from the fact that hypophysectomy does not influence the ACTH content of the brain in the rat (Krieger et al. 1977a). As mentioned before, cells in the mediobasal hypothalamus containing ACTH have been shown to be simultaneously immunoreactive with antisera against many other members of the opiomelanocortin family, including β-LPH, α-MSH, β-MSH, α-END and β-END (Bloch et al. 1979; Bugnon et al. 1979; Sofroniew 1979; Nilaver et al. 1979; Bloch et al. 1978). According to Sofroniew (1979) all perikarya containing ACTH contain β-END as well, and Nilaver et al. (1979) made a similar observation for ACTH and β-LPH. Because of these findings the set of neurons outlined above may be designated as the bed nucleus of the opiomelanocortin peptide group (Joseph 1980). This bed nucleus gives rise to numerous ACTH-containing fibres which are distributed widely over the brain (Joseph 1980; Romagnano and Joseph 1983; Knigge and Joseph 1984). The main features of this projection are summarized below (Fig. 27).

Dense accumulations of ACTH-positive fibres and terminals are found in the anterior, mediobasal and periventricular zones of the hypothalamus. Many of the periventricular fibres penetrate the ependymal lining of the third ventricle. The dorsomedial nucleus and the paraventricular nucleus (both its parvo- and magnocellular portions) are also heavily innervated by ACTH-immunoreactive fibres, and the same holds true for the organum vasculosum of the lamina terminalis. Projections to the neurohypophysis proceed into the internal zone of the median eminence, descend into the pituitary stalk and distribute fairly uniformly throughout the neural lobe (Knigge and Joseph 1981).

Numerous ACTH-positive fibres pass rostrodorsally to innervate the medial preoptic area, the lateral septal nucleus and the bed nucleus of the stria terminalis. Other fibres pass laterally and distribute themselves to the central, basolateral and dorsal portions of the basomedial amygdaloid nuclei. A certain proportion of these fibres course in the ventral amygdalofugal pathway.

Large numbers of ACTH-immunoreactive fibres emerge from the bed nucleus of opio-

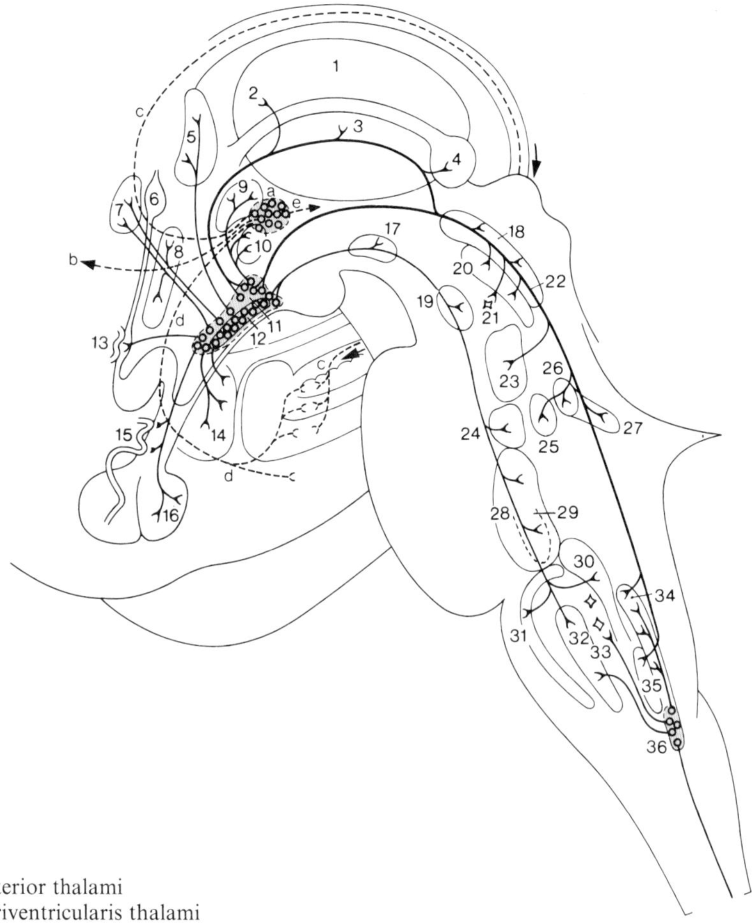

1 Thalamus
2 Nucleus anterior thalami
3 Nucleus periventricularis thalami
4 Nuclei habenulae
5 Nucleus interstitialis striae terminalis
6 Commissura anterior
7 Nucleus septi lateralis
8 Nucleus preopticus medialis
9 Nucleus paraventricularis
10 Nucleus dorsomedialis
11 Hypothalamic bed nucleus of opiomelanocortin cells
12 Nucleus infundibularis
13 Organum vasculosum laminae terminalis
14 Corpus amygdaloideum
15 Eminentia mediana
16 Lobus posterior hypophyseos
17 Area tegmentalis ventralis
18 Griseum centrale mesencephali
19 Cell group A8
20 Nucleus raphes dorsalis
21 Formatio reticularis mesencephali
22 Fasciculus longitudinalis dorsalis
23 Nucleus centralis superior
24 Nucleus raphes pontis
25 Locus coeruleus
26 Nuclei parabrachiales
27 Cell group A4
28 Nucleus raphes magnus
29 Cell group A5
30 Nucleus raphes obscurus
31 Nucleus raphes pallidus
32 Cell group A1
33 Nucleus reticularis gigantocellularis
34 Nucleus solitarius
35 Cell group A2
36 Medullary group of opiomelanocortin cells
a Group of α-MSH-containing cells situated in dorsal hypothalamus and zona incerta
b Projection of group *a* to nucleus caudatus and putamen
c Dorsal projection of group *a* to formatio hippocampi
d Ventral projection of group *a* to formatio hippocampi
e Projection of group *a* to the spinal cord

Fig. 27. Corticotropin- and α-MSH-containing cells and pathways

melanocortin-containing cells and pass dorsocaudally in the subependymal layer of the third ventricle. These periventricular fibres constitute separate anterior and posterior components, which join in the region of the posterior commissure, from where they continue caudally in the fasciculus longitudinalis dorsalis of Schütz. The anterior component gives off fibres which innervate the nucleus periventricularis thalami and the nucleus anterior dorsalis thalami, while in the epithalamic region fibres are present in the zone between the medial and lateral habenular nuclei.

Numerous cell masses in the brain stem are supplied by ACTH-containing fibres. Somewhat schematically, it may be said that these fibres are concentrated into two fibre streams, a larger dorsomedial one and a smaller ventrolateral one. The dorsomedial stream occupies a periventricular position and follows the trajectory of the fasciculus longitudinalis dorsalis; the ventrolateral stream passes caudally through the lateral part of the mesencephalic and rhombencephalic tegmentum. Throughout the brain stem the dorsomedial and ventrolateral streams are interconnected by numerous ACTH-positive fibres. The cell masses supplied by the dorsal stream include the periaqueductal grey, the nucleus cuneiformis, the locus coeruleus, the dorsal and ventral parabrachial nuclei, the A4 and A2 groups of Dahlström and Fuxe (1964) and the nucleus solitarius. Throughout the rostral rhombencephalon, a moderate number of ACTH-containing fibres are present within the ependymal lining of the tegmentum. The ventrolateral stream of ACTH-immunoreactive fibres follows a direct caudal course and distributes to the area tegmentalis ventralis and to the A8 and A5 groups. The raphe nuclei in the brain stem also receive ACTH-positive fibres. The exact source and course of these fibres remains to be determined; however, it seems that the fibres terminating in the nucleus raphes dorsalis and the nucleus centralis superior arise from the dorsomedial fibre stream, whereas the nuclei raphes pontis, raphes magnus, raphes obscurus and raphes pallidus are supplied mainly by the ventrolateral stream.

It is important to note that, apart from the large hypothalamic group, a small second pool of opiomelanocortin cells appears to be present in the caudal, commissural portion of the nucleus solitarius. Within this group, cells immunoreactive with antisera against endorphins (Schwartzberg and Nakane 1981), ACTH (Knigge and Joseph 1981; Romagnano and Joseph 1983; Joseph et al. 1983) and γ-MSH (Kawai et al. 1984) have been observed so far. There is experimental evidence indicating that axons emanating from this group of neurons are distributed to all divisions of the nucleus solitarius, project to the nucleus reticularis gigantocellularis and the A 1 group, and descend throughout the length of the spinal cord in the area surrounding the central canal.

Addendum: Using a retrograde labelling technique in combination with immunohistochemistry, Kitahama et al. (1984) produced evidence suggesting that in the cat, ACTH-immunoreactive cells situated in the arcuate nucleus project to the pontine tegmentum, particularly to the locus coeruleus.

β-Lipotropin

Radioimmunological (Krieger et al. 1977b) and biochemical (Akil et al. 1978) studies have shown that β-lipotropin (Fig. 26) is present in various parts of the brain, and there is ample evidence that many elements in the hypothalamic bed nucleus of opiomelanocortin cells contain this peptide (Zimmerman et al. 1978; Bloch et al. 1978, 1979; Nilaver et al. 1979; Bugnon et al. 1979; Pelletier et al. 1980). Fibres containing β-LPH are scattered throughout the hypothalamus (Zimmerman et al. 1978) and Watson and collaborators (1978) observed that β-LPH-immunoreactive fibres and ACTH-immunoreactive fibres are distributed in a very similar fashion throughout the brain. However, to my knowledge, detailed mapping studies of the β-LPH-containing fibres in the central nervous system have not yet been carried out. The question as to whether in the pituitary β-LPH is actually secreted as a hormone or is present only as a precursor for smaller peptides so far remains unanswered (Grossman and Rees 1983), and mutatis mutandis the same holds true for its role in the central nervous system.

α-Melanocyte-Stimulating Hormone

α-Melanocyte-stimulating hormone (α-MSH) is a tridecapeptide (Fig. 26) secreted by the intermediate lobe cells of the pituitary gland of all groups of vertebrates. In poikilotherm vertebrates, α-MSH is clearly involved in the control of skin color adaptation, but its function as a hormone in homeotherms remains unknown. Radioimmunological (Dubé et al. 1978; Oliver and Porter 1978; Gramsch et al. 1980) and immunohistochemical studies (Dubé et al. 1978; Jacobowitz and O'Donohue 1978; Bloch et al. 1978, 1979; Bugnon et al. 1979; Pelletier et al. 1980; Watson and Akil 1980) have revealed that α-MSH is also present in the mammalian central nervous system. The hypothalamic bed nucleus of opiomelanocortin cells contains numerous α-MSH-positive elements, the distribution of the axons of which in the brain appears to be very similar to that reported for β-LPH and ACTH (Jacobowitz and O'Donohue 1978). However, the hypothalamus contains a second group of α-MSH-positive cells which, with regard to position, biochemical properties and distribution of its efferents, differs from that situated in the mediobasal hypothalamus (Watson and Akil 1980; Guy et al. 1980; Jegou et al. 1983; Köhler et al. 1984b). The elements of this second group are located mainly in the dorsal portion of the intermediate hypothalamic zone and in the adjacent zona incerta (Fig. 27). Within the hypothalamus these elements tend to concentrate in the area between the dorsomedial nucleus and the fornix and in the lateral part of the lateral hypothalamic area (Köhler et al. 1984b). Interestingly, the elements of this dorsal group contain exclusively α-MSH, and no other peptides related to the opiomelanocortin family (Watson and Akil 1980; Guy et al. 1980; Jegou et al. 1983); hence, the biosynthetic route for the production of α-MSH in these cells may well be quite different from that in the cells in the intermediate lobe of the pituitary and in the mediobasal hypothalamus. There is experimental evidence indicating that the dorsal group of α-MSH-positive cells, unlike those of the mediobasal group, projects to the caudate-putamen complex, the neocortex, the entorhinal cortex and the various parts of the hippocampal formation (Guy et al. 1980; Köhler et al. 1984b). Moreover, it has been shown that a certain proportion of these elements project to the spinal cord (Köhler et al. 1984b). The fibres passing from the dorsal group of α-MSH-positive cells to the hippocampus follow two different routes, one dorsal and the other ventral. The fibres which follow the dorsal route course through the septal region and enter the fimbria, while those following the ventral route pass through the amygdala and the piriform cortex.

Addenda:

1. Using a combination of techniques, Shiosaka and Tohyama (1984) demonstrated that α-MSH-immunoreactive neurons situated in the stratum pyramidale and stratum oriens of the cornu Ammonis project to the contralateral hippocampal formation.
2. Köhler and Swanson (1984) have shown that numerous α-MSH-positive cells situated in the lateral hypothalamic area and zona incerta also contain the enzyme AChE but not ChAT. They also demonstrated that at least some of the elements in which α-MSH and AChE are co-localized project to the hippocampal formation and the neocortex.

β-Melanocyte-Stimulating Hormone (β-MSH; Fig. 26)

According to Bloch and colleagues (1978, 1979; Bugnon et al. 1979) the hypothalamic bed nucleus of opiomelanocortin cells contains neurons which are immunoreactive with anti-β-MSH-serum.

γ-Melanocyte-Stimulating Hormone (γ-MSH)

The N-terminus region of POMC contains a structure which, on the basis of its close resemblance to α-MSH, has been designated as γ-MSH (Fig. 26; Nakanishi et al. 1979). Radioimmunological and immunohistochemical studies (Bloom et al. 1980; Shibasaki et al. 1980, 1981; Tanaka et al. 1980; Osamura et al. 1982) have shown that γ-MSH is present in the brain as well as in the pituitary. According to the recent mapping study of Kawai and collaborators (1984), the brain of the rat contains two groups of γ-MSH-immunoreactive cells, a large one in the mediobasal hypothalamus and a small one in the caudal, commissural part of the nucleus solitarius. The distribution of the fibres emanating from the hypothalamic groups of γ-MSH-positive cells is very similar to that reported for ACTH. Thus, γ-MSH-containing fibres have been observed in, among other centres, the hypothalamic periventricular zone, the arcuate, paraventricular and dorsomedial nuclei, the medial and lateral preoptic areas, the ventrolateral part of the septum, the amygdaloid complex, the bed nucleus of the stria terminalis, the periaqueductal grey, the area tegmentalis ventralis, the A8 group, the cuneiform nucleus and the lateral parabrachial nucleus. Kawai and collaborators (1984) also observed γ-MSH-positive fibres in the stria terminalis, the diagonal band of Broca, and the nucleus accumbens, and they emphasized that a cell group known as Barrington's nucleus is densely innervated by γ-MSH-positive fibres. This cell group is located directly rostromedial to the locus coeruleus, and is supposed to represent the pontine micturition reflex centre. In contrast, only a few fibres were seen in the locus coeruleus. The caudal rhombencephalon also appears to contain γ-MSH-immunoreactive fibres. However, the fibres identified in this area were much finer than those seen in the more rostral portions of the brain. The centres containing these fibres include the nucleus solitarius, the reticular formation and cell group A1. Because electrolytic lesions in the mediobasal hypothalamus resulted in an ipsilateral disappearance of γ-MSH-immunoreactive fibres in the telencephalon, diencephalon, mesencephalon and upper rhombencephalon, while the fibres in the lower rhombencephalon remained intact, Kawai and colleagues (1984) suggested that the fibres in this last area originate from the γ-MSH-positive neurons situated in the caudal, commissural part of the nucleus solitarius.

β-Endorphin

β-Endorphin (β-END) was isolated and characterized in 1976 from camel and porcine pituitaries as a 31-amino-acid peptide, identical to the carboxyl terminal sequence of β-LPH, which is considered its biosynthetic precursor (Fig. 26; Bradbury et al. 1976a; Li and Chung 1976). Radioimmunological studies have shown that the brain of the rat (Bradbury et al. 1976b; Rossier et al. 1977) and that of man (Emson et al. 1984) contain significant amounts of β-END. In the rat hypophysectomy does not affect β-END levels in the brain (Cheung and Goldstein 1976; Rossier et al. 1977), a finding which indicates that this peptide is produced by intrinsic neurons. There is physiological evidence suggesting that β-END may act as an inhibitory neuromodulator in the brain (Nicoll et al. 1977). β-Endorphin is an extremely potent opioid agonist having a high stereospecific binding affinity for central nervous opiate receptors (Li and Chung 1976; Bradbury et al. 1976c; Loh et al. 1976; Ferrara et al. 1979).

The presence of β-END-containing perikarya in the mediobasal hypothalamus is well established (Bloom et al. 1978; Bloch et al. 1978, 1979; Bugnon et al. 1979; Sofroniew 1979; Watson and Akil 1980; Finley et al. 1981a), and in the caudal rhombencephalon a second group of β-END-positive elements has been observed (Schwartzberg and Nakane 1981). This caudal group is reportedly situated adjacent to the A2 group of noradrenergic cells (Dahlström and Fuxe 1964) and doubtless corresponds to the small pool of opiomelanocortin cells located in the caudal part of the nucleus solitarius.

With regard to course and termination, the efferents of the group of β-END-positive cells in the mediobasal hypothalamus closely resemble those of the ACTH- and γ-MSH-containing elements in the same area. Finley and colleagues (1981a), who studied these efferents in some detail, observed nerve fibres and terminals with β-END-like immunoreactivity in the following structures: the hypothalamic periventricular area, the paraventricular nucleus, the most lateral part of the median eminence, the medial and lateral preoptic areas, the ventral part of the lateral septal nucleus, the nucleus accumbens, the bed nucleus of the stria terminalis, the central and medial amygdaloid nuclei and parts of the basal and lateral amygdaloid nuclei, some periventricular thalamic nuclei, the periaqueductal grey, the cuneiform nucleus, the nucleus raphes dorsalis, the area tegmentalis ventralis, the locus coeruleus, the lateral parabrachial nucleus, the pontine and medullary reticular formation, the nucleus raphes magnus, the A2 group and the nucleus solitarius. In light of the findings reported for γ-MSH (Kawai et al. 1984), it seems likely that the centres in the lower rhombencephalon mentioned are innervated by the small pool of END-positive cells situated adjacent to the A2 group of noradrenergic cells (Schwartzberg and Nakane 1981).

Comment on the Functional Significance of POMC Derivatives

From the data reviewed above it appears that in the anterior and intermediate lobes of the pituitary gland, as well as in the brain, cells occur in which a large peptide molecule, POMC, is proteolytically cleaved into smaller bioactive peptides, among which ACTH, β-END and three copies of MSH may be especially mentioned. In the brain two pools of cells which process POMC are found, a large one in the mediobasal part of the hypothalamus which largely coincides with the infundibular nucleus, and a small one in the caudal portion of the nucleus solitarius. The large infundibular pool projects to several areas situated within the preoptico-hypothalamic continuum, to a number of limbic telencephalic cell masses, to some periventricular thalamic nuclei and to a variety of mesencephalic and rhombencephalic centres. The projections of the small solitarius pool are confined to some cell groups in the caudal rhombencephalon and to the spinal cord. It is known that the anterior and intermediate lobes of the pituitary process the POMC molecules quite differently. In the anterior lobe ACTH is the main product, whereas in the intermediate lobe α-MSH and β-END predominate. Differences in the processing of POMC and its fragments may also exist between subsets of neurons belonging to the infundibular and solitarius pools, and these differences may result in the production of peptides with widely different biological potencies. Thus, β-END and α-MSH have been shown to possess activities which can be markedly affected by a particular type of acetylation (α-*N*-acetylation). It has now been found that in the amygdaloid complex and in the periaqueductal grey only the non-acetylated forms of β-END and α-MSH are present, whereas in the nucleus accumbens the acetylated derivatives of these two peptides predominate (Dennis et al. 1983). These findings suggest that the infundibular POMC-processing cells which project to the nucleus accumbens contain an acetylating enzyme which is lacking in the elements projecting to the amygdala and the periaqueductal grey.

There is a vast body of literature on the central actions of the various members of the opiomelanocortin groups of peptides. A detailed discussion of this literature falls outside the scope of this book and is beyond the competence of its author. However, an attempt is made to present a brief survey of the main physiological and behavioural effects elicited by these peptides, with emphasis on the relation between these phenomena and the morphological data on the pools of POMC-containing neurons and their efferents discussed earlier in this section. A few general notes preface this survey, touching on some of the problems encountered in the study of the processes and relations to be considered.

Given the fact that the immunohistochemical data available strongly suggest that in the neurons of the 'infundibular' and probably also the 'solitarius' pools of cells various POMC-derived peptides are co-synthesized, co-stored, and co-transported by the axons and co-released by all terminals, the following questions arise: (a) Are these various peptides always released synchronously, or rather differentially, e.g. under the influence of stimulation of different duration or intensity? (b) How do the various peptides, once released, interact? Antagonism between ACTH or ACTH-like peptides and β-END has frequently been reported (e.g. Smock and Fields 1980; Belcher et al. 1982; De Wied 1982), but there is also evidence of synergism or potentiation between peptides derived from POMC (Akil et al. 1984).

It may be difficult to distinguish the actions of the central POMC-processing elements from those of the peripheral (i.e. the pituitary) ones. To mention only one example: pituitary-derived β-END possibly plays a role in the central phenomenon known as stress-induced analgesia (Akil et al. 1984; Basbaum and Fields 1984).

With regard to the origin of α-MSH present in the brain the situation is particularly complex. This peptide may originate from peripheral as well as central POMC-processing cells, but, as we have seen, a special group of α-MSH-producing neurons with a characteristic pattern of efferent projections is present in the dorsal hypothalamus and adjacent zona incerta.

Apart from the POMC-derived endorphins there are two other groups of central opioid peptides, i.e. the enkephalins and the dynorphins. Moreover, a multiplicity of opioid receptors appear to be present in the central nervous system (Akil et al. 1984; Pasternak et al. 1983). All of the endogenous opioids are believed to be involved in the regulation of the organism's responses to stress, but the mode of action and the morphological pattern of the various components in this intricate complex of morphinomimetic ligands and receptors is at present unclear (Akil et al. 1984). This limitation, in combination with uncertainties mentioned above renders it impossible to make a sharp distinction between the actions of POMC-derived peptides on the one hand and endogenous opioid peptides on the other. The following survey of the physiological and behavioural effects of some of the POMC-derived peptides in relation to the structures possibly involved should be viewed against the background of the general aspects just discussed.

De Wied and his associates (e.g. De Wied 1964, 1977, 1982; De Wied and Gispen 1977; Bohus 1979) have claimed that α-MSH, ACTH and certain ACTH fragments affect complex functions such as learning, memory and motivation. Lesion studies (Van Wimersma Greidanus and De Wied 1976; Van Wimersma Greidanus et al. 1979a, b; Bohus and De Wied 1980) suggest that the nucleus parafascicularis thalami, the hippocampal formation and the amygdaloid complex may well be target sites for these peptides. The question as to whether centrally produced POMC-derivatives play a role in the processes mentioned remains unanswered so far. However, it is worthy of note that fibres originating from the infundibular pool of POMC cells project to the amygdala, and that the hippocampus is supplied by fibres from the group of α-MSH-producing cells situated in the dorsal hypothalamus and adjacent zona incerta. ACTH-containing fibres penetrate in several places the ependymal lining of the ventricular system of the brain, and Joseph (1980) has suggested that the secretion of ACTH and/or related peptide sequences from these fibres into the cerebrospinal fluid may represent one mechanism of delivery of these neuropeptides to brain regions associated with the behavioural effects observed by De Wied and his colleagues.

Köhler et al. (1984b) observed that the projections to the hippocampus and the spinal cord arising from the diencephalic groups of α-MSH-containing cells are particularly diffusely organized; they accordingly suggested that these projections may play a role in arousal associated with a variety of motivated behaviours.

β-END is known to be a potent stimulator of the secretion of growth hormone, prolactin and vasopressin (Guillemin 1978). The projections from the infundibular pool of POMC-containing cells to the median eminence and the posterior lobe of the pituitary are probably implicated in these neuroendocrine functions.

The projections of the infundibular POMC-containing cells to the preoptic and anterior hypothalamic areas may be related to the influence of β-END on body temperature observed by Holaday et al. (1978) and Martin et al. (1979).

The periaqueductal grey and the nucleus raphes magnus are important links in a descending pain-control system which ultimately inhibits noxiously evoked activity in spinal dorsal horn neurons (see the following section on enkephalins). This system can be activated by microinjections of morphine in the periaqueductal grey and in the nucleus raphes magnus, as well as by electrical stimulation of the same centres (stimulation-produced analgesia, SPA). Stress may lead to a marked reduction in responsiveness to pain (stress-induced analgesia, SIA), and a suppression of pain perception can also be achieved by acupuncture (acupuncture analgesia, AA). It is assumed that in SIA and AA the descending pain-control system is also involved. Several groups of investigators (e.g. Loh et al. 1976; Tseng et al. 1976) have reported that the administration of β-END, intraventricularly or into the periaqueductal grey, leads to analgesia, and the suggestion has been made that the various types of analgesia mentioned above are all effected by the release of β-END from the fibres which pass from the infundibular pool of POMC-processing cells to the periaqueductal grey and to the nucleus raphes magnus. There is evidence for the release of β-END into plasma, cerebrospinal fluid and brain tissue following acupuncture or stress (Clement-Jones et al. 1980; Pert et al. 1981) and for the development of a naloxone-reversible (and thus opioid-induced) SIA in both animals and man (Willer et al. 1981; Watkins and Mayer 1982). It is also plausible that SPA from electrodes in the periaqueductal grey results from activation of the axons of infundibular POMC cells which produce and release β-END (Akil et al. 1984). However, it is not known which endogenous opioid is actually involved in activation of the descending pain-control system, and it has already been mentioned that peripherally produced β-END may play a role in SIA. Finally, it should be noted that ACTH and γ-MSH are probably also implicated in the regulation of responsiveness to pain (Akil et al. 1984).

Romagnano and Joseph (1983) have suggested a possible relationship between the depression of respiration, blood pressure and heart rate induced by β-END observed in physiological studies, and the presence of fibres containing POMC-derived peptides in cell groups involved in the regulation of respiration (medial parabrachial nucleus, nucleus solitarius, A1 region) and in cardiovascular control (nucleus solitarius, medullary raphe nuclei). Remarkably, the cells of origin of these fibres are situated in the caudal part of the nucleus solitarius, i.e. the centre involved in both regulatory mechanisms.

There is a striking coincidence of the distribution of fibres containing POMC-derived peptides and regions containing dopaminergic (area tegmentalis ventralis, A8 group), noradrenergic (locus coeruleus, A1, A2, A4, A5 groups) and serotoninergic (raphe nuclei) cell groups. Romagnano and Joseph (1983), who remarked upon this coincidence, cited data from the literature indicating that several of these monoaminergic cell groups are involved in analgesia responses.

Pilcher and Jacob (1984) noted that throughout the central nervous system the fibres containing POMC-derived peptides and the perikarya containing CRF show a remarkable concordance of localization. In view of the well-known interaction of CRF with the peripheral (pituitary) pool of POMC-processing cells, they suggested the possibility of a similar relationship between CRF and the central pools of these elements.

Enkephalins

Hughes and co-workers (Hughes 1975; Hughes et al. 1975) isolated and characterized two pentapeptides from the brain which were named methionine-enkephalin (M-ENK) and leucine-enkephalin (L-ENK). Together with other structurally related peptides they form the endorphins, a group of substances which act as endogenous ligands for opiate receptors. The enkephalins stem from a large precursor molecule, called pro-enkephalin. This precursor contains within its structure seven peptides with the M-ENK or L-ENK active core. Four of the seven peptides produced are simply M-ENK; two are carboxyl extended: M-ENK-Arg^6-Phe^7 and -Arg^6-Gly^7-Leu^8. Finally, one copy of L-ENK is produced (Fig. 28; Akil et al. 1984).

Antibodies have been raised against M-ENK and L-ENK coupled to a carrier protein, and the localization of these pentapeptides in the central nervous system has been studied immunohistochemically with the aid of these antibodies (Hökfelt et al. 1977b; Uhl et al. 1979a; Finley et al. 1981c; Arluison et al. 1983; Khachaturian et al. 1983a, b). Enkephalin-containing neurons are distributed widely in the central nervous system, both as local circuit neurons and as projection neurons. In many regions of the brain the localization of enkephalin correlates well with the distribution of opiate receptors, as determined by biochemical and autoradiographic techniques (Wamsley et al. 1982).

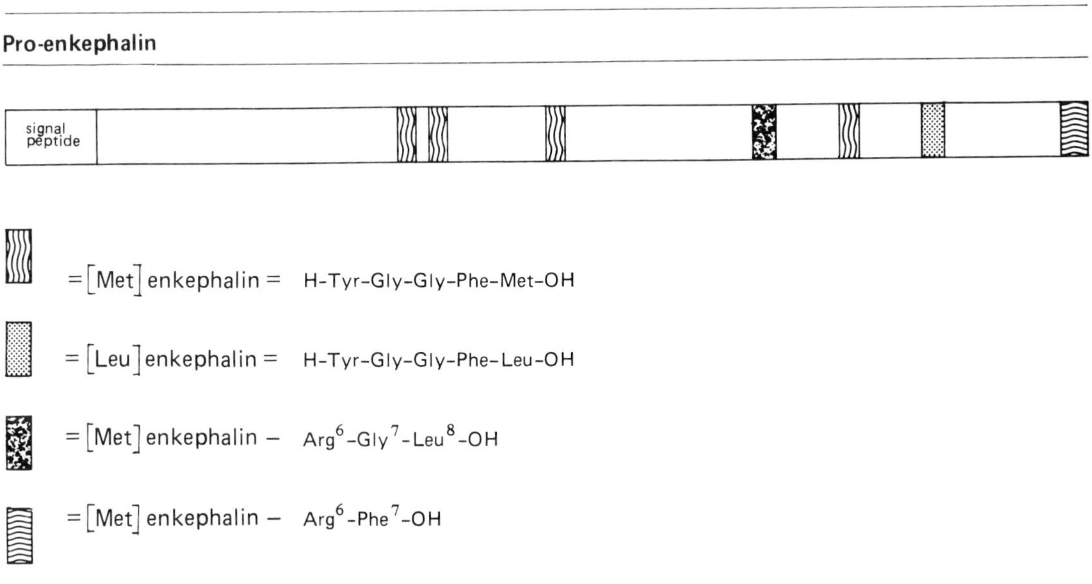

Fig. 28. Schematic representation of the pro-enkephalin precursors molecule. The localization and primary structures of its enkephalin- and 'enkephalinoid'-cleavage products are indicated. The diagram is simplified from Udenfriend and Kilpatrick (1983). For significance of the amino acid code, see Table 7 on p. 198

Several enkephalinergic projection systems have been studied by combining immunohistochemical techniques with the use of retrograde tracers (Hökfelt et al. 1979b; Beitz 1982b) or with stereotaxic injections of neurotoxic agents (colchicine, kainic acid) and microknife deafferentiations (DelFiacco et al. 1982). There is considerable evidence suggesting that the enkephalins may serve as inhibitory neurotransmitters (for review, see North 1979; Iversen et al. 1980). Although in most regions of the neuraxis L-ENK is present in higher concentrations than is M-ENK, and both substances have been found to occur in separate neurons (Larsson et al. 1979), they are treated together in the elementary survey which follows (Fig. 29).

In both the neocortex and the allocortex ENK-containing neurons have been found (Finley et al. 1981c; Khachaturian et al. 1983a). In the neocortex these elements are small, sparsely scattered, and distributed primarily in layer II and to a lesser extent in layer III. In the cingulate and entorhinal areas they tend to be more numerous, occurring likewise mainly in layers II and III. ENK-containing neurons situated in the entorhinal area have been reported to project to the subiculum and, via the 'perforant path', to the stratum moleculare of the fascia dentata (Gall et al. 1981). Another hippocampal enkephalinergic system appears to correspond to the mossy fibre projection from the granule cells in the fascia dentata to the cornu Ammonis (Gall et al. 1981).

A certain proportion of the periglomerular and granule cells in the olfactory bulb and of the elements situated in the anterior olfactory nucleus have been reported to contain enkephalin (Finley et al. 1981c; Khachaturian et al. 1983a). The lateral and, to a lesser extent, the medial septal nucleus contain enkephalinergic perikarya (Hökfelt et al. 1977b; Khachaturian et al. 1983a).

The central nucleus of the amygdaloid complex is densely populated with ENK-immunoreactive cells (Finley et al. 1981c; Arluison et al. 1983; Khachaturian et al. 1983a), and several other nuclei of that complex (nucleus corticalis, nucleus medialis, nucleus lateralis) have also been found to contain ENK-positive perikarya. Enkephalinergic fibres, most of which originate from the central amygdaloid nucleus, course in the stria terminalis to terminate primarily in the bed nucleus of that pathway (Uhl et al. 1978; Finley et al. 1981c). The centre last mentioned contains a dense plexus of enkephalinergic fibres and terminals (Haber and Elde 1981). The other large projection system connecting the amygdaloid complex with the basal forebrain, i.e. the ventral amygdaloid pathway, also contains ENK-immunoreactive fibres (Finley et al. 1981c).

The neostriatum, i.e. nucleus caudatus and the putamen, contains numerous enkephalinergic neurons which give rise to a massive striopallidal projection (Cuello and Paxinos 1978; Brann and Emson 1980; Cuello et al. 1981; Del Fiacco et al. 1982). The cells of origin of this projection are medium-sized spiny elements which constitute 15%–20% of the total neurons in the neostriatum (Pickel et al. 1980; DiFiglia et al. 1982a). The greatest accumulation of ENK-positive cells is found in the ventral regions of the caudate nucleus and the putamen (Pickel et al. 1980; DiFiglia et al. 1982a; Arluison et al. 1983). The enkephalinergic fibres passing from the neostriatum to the globus pallidus are radially organized and establish a topical correspondence between these two structures (Del-Fiacco et al. 1982). The great majority of these fibres terminate in the lateral segment of the globus pallidus; in the inner portion of its medial segment the density of enkephalinergic fibres is moderate, while such fibres are sparse in the outer portion of the medial segment (Haber and Elde 1981). Because the lateral segment of the globus pallidus projects almost exclusively to the subthalamic nucleus, which, in turn, projects back to the globus pallidus, it seems likely that the enkephalinergic striopallidal pathway strongly influences the pallido-subthalamo-pallidal loop system (Haber and Elde 1981; Graybiel and Ragsdale 1983). Zech et al. (1983) reported that in patients suffering from Huntington's

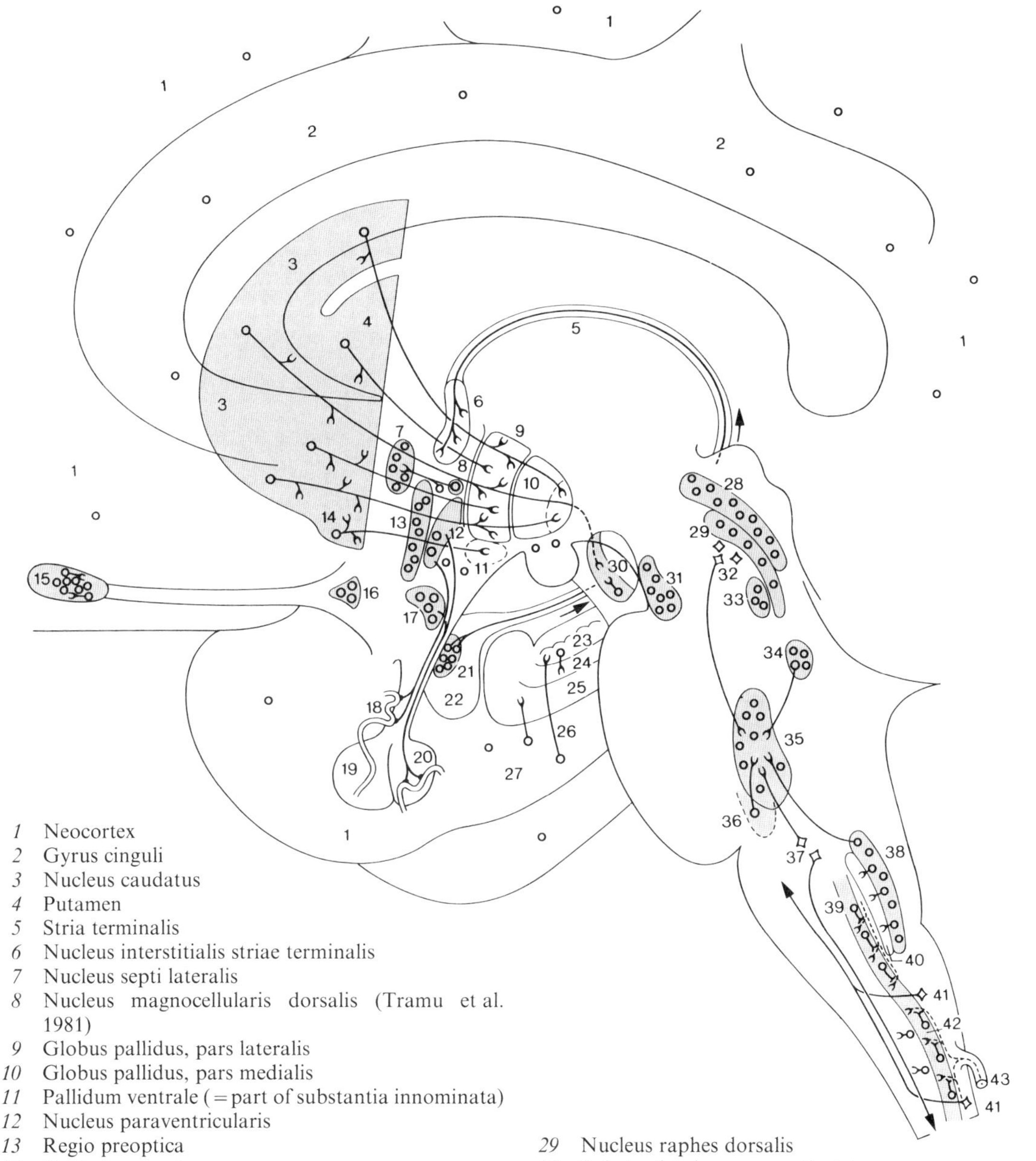

1 Neocortex
2 Gyrus cinguli
3 Nucleus caudatus
4 Putamen
5 Stria terminalis
6 Nucleus interstitialis striae terminalis
7 Nucleus septi lateralis
8 Nucleus magnocellularis dorsalis (Tramu et al. 1981)
9 Globus pallidus, pars lateralis
10 Globus pallidus, pars medialis
11 Pallidum ventrale (= part of substantia innominata)
12 Nucleus paraventricularis
13 Regio preoptica
14 Nucleus accumbens
15 Bulbus olfactorius
16 Nucleus olfactorius anterior
17 Nucleus supraopticus
18 Eminentia mediana
19 Lobus anterior hypophyseos
20 Lobus posterior hypophyseos
21 Nucleus centralis amygdalae
22 Corpus amygdaloideum
23 Fascia dentata
24 Cornu Ammonis
25 Subiculum
26 'Perforant path'
27 Gyrus parahippocampalis
28 Griseum centrale mesencephali
29 Nucleus raphes dorsalis
30 Substantia nigra, pars reticulata
31 Nucleus interpeduncularis
32 Nucleus cuneiformis
33 Nucleus tegmentalis dorsalis
34 Nuclei parabrachiales
35 Nucleus raphes magnus
36 Cell group A5
37 Formatio reticularis rhombencephali
38 Nucleus solitarius
39 Substantia gelatinosa nuclei spinalis nervi trigemini, pars caudalis
40 Nucleus dorsalis nervi vagi
41 Cellulae marginales
42 Substantia gelatinosa spinalis
43 Radix dorsalis nervi spinalis

Fig. 29. Enkephalin-containing cell groups and pathways

chorea the plexus of ENK-containing fibres in the lateral pallidal segment disappears almost completely.

The highest concentrations of enkephalins in the central nervous system are found in the neostriatum. In addition to the cell bodies mentioned above, this centre contains a very dense network of ENK-immunoreactive fibres and terminals. This network is particularly dense in the more ventral portions of the neostriatum, where it has a patchy appearance (Haber and Elde 1981; Arluison et al. 1983). It is also worthy of note that the neostriatum is particularly rich in opiate receptors (Kuhar et al. 1973; Wamsley et al. 1982). Because the medium-sized spiny striopallidal projection neurons are known to have axon collaterals intrinsic to the caudate nucleus and putamen (DiFiglia et al. 1982a), it is likely that the enkephalinergic fibre plexuses in both the neostriatum and the globus pallidus originate from these cells.

Not only the ventral parts of the neostriatum but also the adjacent nucleus accumbens contain numerous ENK-positive cells and a particularly dense network of enkephalinergic fibres and terminals (Khachaturian et al. 1983a; Arluison et al. 1983; Haber and Elde 1981). The so-called ventral pallidum, i.e. a certain portion of the substantia innominata to which the nucleus accumbens is known to project, shows the most intensive enkephalin staining of the entire brain (Haber and Elde 1981).

In primates the medial portion of the pars reticulata of the substantia nigra contains a dense plexus of enkephalinergic fibres (Haber and Elde 1981), and it has been suggested that these fibres originate from the neostriatum (Pickel et al. 1980; DelFiacco et al. 1982; DiFiglia et al. 1982a). Neuropathological findings lend support to the existence of such an enkephalinergic strionigral projection. Thus, it has been observed that there is a marked reduction of ENK immunoreactivity in the substantia nigra of patients with Huntington's chorea (Emson et al. 1980a), as well as in patients with striopallidal infarction (Pioro et al. 1984). However, it is also possible that at least some of the enkephalin found in the substantia nigra is produced by intrinsic neurons: ENK-positive perikarya have been observed in the medial part of the substantia nigra of the cat (Conrath-Verrier et al. 1983). A significant decrease in M-ENK has been observed in the substantia nigra of patients with Parkinson's disease, and it has been suggested that in the substantia nigra not only dopaminergic elements but also enkephalinergic local circuit neurons may degenerate in this disease (Taquet et al. 1982).

Enkephalins have a profound influence on the hypothalamo-hypophyseal axis. It has been shown that these pentapeptides can inhibit the release of luteinizing hormone, follicle-stimulating hormone and thyroid-stimulating hormone, and can also enhance the release of growth hormone, prolactin and vasopressin (see Micevych and Elde 1980). An extensive enkephalinergic neuroendocrine neuronal system in the cat has been described; it comprises large perikarya in the nucleus supraopticus and nucleus paraventricularis, giving rise to fibres and terminals in the median eminence, hypophyseal stalk and posterior pituitary, and thus closely resembling the well-known vasopressinergic and oxytocinergic neuronal systems (Micevych and Elde 1980; Haber and Elde 1981). Magnocellular enkephalin-containing neurons have also been found in the supraoptic and paraventricular nuclei of the rat by one group of authors (Finley et al. 1981c), but another group (Khachaturian et al. 1983b) was unable to confirm this observation. DiFiglia and Aronin (1984b) presented immunohistochemical and ultrastructural evidence suggesting that in the monkey, at least some of the ENK-containing neurons in the paraventricular nucleus belong to the population of magnocellular neurosecretory cells.

The structures constituting together the so-called limbic axis contain numerous enkephalinergic neurons. The preoptic region, in particular the medial preoptic nucleus; the various parvocellular hypothalamic nuclei, including the dorsomedial and ventromedial nuclei and the lateral hypothalamic area; the dorsal tegmental nucleus; and the interpeduncular nucleus should be mentioned in this context (Hökfelt et al. 1977b; Finley et al. 1981c; Khachaturian et al. 1983b; Hamill et al. 1984; DiFiglia and Aronin 1984b). Evidence suggesting the presence of an enkephalin-containing pathway passing from the area just ventrolateral to the anterior hypothalamic nucleus to the lateral septal nucleus has been presented (Sakanaka et al. 1982a). In the hypothalamus of the guinea pig, enkephalin-containing perikarya are particularly numerous in the perifornical area, where they form a separate nucleus adjoining the paraventricular nucleus, termed the 'magnocellular dorsal nucleus' (Tramu et al. 1981). This nucleus has been shown to project bilaterally to the lateral septal nucleus (Poulain et al. 1984).

Enkephalinergic neurons are found in several centres known or supposed to exert a modulatory influence on the conduction of nociceptive impulses. These centres include the griseum centrale mesencephali, the dorsal raphe nucleus, the cuneiform nucleus, the nucleus raphes magnus, the rhombencephalic medial reticular formation, and the substantia gelatinosa in the spinal cord and in the caudal part of the spinal trigeminal nucleus.

There is considerable evidence that the *griseum centrale mesencephali* (or periaqueductal grey) represents a key structure in a chain of descending neurons, which ultimately exerts its influence on enkephalinergic local circuit neurons situated in the substantia gelatinosa of the spinal trigeminal nucleus and the spinal dorsal horn. These enkephalinergic interneurons are believed to form axon-axonic synapses with pain afferents on which they exert a powerful inhibitory influence (for review, see Basbaum and Fields 1978; Fields and Basbaum 1978). The nucleus raphes magnus is considered to be an important medullary relay station in the chain of descending neurons just mentioned. However, the following synopsis will show that the neuronal organization of this descending pain-inhibiting or pain-suppressing system is presumably very complex and by no means fully clarified.

It is known that the periaqueductal grey can generate a potent analgesia in response to either electrical stimulation or opiate microinjection. This region contains a high concentration of opiate receptors (Wamsley et al. 1982) and the analgesic action of systemically or locally administered morphine is supposed to rest on occupation of these receptors. Thus, morphine would mimic the pain-inhibiting effect of endogenous substances like enkephalin. In addition to opiate receptors, the periaqueductal grey contains throughout its rostrocaudal extent a number of discrete populations of ENK-immunoreactive neurons (Moss et al. 1983; Conrath-Verrier et al. 1983) and a terminal field of enkephalinergic fibres of varying density (Haber and Elde 1981; Moss et al. 1983). Because the periaqueductal grey is not known to give rise to any substantial enkephalinergic efferent projections, the ENK-containing cells located in this centre presumably represent local circuit neurons. In relation to the organization of the descending pain-inhibiting system it is of interest that the periaqueductal grey projects to the nucleus raphes magnus. However, this projection is not enkephalinergic, but rather contains a major neurotensin and serotonin component (Beitz 1982a).

The *nucleus raphes dorsalis* contains, in addition to serotoninergic neurons, numerous ENK-immunoreactive cells throughout its rostrocaudal extent (Moss et al. 1981; Khachaturian et al. 1983b). There is experimental evidence suggesting that this centre contributes substantially to opiate and stimulus-produced analgesia (Oliveras et al. 1979). It has been hypothesized that the serotonin- and enkephalin-containing cells of the nucleus raphe dorsalis feed into the descending analgesia system, and that they contribute additionally to the modulation of the affective component of pain through the raphe dorsalis projections to limbic and other forebrain structures (Moss et al. 1981).

The *nucleus cuneiformis,* which forms part of the mesencephalic reticular formation, contains enkephalinergic cells which project to the nucleus raphes magnus. Substance P-immunoreactive neurons contribute to the same projection (Beitz 1982a). According to Beitz (1982a), this finding is consistent with the idea of two separate, pharmocologically distinct descending inhibitory mechanism arising from the midbrain as proposed by Carstens et al. (1980, 1981). Thus, the influence on the nucleus raphes magnus exerted by the enkephalin- and substance P-containing fibres from the cuneiform nucleus may be quite different from that exerted by the neurotensinergic and serotoninergic fibres arising from the periaqueductal grey.

The *nucleus raphes magnus* contains numerous enkephalinergic perikarya (Hökfelt et al. 1977a; Finley et al. 1981c; Khachaturian et al. 1983b; Conrath-Verrier et al. 1983). In addition to the cuneiform nucleus already mentioned, some other centres, including the nucleus solitarius and the nucleus reticularis paragigantocellularis, provide both enkephalin and substance P inputs to the nucleus raphes magnus. Moreover, enkephalinergic projections to this centre were found to originate from the lateral parabrachial nucleus, from a certain portion of the nucleus reticularis gigantocellularis, and from an area corresponding to the A5 group of Dahlström and Fuxe (Beitz 1982a). The nucleus raphes magnus gives rise to a descending pathway which courses in the spinal dorsolateral funiculus and terminates in laminae I, II and V of the spinal dorsal horn (Basbaum and Fields 1979). This pathway was originally thought to consist entirely of serotoninergic fibres. However, serotoninergic neurons form only a minority within the nucleus raphes magnus (Steinbusch and Nieuwenhuys 1983), and the experimental work of Bowker et al. (1981b) has shown that most (55%) of the raphespinal projection neurons in the nucleus raphes magnus are non-serotoninergic. The role of the enkephalinergic neurons in the nucleus raphes magnus is not known at present. It is noteworthy that a considerable number of serotoninergic cells situated in the mesencephalic tegmental area adjacent to the nucleus centralis superior project upon the dorsal horn of the spinal cord (Bowker et al. 1981b), and thus may well form part of the descending pain-control system.

The *nucleus reticularis gigantocellularis* contains a group of enkephalinergic cells which project to the spinal cord (Hökfelt et al. 1979b). Whether these neurons form part of the descending pain-inhibiting system remains to be elucidated. Other parts of the rhombencephalic reticular formation have also been reported to contain ENK-positive cells (Khachaturian et al. 1983b; Conrath-Verrier et al. 1983).

A dense plexus of ENK-immunoreactive fibres is found in the *superficial dorsal horn of the spinal cord* and the *spinal trigeminal nucleus* (Hökfelt et al. 1977a; Haber and Elde 1981; Conrath-Verrier et al. 1983; LaMotte and De Lanerolle 1983a), and these structures are also known to be rich in opiate receptors (Wamsley et al. 1982). The superficial laminae of the spinal dorsal horn and the caudal part of the spinal trigeminal nucleus contain numerous small enkephalinergic neurons (Hökfelt et al. 1977a; Uhl et al. 1979a; Aronin et al. 1981; LaMotte and De Lanerolle 1981; Conrath-Verrier et al. 1983) which are thought to give rise to the fibre plexus mentioned above. As already indicated, these elements are believed to constitute the final link in the descending pain-suppression system (Hökfelt et al. 1977a; Basbaum and Fields 1978; Fields and Basbaum 1978). Their terminals probably act directly on the sensory input of the dorsal horn and the spinal trigeminal nucleus by presynaptically inhibiting the release of substance P from primary afferent fibres (see, e.g. Jessell and Iversen 1977).

Enkephalin-containing neurons in the spinal cord are by no means confined to the substantia gelatinosa (Aronin et al. 1981; Glazer and Basbaum 1981; Khachaturian et al. 1983b), and it is noteworthy that elements containing these pentapeptides probably also play a role in the ascending nociceptive system. Thus, numerous large neurons present in the *marginal zone of the spinal grey* (lamina I), elements which are known to give rise to long axons ascending to the brain stem and/or thalamus, appear to be enkephalinergic (Glazer and Basbaum 1981). Moreover, it has been reported that ENK-immunoreactive axonal endings make direct synaptic contacts with dorsal horn spinothalamic projection neurons (Ruda 1982).

The nucleus solitarius and the parabrachial nuclei, centres known to be involved in autonomic regulation, contain ENK-positive cells and dense plexuses of enkephalinergic fibres and terminals (Uhl et al. 1979a; Finley et al. 1981c; Haber and Elde 1981; Khachaturian 1983b; Kalia et al. 1984b). A dense network of ENK-immunoreactive fibres has also been observed in the dorsal motor nucleus of the vagus (Khachaturian et al. 1983b; Haber and Elde 1981; Kalia et al. 1984b).

There is experimental evidence suggesting that at least some sympathetic preganglionic neurons located in the upper lumbar segments of the spinal cord, contain enkephalin (Dalsgaard et al. 1982b). Whether enkephalin co-exists with acetylcholine (the classical neurotransmitter of preganglionic neurons) in these neurons is unknown.

It has been reported that the efferent olivocochlear bundle contains enkephalinergic fibres (Fex and Altschuler 1981).

In the rat the ventral horn of the spinal cord and some motor nuclei in the lower brain stem (the nucleus ambiguus and the hypoglossal, facial and motor trigeminal nuclei) have been found to contain plexuses of enkephalinergic fibres (Senba et al. 1982). It has also been experimentally shown that ENK-positive cells located in the ventrolateral part of the caudal rhombencephalic reticular formation project ipsilaterally to the facial nucleus (Senba and Tohyama 1983).

Addenda:

1. Finley et al. (1981c) mentioned the presence of ENK-positive perikarya and fibres in numerous regions which have not been included in the preceding text or in Fig. 29 – telencephalon: cells in the periventricular preoptic nucleus, the olfactory tubercle and (sparsely scattered) in the fascia dentata, cornu Ammonis and subiculum; numerous fibres in the diagonal band of Broca; diencephalon: cells in the anterior, suprachiasmatic and arcuate nuclei; scattered neurons in some periventricular thalamic nuclei and in the ventral nucleus of the lateral geniculate body; neuronal processes in the organum vasculosum of the lamina terminalis, the ventral amygdalofugal pathway, the lateral and medial habenular nuclei; scattered fibres in several thalamic nuclei, including the periventricular, parafascicular medial, and ventral nuclei, and the ventral part of the lateral geniculate body; mesencephalon: fibres in the colliculus superior, particularly its medioventral part, and in various regions of the mesencephalic tegmentum; rhombencephalon: perikarya in the central grey, the locus coeruleus, the subcoeruleus area, the dorsal cochlear nucleus, the granular layer of the cortex cerebelli (numerous elements, resembling Golgi cells in both morphology and distribution), the medial and spinal vestibular nuclei, the area postrema (numerous cells); fibre plexuses in a variety of structures, including the periventricular region, the medial and lateral parts of the reticular formation, the raphe nuclei and the parabrachial nuclei.

2. Zamir et al. (1984b) presented a brief survey of the modes of generation of L-ENK. They pointed out that the amino acid sequence of L-ENK is found within several larger peptides which are generated from the precursors pro-enkephalin and pro-dynorphin, and they presented evidence indicating that the relatively high levels of L-ENK found in the substantia nigra of the rat are supplied by striatonigral axons and generated from the precursor prodynorphin.

3. It is known that the efferent olivocochlear bundle arises from two cell groups situated in the vicinity of the superior olivary complex, one medial, the other lateral. Using antisera against enkephalin and CHAT, Altschuler et al. (1984) demonstrated that in cells in the lateral group, enkephalin immunoreactivity and CHAT immunoreactivity

are co-localized. In the medial group only CHAT immunoreactivity was found. These observations indicate that the olivocochlear bundle is at least partly cholinergic (cf. Fig. 2), and that in some of its fibres both acetylcholine and enkephalin occur.

Dynorphins

Apart from the enkephalins and the endorphins, already discussed, there is a third family of endogenous opioid peptides, the dynorphins. The members of this family stem from a precursor known as a pro-dynorphin or pro-neo-endorphin-dynorphin (Fig. 30), which produces the following peptides (the number of amino acids found in each is added in parentheses): α-neo-endorphin (10), β-neo-endorphin (9), dynorphin A (17), dynorphin A (1-8; 8) and dynorphin B (13) (Goldstein et al. 1981; Kangawa et al. 1981; Seizinger et al. 1981; Kakidani et al. 1982; Tachibana et al. 1982). All of these peptides contain the amino acid sequence of L-ENK at their N-terminal (Fig. 30), and all have been isolated as separate products from mammalian brain tissue. The exact processing of the precursor into its biologically active end products has not yet been determined. Watson et al. (1983) demonstrated that dynorphin A, dynorphin B and α-neo-endorphin are present in the same neurons of several nuclear groups, but there is also evidence that the pro-dynorphin molecule may be processed differently in different regions of the brain (Weber et al. 1982; Cone et al. 1983). In the following survey I will confine myself to dynorphin A, which will be designated as dynorphin (DYN) without further specification (Fig. 31).

Dynorphin was first isolated from porcine pituitary glands (Goldstein et al. 1979, 1981), but later it was also reported to be present in the gut (Watson et al. 1981; Tachibana et al. 1982) and in the peripheral (Watson et al. 1981; Botticelli et al. 1981; Vincent et al. 1984) and central (Goldstein and Ghazarossian 1980; Watson et al. 1981, 1982b; Vincent et al. 1982; Khachaturian et al. 1982; Molineaux et al. 1982; Watson et al. 1983; Cone et al. 1983; Zamir et al. 1983) nervous system. This peptide is an extraordinarily potent opiate agonist (Goldstein et al. 1979); moreover, it appears to be the selective ligand of a particular subclass of opioid receptors, the so-called kappa receptors (Chavkin et al. 1982; Young et al. 1983). The studies quoted above have shown that there is no clear parallel between the distribution of dynorphin and those of the other endogenous opioid peptides, i.e. the enkephalins and endorphin.

Biochemical and radioimmunological studies (Molineaux et al. 1982; Zamir et al. 1983, 1984a; Cone et al. 1983) have revealed that DYN occurs in all parts of the central nervous system but is unevenly distributed. The highest concentrations by far have been found in the neural lobe of the pituitary. Within the brain *sensu strictiori* the highest levels have been observed in the hypothalamus. In extrahypothalamic areas DYN levels appear to be high in the lateral preoptic area, the nuclei of the diagonal band of Broca, the bed nucleus of the stria terminalis and the substantia nigra. Relatively high concentrations of DYN have been found in the amygdaloid complex, the striatum, the hippocampus, the nucleus spinalis of the trigeminal nerve, the nucleus solitarius and the area postrema.

The nucleus supraopticus and the magnocellular part of the nucleus paraventricularis contain numerous DYN-immunoreactive neurons (Watson et al. 1981). It has been shown experimentally that the axons of these cells pass to the median eminence and to the neural lobe of the pituitary (Millan et al. 1983, 1984), and it has, moreover, been demonstrated that in these hypothalamo-hypophyseal projection neurons DYN co-exists with vasopressin (Watson et al. 1982a). The suprachiasmatic nucleus also contains numerous DYN-positive cells, but these do not project to the neural lobe of the pituitary.

Scattered DYN-positive perikarya have been observed in the following centres and areas

Pro-dynorphin and its derivatives

 = [Leu]enkephalin = H-Tyr-Gly-Gly-Phe-Leu-OH

α-Neo-endorphin [α-NE]

H-Tyr^{1}-Gly-Gly-Phe-Leu-Arg-Lys-Tyr-Pro-Lys^{10}-OH

β-Neo-endorphin [β-NE] = α-NE (1-9)

Dynorphin A (1-17) [Dyn A (1-17)]

H-Tyr^{1}-Gly-Gly-Phe-Leu-Arg-Arg-Ile-Arg-Pro^{10}-Lys-Leu-Lys-Trp-Asp-Asn-Gln^{17}-OH

Dynorphin B [Dyn B] = Rimorphin

H-Tyr^{1}-Gly-Gly-Phe-Leu-Arg-Arg-Gln-Phe-Lys^{10}-Val-Val-Thr^{13}-OH

Dynorphin A (1-8) [Dyn A (1-8)]

Fig. 30. Schematic representation of the pro-dynorphin or pro-neo-endorphin-dynorphin precursor molecule. The molecule contains three (Leu)enkephalin sequences, which become located at the N-terminal of the cleavage products indicated. The primary structure of these cleavage products is shown (*below*). For significance of the amino acid code, see Table 7 on p. 198

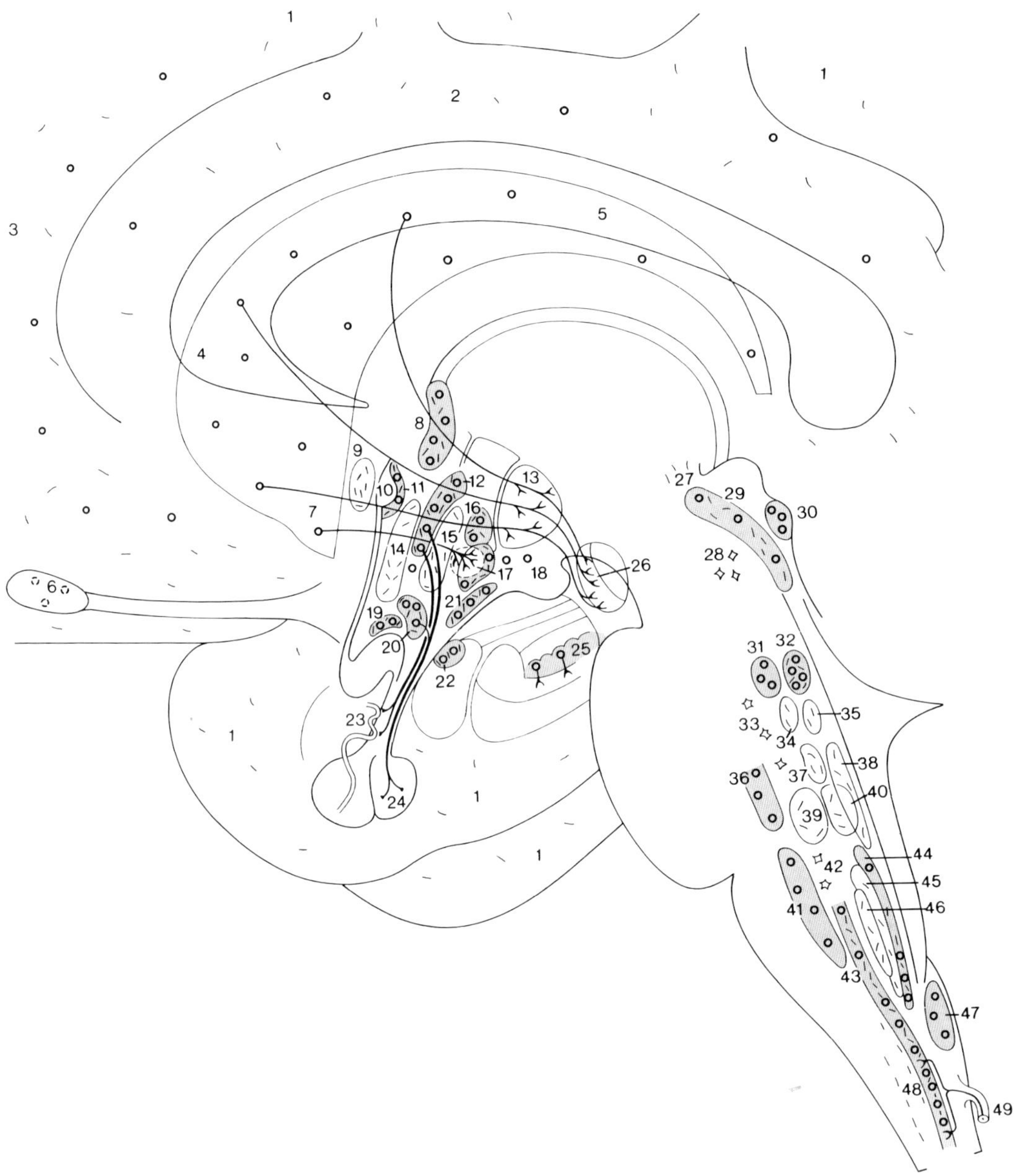

Fig. 31. Dynorphin-containing cells and fibres

Explanations to Figure 31

1 Neocortex
2 Gyrus cinguli
3 Cortex prefrontalis
4 Caput nuclei caudati
5 Cauda nuclei caudati
6 Bulbus olfactorius
7 Nucleus accumbens
8 Nucleus interstitialis striae terminalis
9 Nucleus septi lateralis
10 Commissura anterior
11 Nucleus interstitialis commissurae anterioris
12 Nucleus paraventricularis, pars magnocellularis
13 Globus pallidus, pars medialis
14 Nucleus preopticus medialis
15 Nucleus anterior hypothalami
16 Nucleus dorsomedialis
17 Pallidum ventrale (=part of substantia innominata)
18 Area hypothalamica lateralis
19 Nucleus suprachiasmaticus
20 Nucleus supraopticus
21 Nucleus ventromedialis
22 Nucleus centralis amygdalae
23 Eminentia mediana
24 Lobus posterior hypophyseos
25 Fascia dentata
26 Substantia nigra, pars reticulata
27 Area pretectalis, pars medialis
28 Nucleus cuneiformis
29 Griseum centrale mesencephali
30 Colliculus inferior
31 Nuclei lemnisci lateralis
32 Nucleus parabrachialis lateralis
33 Formatio reticularis mesencephali
34 Nucleus motorius nervi trigemini
35 Nucleus sensorius principalis nervi trigemini
36 Cell group A5
37 Nucleus nervi facialis
38 Nucleus vestibularis medialis + spinalis
39 Nucleus cochlearis ventralis
40 Nucleus cochlearis dorsalis
41 Cell group A1
42 Formatio reticularis myelencephali
43 Nucleus spinalis nervi trigemini
44 Nucleus solitarius
45 Nucleus dorsalis nervi vagi
46 Cell group A2
47 Nucleus cuneatus medialis
48 Substantia gelatinosa
49 Radix dorsalis nervi spinalis

of the brain: nucleus infundibularis, lateral hypothalamic area (numerous large elements), bed nuclei of the stria terminalis and anterior commissure, nucleus centralis amygdalae, nucleus caudatus plus putamen (many small cells, particularly in the medial part of the head of the caudate nucleus and throughout the tail), dentate gyrus (a small number of perikarya which give rise to mossy fibres that project to hippocampal fields CA2 and CA3), cingulate and prefrontal cortex, griseum centrale mesencephali, cuneiform nucleus, inferior colliculus, nuclei of the lateral lemniscus, parabrachial nuclei (particularly the lateral nucleus), rhombencephalic medial reticular formation, the noradrenergic cell groups A1, A2 and A5, the superficial zone of the nucleus spinalis nervi trigemini, the nucleus solitarius and the nucleus cuneatus medialis. DYN-containing fibres and terminals appear to be present in a variety of brain centres, including the supraoptic and paraventricular nuclei, the nucleus infundibularis, the dorsomedial and ventromedial hypothalamic nuclei, the anterior hypothalamic nucleus, the medial preoptic nucleus, the nucleus suprachiasmaticus, the lateral hypothalamic area, the median eminence, the neural lobe of the pituitary, the glomerular zone of the olfactory bulb (the fibres occur in some, but not all glomeruli; Khachaturian et al. 1982), the lateral septal nucleus, the bed nucleus of the stria terminalis, the nucleus centralis amygdalae, the nucleus caudatus and the putamen, the nucleus accumbens, the entopeduncular nucleus (i.e. the homologue of the internal segment of the primate globus pallidus), the ventral pallidum (which contains a very dense network of DYN-positive fibres), the hippocampal formation (mossy fibres), the neocortex (scattered fibres in all areas and layers), the griseum centrale mesencephali (particularly its dorsal portion), the medial pretectal area, the cuneiform nucleus, the pars reticulata of the substantia nigra (which, like the ventral pallidum, is very densely innervated), the lateral parabrachial nucleus, the dorsal and the ventral cochlear nuclei, the medial and spinal vestibular nuclei, the nucleus princeps and the nucleus spinalis of the trigeminal nerve, the nucleus solitarius (particularly the commissural part), the motor trigeminal nucleus, the facial nucleus and the dorsal vagal nucleus (Watson et al. 1981, 1982b; Vincent et al. 1982; Khachaturian et al. 1982).

It is known that vasopressinergic and oxytocinergic fibres arising from the parvocellular portion of the paraventricular nucleus descend to the brain stem. Millan et al. (1984) have presented experimental evidence suggesting that, contrary to the tightly coupled co-localization of vasopressin and DYN in the hypothalamo-hypophyseal tract, in the brain stem these two peptides occur only independently, i.e. in separate fibres.

It is important to note that in the periaqueductal grey, enkephalin- as well as DYN-containing neurons occur. However, it has been established that the former are located more dorsally than the latter (Watson et al. 1982b). Because in the periaqueductal grey endorphin-containing fibres occur as well as enkephalinergic and dynorphinergic neurons, the influence exerted in this centre upon the descending pain-control system could result from the release of any or all of the three endogenous opioid peptides (Basbaum and Fields 1984).

Some neurons in the nucleus raphes magnus, the raphes pallidus and the adjacent nucleus reticularis magnocellularis contain serotonin as well as DYN (Basbaum and Fields 1984).

The spinal ganglia contain DYN-positive cells, and within the spinal cord DYN-immunoreactive fibres are concentrated in laminae I, IIa and V of the dorsal horn, as well as in the central grey area around the central canal. Moreover, numerous DYN-containing cell bodies have been observed in laminae I and IIa of the spinal cord (Botticelli et al. 1981; Watson et al. 1982b; Vincent et al. 1982).

Little is known about the physiological role of DYN in the various parts of the central nervous system. However, the presence of DYN-positive perikarya in the supraoptic, paraventricular, suprachiasmatic and infundibular nuclei of the hypothalamus suggests that DYN could affect several neuroendocrine functions. Sensory functions might be influenced by DYN at various levels of the central nervous system: nociception (substantia gelatinosa and lamina V of the spinal cord, the superficial zone of the spinal trigeminal nucleus, the mesencephalic periaqueductal grey); proprioception and tactile sensibility (nucleus cuneatus medialis, nucleus princeps nervi trigemini); regulation of input from baro- and chemoreceptive afferents (nucleus solitarius, particularly its caudal, commissural part); regulation of vestibular input (medial and spinal vestibular nuclei); processing of auditory information (dorsal and ventral cochlear nuclei, nuclei of the lateral lemniscus, colliculus inferior); olfaction (primary olfactory fibres terminating in some olfactory glomeruli). Finally, it should be mentioned that the presence of numerous DYN-positive perikarya in the caudatus-putamen complex and of dense networks of DYN-immunoreactive fibres in the medial segment of the globus pallidus and in the pars reticulata of the substantia nigra strongly suggests the presence of a dynorphinergic strio-pallidonigral projection. The cells of origin of the terminal plexus of DYN-positive fibres found in the pallidum ventrale are probably located in the nucleus accumbens.

Angiotensin II

The so-called renin-angiotensin system is a peptide hormone-generating system, the actions of which are all aimed at the maintenance of fluid homeostasis (Phillips 1978; Simpson 1981).

Renin is a proteolytic enzyme secreted by juxtaglomerular cells in the kidney. Blood-borne renin acts on angiotensinogen, a peptide produced by the liver, to form the decapeptide angiotensin I. This decapeptide, in turn, is cleaved by angiotensin-converting enzyme, produced in the lungs, to form the octapeptide angiotensin II (ANG) within the circula-

Angiotensinogen (Renin substrate)	H-Asp^{1}-Arg-Val-Tyr-Ile-His-Pro-Phe-His-Leu^{10}-Leu-Val-Tyr-Ser^{14}-OH
	↓ *Renin*
Angiotensin I [ANG I]	H-Asp^{1}-Arg-Val-Tyr-Ile-His-Pro-Phe-His-Leu^{10}-OH
	↓ *Angiotensin-converting enzyme*
Angiotensin II [ANG II]	H-Asp^{1}-Arg-Val-Tyr-Ile-His-Pro-Phe-OH

Fig. 32. Biosynthesis of angiotensin II

tion (Fig. 32). It is this octapeptide which mediates the principal actions of the renin-angiotensin system, i.e. vasoconstriction, sodium retention, antidiuresis and drinking behaviour. These effects are elicited by (a) direct action on peripheral structures, (b) action on some of the circumventricular organs, and (c) action on certain centres within the central nervous system *sensu strictiori.*

Peripherally, ANG acts on arterial smooth muscle to cause vasoconstriction, with a resultant increase in blood pressure, and provokes aldosterone secretion from the adrenal cortex, with consequent sodium retention.

Circumventricular organs are small receptive loci which lie in regions of the central nervous system devoid of a blood-brain barrier and, hence, are accessible to circulating peptides and other compounds. The circumventricular organs involved in the action of ANG are the area postrema, the subfornical organ and the organum vasculosum of the lamina terminalis (see Fig. 33).

The area postrema is situated in the subventricular zone of the most caudal part of the rhombencephalon. Action of ANG on this structure provokes a distinct central pressor response in various mammals (Joy and Lowe 1970; Ueda et al. 1972), but not in the rat (Haywood et al. 1980; Brooks et al. 1983).

The subfornical organ (SFO) and the organum vasculosum of the lamina terminalis (OVLT) are both situated in the wall of the third ventricle, the former rostrodorsally, just between the two interventricular foramina, the latter rostroventrally, immediately dorsal to the optic chiasma. Both organs contain considerable numbers of ANG receptors (Van Houten et al. 1980), and both have been shown to contain neurons that are exquisitely sensitive to ANG (Phillips and Felix 1976; Hoffman and Phillips 1976; Felix and Phillips 1978; Knowles and Phillips 1980). Local application of ANG to these organs evokes an increase in blood pressure, secretion of antidiuretic hormone and a short-latency drinking behaviour, even in animals that are intravenously overloaded with water (Hoffman and Phillips 1976; Phillips 1976; Simpson 1981). The pressor response is due both to sympathetic activation and to the release of antidiuretic hormone. It has been suggested that the SFO is a receptor only for plasma-borne ANG, whereas the OVLT is available to both plasma-borne and CSF-borne ANG (Phillips 1978). The inner surface of the OVLT protrudes into the third ventricle and has been found to be densely studded with ANG receptors (Landas et al. 1980).

In addition to the organs discussed above, some areas in the brain not situated where the blood-brain barrier is impaired have been shown to be sensitive to ANG. These include the medial preoptic area and the paraventricular nucleus. The preoptic region contains a rather high concentration of ANG receptors (Sirett et al. 1977), and its cells have been found to increase their firing when ANG is applied microiontophoretically (Gronan and York 1978). Moreover, this area appears to be a receptor site for the dipsogenic effect of ANG (Swanson et al. 1978; Richardson and Mogenson 1981). The paraventricular nucleus contains cells which release antidiuretic hormone in the posterior pituitary, and this centre is also implicated in various autonomic neural mechanisms. It is not clear how ANG gains access to the two centres in question; it has been suggested that small amounts of this peptide hormone may cross the blood-brain barrier and thus are able to act directly on the cells of these centres (Swanson and Mogenson 1981), but it is also possible that centrally produced ANG plays a role in their stimulation (see below).

The diencephalic circumventricular organs, the medial preoptic area and the paraventricular nucleus, together with their respective efferents, form part of the neuroanatomical correlate of the synergistic endocrine, autonomic and somatomotor reactions involved in the maintenance of fluid homeostasis (Swanson and Mogenson 1981). It has been experimentally established that the SFO projects to the OVLT, the medial preoptic area, and the paraventricular nucleus (Miselis et al. 1979; Camacho and Phillips 1981; Swanson and Mogenson 1981). The OVLT sends fibres to the supraoptic nucleus, a cell mass which, like the paraventricular nucleus, contains antidiuretic hormone-producing cells (Camacho and Phillips 1981). The medial preoptic area distributes fibres to the paraventricular nucleus and several other hypothalamic centres, to the periaqueductal grey and to the area tegmentalis ventralis (Swanson 1976b; Swanson et al. 1978). The cell mass last mentioned projects in turn to the nucleus accumbens, one of the regions presumably involved in the somatomotor control of drinking behaviour (Swanson and Mogenson 1981). The paraventricular nucleus integrates endocrine and autonomic responses related to the action of ANG. It sends neurosecretory fibres to the posterior pituitary gland, where antidiuretic hormone is released; moreover, it projects, directly to autonomic centres in the lower brain stem and spinal cord which are involved in the regulation of blood pressure (Swanson and Mogenson 1981).

So far, the general or peripheral renin-angiotensin system and the mode in which plasma-borne ANG acts on the central nervous system have been considered. However, evidence is accumulating that the central nervous system possesses an endogenous ANG-forming system. All components of the so-called renin-angiotensin cascade, i.e. renin (Hirose et al. 1978), angiotensinogen (Ganten et al. 1971; Lewicki et al. 1978; Sernia and Reid 1980), angiotensin I (Fischer-Ferraro et al. 1981), angiotensin-converting enzyme (Yang and Neff 1972; Printz et al. 1982) and angiotensin II (Fischer-Ferraro et al. 1981; Phillips et al. 1980; Sirett et al. 1981; see, however, Meyer et al. 1982) have been reported to be present in the brain. Moreover, the central nervous system possesses abundant ANG receptor binding sites, indeed more than any other tissue in the body (Bennett and Snyder 1976; Sirett et al. 1977; Mendelsohn et al. 1984).

The presence of renin in various parts of the central nervous system has been quantified by biochemical analysis (Michelakis et al. 1974). The highest concentrations of renin activity were found in the pineal gland, the anterior pituitary and the choroid plexus, followed by the hypothalamus, the cerebellum and the amygdala. Lower levels were ob-

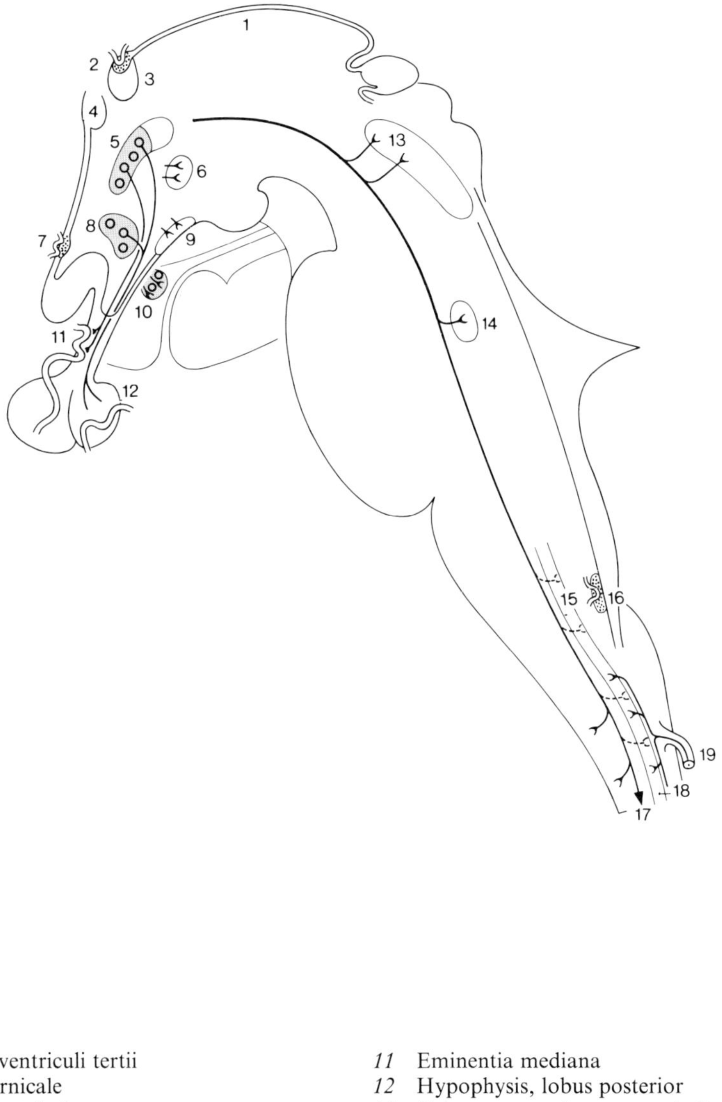

1 Tela choroidea ventriculi tertii
2 Organum subfornicale
3 Foramen interventriculare
4 Commissura anterior
5 Nucleus paraventricularis, pars magnocellularis
6 Nucleus dorsomedialis
7 Organum vasculosum laminae terminalis
8 Nucleus supraopticus
9 Nucleus infundibularis
10 Nucleus medialis + centralis amygdalae
11 Eminentia mediana
12 Hypophysis, lobus posterior
13 Griseum centrale mesencephali
14 Locus coeruleus
15 Nucleus spinalis nervi trigemini, pars caudalis
16 Area postrema
17 Nucleus intermediolateralis
18 Substantia gelatinosa
19 Radix dorsalis nervi spinalis

Fig. 33. Angiotensin II-containing cells and fibres

tained in the cerebral cortex, hippocampus, thalamus, rhombencephalon and spinal cord. An immunohistochemical study (Fuxe et al. 1980) revealed the presence of renin-like immunoreactivity in the supraoptic and paraventricular nuclei and in the cerebellar Purkinje cells.

Sirett et al. (1981) determined the regional distribution of ANG by radioimmunoassay. The hippocampus appeared to have the highest concentration and the cortex the lowest. The concentrations relative to those of the cortex were: hippocampus, 8; striatum, 5; cerebellum, 4; several other regions, including septum, thalamus, hypothalamus, midbrain and myelencephalon, 3.

The presence of ANG-like immunoreactivity in neuronal perikarya and terminals has been reported by several groups of investigators (Fuxe et al. 1976, 1982; Changaris et al. 1978; Phillips et al. 1979; Hoffman et al. 1982; cf. Fig. 33). ANG-positive cell bodies have been found in the supraoptic nucleus and in the magnocellular component of the paraventricular nucleus (Phillips et al. 1979; Fuxe et al. 1982; Hoffman et al. 1982). Some scattered ANG-containing perikarya appear to be present in the perifornical area and in the medial part of the amygdaloid complex (Fuxe et al. 1982). With regard to the localization of renin, ANG, vasopressin and oxytocin in the cells of the supraoptic and paraventricular nuclei, differing observations have been made. Phillips et al. (1979) reported that ANG and vasopressin are stored in separate cells, but according to Hoffman and colleagues (1982), these two peptide hormones occur together in the same neurons. Fuxe et al. (1982) confirmed the co-existence of ANG and vasopressin, but added the remarkable observation that renin occurs exclusively in cells which also contain oxytocin.

The ANG-containing cells in the supraoptic and paraventricular nuclei send fibres to the median eminence and the posterior pituitary gland. The medial part of the median eminence contains a dense plexus of ANG-positive fibres, which is concentrated in the border zone of the external and internal laminae of that structure (Changaris et al. 1978; Phillips et al. 1979).

Attention may now be given to the ANG immunoreactivity, and in particular to the ANG-positive terminals, found in various parts of the brain and spinal cord. Because the bulk of the ANG-positive perikarya are situated in the magnocellular supraoptic and paraventricular nuclei, and because the first of these nuclei is known to project mainly to the posterior pituitary gland, Fuxe et al. (1982) surmised that most of these terminals belong to magnocellular neurons in the paraventricular nucleus. This nucleus does give rise to descending fibres which distribute to several centres in the brain stem and spinal cord; however, it has been experimentally established that these descending projections originate from the parvocellular, rather than from the magnocellular portion of the nucleus (see the section on vasopressin and oxytocin). Therefore, the sources of the ANG immunoreactivity to be discussed below remain to be established and, accordingly, the relations indicated in Fig. 33 should be considered highly speculative.

Within the telencephalon the nucleus centralis amygdalae contains high densities of ANG-immunoreactive terminals (Fuxe et al. 1976). As mentioned before, the hippocampus and the striatum contain a relatively very high concentration of ANG immunoreactivity. Taking the distribution of angiotensin-converting enzyme and of the specific ANG receptor binding into consideration, Sirett et al. (1981) considered it likely that the hippocampus and the striatum (including the nucleus accumbens) harbour ANG-containing neurons which project largely to the septum and the hypothalamus.

In the hypothalamus, beyond the supraoptic and paraventricular nuclei, the highest concentrations of ANG-positive terminals are found in the dorsomedial nucleus and in the basal hypothalamus, with scattered terminals in most hypothalamic nuclei (Hökfelt et al. 1978b; Fuxe et al. 1982).

Scattered ANG-positive fibres and terminals

occur in many areas of the brain stem. High densities of ANG-immunoreactive structures are found in the spinal nucleus of the trigeminal tract. Moderate densities are present in the periaqueductal grey, especially in the rostral part, and in the locus coeruleus (Fuxe et al. 1976).

In the spinal cord, high densities of ANG-immunoreactive fibres and terminals are observed in the substantia gelatinosa and in the upper thoracic part of the intermediolateral column; the latter structure is composed of sympathetic preganglionic neurons (Fuxe et al. 1976). Two zones of ANG innervation have been reported to exist within the substantia gelatinosa, one in the most superficial portion of this structure, i.e. lamina I of the spinal grey, and another situated deeper, in the inner part of lamina II (Fuxe et al. 1982). Some cell bodies in the spinal ganglia appear to be ANG positive (Fuxe et al. 1982); the observation that following dorsal rhizotomy there is a considerable reduction of the ANG immunoreactivity in the deeper zone of the substantia gelatinosa indicates that this zone is formed largely by the central branches of primary afferent neurons. In contrast, the ANG-immunoreactive terminals in the intermediolateral column and anterior horn disappear completely after sectioning of the spinal cord, and thus presumably belong to descending ANG-containing pathways (Hökfelt et al. 1978a; Fuxe et al. 1982).

In summary, it may be stated that there is evidence for the existence of a central renin-angiotensin system, i.e. a set of ANG-producing neurons situated within the confines of the blood-brain barrier, but that our knowledge of the morphology of this system is still very fragmentary. However, the data available suggest that ANG-producing neurons are involved in vasopressin release and in sympathetic activation, and thus may participate in the regulation of fluid balance and blood pressure. Interestingly, evidence is accumulating that overactivity of the brain renin-angiotensin system may play a role in the development and maintenance of hypertension (Weyhenmeyer and Phillips 1982; Phillips 1983). On the other hand, it cannot be excluded that certain sets of ANG-releasing neurons are involved in functions entirely unrelated to blood pressure and body fluid regulation. If that were the case, the term 'central renin-angiotensin *system*' would have to be discarded.

CHAPTER 4

Conclusions and Comments

In the preceding chapter a survey of a number of chemically defined cell groups and pathways has been presented. In this chapter I shall attempt (a) to draw certain general conclusions from the data gathered, and (b) to explore the relations between 'classical' and 'chemical' neuroanatomy. Before embarking on the enterprise just outlined it seems proper to indicate the limitations of this study and to articulate a number of caveats.

1. Our knowledge of chemical neuroanatomy is still incomplete and we lack insight into the question of how incomplete this knowledge actually is. It is to be expected that new neuroactive principles will be discovererd in the coming years and that more sensitive techniques will reveal the neuronal networks containing many of the known (putative) transmitters to be more extensive than is described here. It also seems likely that a considerable number of the data compiled will appear to be erroneous in light of work with new and more specific antisera.

2. The title of this monograph is too ambitious, because I have been unable to cover the entire literature on chemically defined neuron populations. This holds in particular for the neuropeptides. Of the some 35 members of this group described so far, only 16 have been dealt with here. However, all known general categories of neuropeptides are represented in this sample.

3. It is impossible to define neuroactive principles on the basis of their chemical structure per se as neurotransmitters, neuromodulators or neurohormones. Many of the compounds discussed in the previous chapter presumably play multiple roles, depending on their locus of action. It is for this reason that the neutral term *neuromediator* will be frequently employed in the discussions that follow.

4. In the pictorial surveys (Figs. 2, 5 etc.) data derived from various mammals (primarily the rat) and man have been transferred to a schematic diagram of the human brain. Because, with regard to chemical neuroanatomy, the similarities between the various mammalian species investigated so far are much more striking than the differences, it is reasonable to assume that these pictures broadly reflect the distribution of the various neuromediators in the human brain (with the limitations indicated under 1, above). However, a number of striking species differences have been reported; therefore, it should be cautioned that, strictly speaking, in the pictorial surveys certain morphological features of the human brain have been used as non-linear coordinate system on which data derived from various mammals have been projected.

1. The various neuromediator-specified neuronal populations do not obey simple structural rules. They vary considerably with regard to (a) the number of centres in which their cells are found, (b) the extent of the area occupied by these centres, and (c) the area of distribution of their fibres and terminals. Although some of the populations studied show a certain amount of structural parallelism, most appear to have their own characteristic pattern of distribution.

Cells containing, for example, GABA (Fig. 13), cholecystokinin (Fig. 18), somatostatin (Fig. 23) and enkephalin (Fig. 29) are found in many different centres in the central nervous system, spread throughout most parts of the neuraxis. Cells containing noradrenaline (Figs. 6 and 7), serotonin (Fig. 9), ACTH (Fig. 27) or angiotensin II (Fig. 33) occur in a limited number of centres, which are confined to a relatively small area of the brain. Finally, the histamine-containing neurons (Fig. 11) are all concentrated in a single cluster within the caudal hypothalamus.

With regard to the spread of fibres, it has been found that noradrenergic (Figs. 6, 7) and serotoninergic axons (Fig. 9) are distributed to virtually all areas of the central nervous system. Fibres containing adrenaline (Fig. 8), VIP (Fig. 17), vasopressin, oxytocin (Fig. 25) and angiotensin II (Fig. 33), on the other hand, project to only a very limited number of centres. The remaining neuromediator-specified neuron populations occupy an intermediate position between these two extremes.

Numerous centres are composed of several sets of perikarya, each set containing a different neuromediator (see below); at the same time, numerous fibre tracts are also composed of several axonal sets, each containing a different neuromediator (see below). Therefore the term 'structural parallelism' should be reserved for those instances in which two (or more) neuromediator-specific neuron populations share a considerable number of centres and/or projections. The substance P (Fig. 16) and enkephalin (Fig. 29) populations show a striking, though partial, parallelism. It is also worth mentioning in this context that adrenergic fibres are distributed almost exclusively to centres which also receive a noradrenergic innervation, although both types of fibres originate from different sets of centres (Figs. 6–8). It must also be cautioned, however, that if the distribution of a newly discovered peptide B shows extensive parallels with that of a previously investigated peptide A, and both peptides share certain immunogenic amino acid sequences, cross reactivity of the anti-B serum with peptide A is the first possibility to be considered.

2. Unifying concepts concerning the overall functional significance of the various neuromediator-specified neuronal populations have not yet been formulated.

Generalizations concerning the function(s) exerted by the various neuromediators appear to be hard to make. Most authors who have studied the distribution of a given neuromediator are inclined to place their findings in the perspective of what is known concerning the functional significance of the various centres, pathways and areas of termination in which that neuromediator is found. Because most of the neuromediator-specified neuronal networks extend over areas bearing different 'functional labels', sentences like: "The present study demonstrated a wide distribution of substance X-containing structures in the central nervous system, suggesting that substance X may be involved in a variety of functions" are frequently encountered in the neuroimmunohistochemical literature. On the other hand, the notion that a certain population of transmitter-specified neurons may subserve one function or a set of coherent functions is also repeatedly found, mostly only implicit and unspecified, e.g. by denoting that population as 'the substance X *system*'.

It seems well to consider the question concerning functional homogeneity or functional heterogeneity separately for each neuromediator or group of neuromediators. In the light of our present knowledge it seems highly unlikely that all of the actions of the populations of cholinergic or dopaminergic neurons will ever be integrated into a single functional concept (which, as a matter of course, does not exclude the possibility that certain subsets of these populations do fulfil a clearly definable functional role, and thus fully deserve the designation 'system'). Nevertheless it seems possible that for the actions of the remarkable, widely distributed noradrenergic and serotoninergic networks, common functional denominators will ultimately be found (see below). Turning to the neuroactive amino acids, almost all members of the neuronal populations containing GABA or glycine presumably exert an inhibitory action on other neurons, and all members of the neuronal populations containing glutamate and aspartate are most probably excitatory in nature. However, the neurons of these four populations are known to exert the basic actions indicated in a variety of functional contexts, and attempts to find higher-order functional homogeneities behind these functional heterogeneities may be expected to remain unsuccessful.

As regards the neuropeptides, it is important to note that long before their role as central chemical messengers was discovered many of them were known as neurohormones, peripheral hormones or hypophysiotropic factors. Most members of these three categories of substances exert either a single function or a number of different actions which are all aimed at a single ultimate goal. Because it seems unlikely that the occurrence of these peptides in the periphery as well as in the central nervous system is a mere coincidence, it has been repeatedly suggested that the peripheral and central elements producing one and the same peptide may well affect closely related aspects of physiology and behaviour or, in other words, that they are different components of one functional system. To give a few examples: (a) There is evidence that the intestinal hormone cholecystokinin and the central neuromediator cholecystokinin both inhibit feeding behaviour (for references, see Iversen 1983). (b) It is reasonable to assume that the hypophysiotropic hormone LHRH and certain portions of the central LHRH neuronal network both play a role in the regulation of sexual behaviour and reproduction (Witkin et al. 1982; King et al. 1984). (c) It has been suggested that the neurohypophyseal peptides vasopressin and oxytocin and the fibres containing the same peptides which project to the spinal cord may well be involved in the regulation of the same processes, for instance, the control of blood pressure and lactation (Buijs et al. 1983). (d) Both peripherally and centrally produced angiotensin II may be implicated in the maintenance of fluid homeostasis. Iversen (1983) has suggested that such parallel actions of peptides in periphery and brain, leading to similar end results, may well begin to tell us something about the general integration of brain and body. However, it should be cautioned that, in the examples given, there is at best a coupling between the actions of groups of cells producing a given hormone or hypophysiotropic factor and the possible activities of *part* of the central neuronal network using the same substance as a neuromediator. Moreover, it should be added that for many other peptides which are known to be active in both the periphery and the brain, any indication of a relationship between their peripheral and central actions is lacking. It is, for instance, hard to understand what the actions of somatostatin or thyrotropin-releasing hormone on the hypophysis could have to do with their activities as central neuromediators. Clearly, we are still far removed from any general or unifying neuropeptide concept.

3. Different regions of the central nervous system may contain quite different numbers of neuromediators and may show considerable differences in their neuromediator profiles.

For some data illustrating this aspect refer to Table 1. It is remarkable that in the preopticohypothalamic region and in the basal telencephalon (i.e. the continuum formed by the bed nucleus of the stria terminalis, the septum, the nuclei of the diagonal band and the amygdala) all or nearly all of the neuromediators included in this study are present. In the olfactory bulb and the superior colliculus only limited numbers of neuromediators are found, but the least have been registered for the cerebellum, where the amino acids GABA and glutamate/aspartate are, as far as we know, the most important neuromediators. It is also noteworthy that in the cerebellum only two of the 16 neuropeptides included in this study are found. In the thalamus acetylcholine, noradrenaline, serotonin, GABA and glutamate/aspartate appear to be the principal neuromediators. The neuropeptides, though relatively large in number, play presumably only a minor role in the thalamus, because most of them occur in a limited number of structures and only in one or two of the cell masses within that nuclear complex.

Table 1. Neuromediator profiles and total numbers of neuromediators present in some regions of the central nervous system

	acetylcholine	dopamine	noradrenaline	adrenaline	serotonin	histamine	GABA	glutam. + aspart.	glycine	substance P	VIP	CCK	neurotensin	CRF	LHRH	somatostatin	TRH	vasopr. + oxytoc.	ACTH	αMSH	γMSH	endorphin	enkephalin	dynorphin	angiotensin II	total number
olfactory bulb	+	+	+		+		+	+		+					+								+	+		10
basal telencephalon	+	+	+		+	+	**+**	+		**+**	**+**	**+**	**+**	**+**	+	**+**	+	+	**+**	+	**+**	+	**+**	+	+	23
neostriatum	**+**	**+**	+		+	+	**+**	**+**	+	+	+	**+**	+			+				+			**+**	+	+	17
hippocampal form.	**+**	+	+		+		**+**				**+**	**+**		+		**+**		+		**+**			**+**	**+**		13
neocortex	**+**	+	+		+	+	+	**+**	+	+	**+**	**+**	+	**+**		+		+		+			+	+		18
thalamus	**+**	+	**+**	+	**+**	+	**+**	**+**		+	+	+	+	+		+		+	+			+	+			18
preoptico-hypothalamic r.	+	**+**	**+**	**+**	**+**	**+**	+	+	+	**+**	**+**	**+**	**+**	**+**	**+**	**+**	**+**	**+**	**+**	+	**+**	**+**	**+**	**+**	+	25
superior colliculus	+		+		+		**+**	+		+	+				+	+							+			10
cerebellum			+		+		**+**	**+**								+							+			6
medial reticular form.	**+**		**+**		+	+	+			**+**		+	+	**+**	+	**+**	**+**		**+**		**+**	**+**	**+**	**+**		17
spinal dorsal horn	**+**	+	+		**+**		+	+		**+**	+	+	**+**			+		+					**+**	**+**	**+**	15

+, present; **+**, present in a large number of structures, i.e. somata, fibres, terminals

4. Although within a number of nuclei all of the neuromediator-specified cells contain one and the same neuromediator, most nuclei appear to be composed of subpopulations, each containing a different neuromediator. The neuromediator profiles of the various cell masses may differ considerably.

The total numbers of known neuromediators and the neuromediator profiles of the cells in a number of representative nuclei are listed in Table 2. The fragmentary character of our current knowledge results from the fact that in a considerable number of cell masses the identity of the neuromediator(s) present in their constituent perikarya is not known. Judging from the literature I have studied this is the case for several thalamic nuclei, the nucleus interstitialis of Cajal, the nucleus ruber and the pontine nuclei, among other grisea. In some cell masses, such as the nucleus basalis of Meynert, the nucleus reticularis thalami and the somatomotor and branchiomotor nuclei, the whole population of cells may contain the same single neuromediator. However, in many other nuclei in the perikarya of which only a single neuromediator has yet been found, only some cells contain that neuromediator, e.g. the subthalamic nucleus or the substantia nigra, pars reticulata. In centres in which subpopulations of cells each containing a different neuromediator have been demonstrated, it is in most cases very hard to prove that the identities of the neuromediators of all of its constituent cells have been determined. Reference to Table 2 shows that the numbers of neuromediator-specified subpopulations and neuromediator profiles may differ considerably from cell mass to cell mass. Some nuclei, such as the

Table 2. The total number of neuromediators and the neuromediator profiles of a number of cell masses. Only the neuromediators occurring in the perikarya are listed

	acetylcholine	dopamine	noradrenaline	adrenaline	serotonin	histamine	GABA	glutam. + aspart.	glycine	substance P	VIP	CCK	neurotensin	CRF	LHRH	somatostatin	TRH	vasopr. + oxytoc.	ACTH	αMSH	γMSH	endorphin	enkephalin	dynorphin	angiotensin II	total number
n. centralis amygdalae	o						o				o		●	●		●							●	o		8
n. septi medialis	o						●					o	o		o								o			6
bed n. stria terminalis											o		o	●	●	o		●					o	o		8
n. basalis of Meynert	●																									1
n. habenulae medialis	o									o																2
n. reticularis thalami							●																			1
n. preopticus medialis										o		o	●	●	●											6
n. suprachiasmaticus																	o	o					o			6
n. ventromedialis hypoth.										o							o						o			3
n. infundibularis	o	o								o			●	o	o	o			●		o	o	o	o		12
n. nervi oculomotorii	●																									1
n. of Edinger-Westphal	●									●		●														3
s. nigra, p. compacta		●							o														o			3
s. nigra, p. reticulata							●																			1
n. parabrachialis lat.	o											o	o	o									●	o		6
locus coeruleus			●										o	o				o					o			5
n. solitarius	o			o						o		o		o		o			o		o	o		o		13
n. ambiguus	●															o										2
n. nervi hypoglossi	●																									1
n. intermediolateralis	●															o							o			3

o, present; ●, present in a considerable number of cell bodies

nucleus centralis amygdalae, the nucleus infundibularis and the solitary nucleus, contain a particularly large number of neuromediator-specified subpopulations and may be aptly designated as 'multiple neuromediator complexes'.

5. The question as to whether or not the neuromediator-specified subpopulations present within a griseum correspond to particular morphological classes, is often left unanswered.

In fact, information of this kind is mainly confined to cells which, on account of their characteristic size, shape, or position, can be easily recognized in both Nissl (or Golgi) and histochemically stained material. Thus, it is known that the highly characteristic pyramidal cells in the allocortex and neocortex use glutamate as their neurotransmitter, and that Purkinje and basket cells in the cerebellar cortex contain GABA, whereas the small cerebellar granule cells are presumably glutamatergic (Fig. 48). However, for cell populations which cannot be as easily classified on morphological grounds, histochemical results concerning neuromediator specification are often presented without any reference to the structural characteristics of the cells found. This statement holds in particular for studies dealing with the distribution of a particular neuromediator throughout the central nervous system. In the accompanying illustrations of such studies the relevant cells are usually indicated simply as dots. It should be noted that this mode of presentation can result in the creation of serious gaps between classical neuroanatomy and chemical neuroanatomy.

6. Neuromediator-specified neurons occurring within the confines of given grisea are often not equally dispersed in these grisea.

The neurons containing a particular neuromediator may be concentrated in or even confined to one or more parts of the centres studied. In the preceding section numerous examples of the conditions just indicated are mentioned. Thus, CCK-containing cells are concentrated in the ventrocaudal part of the caudatoputamen (Takagi et al. 1984), whereas VIP-positive cells are confined to the dorsomedial part of the same structure (Roberts et al. 1980a). SST-containing neurons are confined to the ventral and lateral portions of the periaqueductal grey (Finley et al. 1981b). Agnati et al. (1982, 1984), who developed a quantitative method which makes it possible to assess whether or not a given cluster of neuromediator-identified neurons consists of subgroups, established that the B7 group of serotoninergic cells comprises at least two separate subpopulations.

Transmitter specifications of cells present within a given griseum may lead to a chemoarchitectonic subdivision which may or may not correspond to a cytoarchitectonic subdivision. For example, in the golden hamster the suprachiasmatic nucleus can be divided on cytoarchitectonic grounds into dorsomedial and ventrolateral subdivisions. Vasopressin- and SST-containing neurons are localized within the dorsomedial subdivision, whereas VIP-immunoreactive neurons are concentrated in the ventrolateral subdivision (Card and Moore 1984). In man, a similar differential distribution of vasopressin- and VIP-containing neurons has been found in the suprachiasmatic nucleus (Stopa et al.

1984), but it has not been established if there is a corresponding cytoarchitectonic subdivision. The interpeduncular nucleus has been the subject of a number of recent immunohistochemical studies (Hemmendinger and Moore 1984: rat; Hamill et al. 1984; rat; Karpadia and De Lanerolle 1984: cat; for nomenclature, see Lenn and Hamill 1984). From these studies it may be concluded that the patterns of distribution of serotonin-, GABA-, substance P-, SST- and L-ENK-containing cells conform partly, but not entirely to the cytoarchitectonic subdivisions of this nucleus. A similar conclusion seems to be warranted for the substance P-, SST-, neurotensin- and enkephalin-containing neurons in the septal region (see the figures in Gall and Moore 1984 and Köhler and Eriksson 1984). Finally, it may be mentioned that in some cytoarchitectonic entities a particular neuromediator is contained in cells which at first sight seem to be evenly distributed, but on closer examination are seen to be arranged in a certain pattern. This phenomenon may be exemplified by some findings on the nucleus caudatus and the putamen. Histochemical studies (e.g. Graybiel and Ragsdale 1979, 1983) have shown that these structures consist of two interdigitating compartments, a discontinuous one consisting of small patches, which is embedded in a continuous one known as 'the matrix'. The investigations of Beach and McGeer (1984) have shown that in primates substance P-containing perikarya are distributed evenly throughout the nucleus caudatus and the putamen, but yet are clustered within the patches mentioned.

7. Subpopulations of neuromediator-specified neurons frequently extend beyond the morphologically defined boundaries of grisea.

This is certainly one of the most important general results of (immuno-)histochemical studies on the central nervous system. The significance of this finding lies in the fact that in classical neuroanatomy the centres delineated on cytoarchitectonic grounds are commonly regarded not merely as morphological entities, but also as functional units. As regards this 'non-conformance' phenomenon, I confine myself to the following three examples:

(a) The large cholinergic neurons present in the basomedial telencephalon are unevenly distributed over a continuum formed by the nucleus basalis of Meynert, the nuclei of the diagonal band of Broca and the medial septal nucleus (Fig. 2).

(b) A group of ACTH-containing neurons coincides largely with the infundibular nucleus. However this group extends rostrally, dorsally as well as caudally, beyond the boundaries of the nucleus (Fig. 27).

(c) Throughout most of the brain stem significant numbers of serotoninergic neurons are located outside the raphe nuclei.

8. Numerous populations or subpopulations of neuromediator-specified cells are not organized along the lines of defined cytoarchitectonic subdivision of the central nervous system.

In the preoptico-hypothalamic region several groups of neuromediator-specified neurons – for example, those containing neurotensin (Fig. 19), LHRH (Fig. 22), SST (Fig. 23) and TRH – are not related to any cytoarchitectonic boundaries. The noradrenergic cell groups in the brain stem – except for the locus coeruleus – do not form distinct nuclei, but are dispersed among non-noradrenergic cells (Fig. 7), and the same holds for the adrenergic cell groups (Fig. 8). For these cell groups in the brain stem numerical nomenclatures (A1–A5, C1–C3) have been proposed, but such numerical designations can be warranted only if they do not correspond at all to known cytoarchitectonic entities.

9. Neurons may contain more than one neuromediator.

This remarkable phenomenon, commonly designated as the co-existence or co-localization of neuromediators, has already been briefly discussed in the introductory section of this monograph. Pertinent data have been compiled from the literature and are assembled in Table 3. Neurons containing two or more neuromediators which are all derived from one and the same precursor molecule (e.g. various pro-opiomelanocortin derivatives) have been left out of consideration. The data gathered warrant the following tentative conclusions regarding the co-existence of neuromediators within the same neurons:

1. Co-existence of neuromediators occurs in many different grisea of the brain; among the centres showing this phenomenon the medullary raphe nuclei take pride of place, with no less than five different neuromediator combinations.
2. As far as is known, generally two, or sometimes three different neuromediators coexist within 'multiple neuromediator' neurons.
3. Representatives of any known classes of neuromediators, i.e. acetylcholine, monoamines, amino acids and neuropeptides, may be involved. The first three classes mentioned are commonly designated as classical neurotransmitters (see Lundberg and Hökfelt 1983; Hökfelt et al. 1984b); they will be denoted here as classical neuromediators.
4. In the central nervous system co-existence of classical neuromediators has not been reported so far.
5. 'Multiple neuromediator' neurons commonly synthesize a classical neuromediator in combination with one, or sometimes two neuropeptides.
6. For some centres co-existence of two different neuropeptides has been reported. Whether such 'peptidergic' cells also contain a classical neuromediator is not known at present.
7. Most frequently one of the monoamines co-exists with one of the so-called gut-brain peptides.
8. Neuropeptides derived from the pro-opiomelanocortin molecule have not yet been reported to be co-localized with either classical neuromediators or other neuropeptides.

The occurrence of two or more neuromediators in the same neurons raises numerous questions, among which the following three may be especially mentioned:

1. How and where are the molecules of the different neuromediators stored in the terminals of the pertinent neurons?
2. How and when are the molecules of the different neuromediators released (provided that they *are* released)?
3. What is the possible functional significance of the co-existence phenomenon?

These questions have been discussed in detail in numerous publications by Tomas Hökfelt and his colleagues (e.g. see Hökfelt et al. 1980a, 1984b; Lundberg et al. 1981; Lundberg and Hökfelt 1983), and the following brief comments are based mainly on these papers.

As regards storage, Hökfelt et al. (1980a) have addressed the question as to whether the different neuromediators are stored in the same or in separate synaptic vesicles. Given the observation that peptides are usually present in large vesicles (diameter about

Table 3. Co-existence of neuromediators in various centres

"classical" neuromediator	peptide	peptide	brain region	source
acetylcholine	substance P		n. laterodorsalis tegmenti	Vincent et al. 1983
"	VIP		cerebral cortex	Eckenstein and Baughman 1984
"	enkephalin		oliva superior complex	Altschuler et al. 1984
dopamine	CCK		area tegmentalis ventralis	Hökfelt et al. 1980a,b
"	neurotensin		" " "	Hökfelt et al. 1984
"	"		n. infundibularis	" " " "
noradrenaline	"		n. solitarius	" " " "
"	enkephalin		locus coeruleus	Charnay et al. 1982
adrenaline	neurotensin		n. solitarius	Hökfelt et al. 1984
serotonin	substance P		medullary raphe nuclei	Hökfelt et al. 1978
"	"	TRH	" " "	Johansson et al. 1981
"	CCK		" " "	Mantyh and Hunt 1984
"	enkephalin		" " "	Glazer et al. 1981
"	dynorphin		" " "	Basbaum and Fields 1984
"	"		n. reticularis magnocellularis	" " " "
GABA	CCK		hippocampus	Vickrey et al. 1983
"	SST		"	Jirikowski et al. 1984
"	"		n. reticularis thalami	Oertel et al. 1983
"	enkephalin		n. caudatus + putamen	Aronin et al. 1984
	substance P	CCK	griseum centr. mesencephali	Skirboll et al. 1982
	somatostatin	APP	n. caudatus + putamen	Vincent et al. 1983a,b
	vasopressin	dynorphin	n. supraopt. + n. paraventr.magn.	Watson et al. 1982
	"	angiotensin II	" " "	Hoffman et al. 1982

APP, avian pancreatic polypeptide (cf. Fig. 15)

100 nm), whereas monoamines may be present in large as well as small vesicles (diameter about 50 nm), they speculated that in presynaptic endings containing both large and small vesicles and a monoamine as well as a neuropeptide, the monoamine and the neuropeptide may co-exist in the large vesicles, whereas the small vesicles may contain only the monoamine. On the basis of subcellular fractionation studies Hökfelt et al. (1984b) reported that the situation just indicated actually exists in the terminals of postganglionic autonomic neurons innervating the vas deferens. In these terminals noradrenaline and a neuropeptide (neuropeptide Y) are stored differentially, whereby the peptide is preferentially stored in large dense-core vesicles, and noradrenaline is present both in these vesicles and in small synaptic vesicles. A somewhat different mode of neuromediator storage is probably present in another set of autonomic postganglionic neurons, i.e. those innervating the submandibular salivary gland. These elements contain the classical neuromediator acetylcholine and the peptide VIP. In their terminals there is a dominance of small clear vesicles, which presumably store most of the acetylcholine, and some large dense-core vesicles, which most likely contain VIP (Lundberg and Hökfelt 1983). Data on the mode of co-localization of neuromediators in the central neurons are extremely scant. Pelletier et al. (1981) have produced evidence suggesting that the same dense-core vesicles found in certain nerve endings in the spinal ventral horn store both serotonin and substance P.

The release of neuromediators, let alone the release of combinations of neuromediators, cannot be studied directly at the ultrastructural level. However, there is evidence that in certain sets of neurons in the peripheral nervous system which contain both a classical neuromediator and a peptide a differential release of these two compounds may take place, because the secretion of the peptide requires a higher stimulation frequency than does the classical neuromediator (Lundberg and Hökfelt 1983). It may well be that comparable mechanisms are operative in the central nervous system.

The functions mediated by two or more neuromediators that are released by the same neuron are not fully established. Theoretically, the potential to release several messengers greatly increases the possible number of different chemical signals that a neuron can use in communicating with other elements, particularly because the elements innervated may be equipped with different sets of receptors (Iversen 1983; Swanson 1983; Hökfelt et al. 1984b). Experimental data, primarily obtained from studies of the peripheral nervous system, yielded the following two ideas concerning the possible significance of co-localization of a classical neuromediator and a neuropeptide (see Lundberg and Hökfelt 1983; Hökfelt et al. 1984b): (a) Often the neuropeptide seems to be responsible for a component of the response, characterized by slow onset and long duration, while the classical neuromediator causes rapid effects of short duration. (b) Often the neuropeptide seems to play an auxiliary role as co-messenger in support of the classical neuromediator. Such synergistic actions of neuropeptides may be effected in different ways, e.g. by blocking inhibitory presynaptic autoreceptors for the classical neuromediator, or by cooperating with the classical neuromediator at postsynaptic sites (Hökfelt et al. 1984b).

Before leaving the intriguing subject of co-localization of neuromediators, we should note that Osborne (1979) has pointed out that the presence of a neuromediator in very small quantities in a neuron which also contains another neuromediator does not necessarily have a functional significance. As he has put it: "Since, from the evolutionary point of view, all neurons are expected to possess the same genetic machinery, it is possible that the synthesis of certain substances is not completely suppressed. Thus, although the substance is present in minute amounts, it may nevertheless be without function."

10. Morphologically homogeneous neuronal populations (sets of neurons) may be chemically heterogeneous in the sense that different elements of a set may either contain different neuromediators or different combinations of neuromediators.

There are several examples of the condition first mentioned. (a) First are the periglomerular cells in the olfactory bulb, a set of superficially situated interneurons (Fig. 49) which comprises two chemical subsets, one containing GABA, the other dopamine (Ribak et al. 1977; Ljungdahl et al. 1977; Priestley et al. 1979; Mugnaini et al. 1984a). Mugnaini et al. (1984b) speculated that the GABA-ergic and the dopaminergic periglomerular cells may be primarily involved in short- and long-lasting modulation respectively. (b) GABA is generally considered to be the neurotransmitter of the cerebellar Purkinje cells (Fig. 48). However, Chan-Palay et al. (1981) found that a large proportion of these elements remain unlabelled by the autoradiographic and immunohistochemical techniques for GABA. They presented evidence suggesting that many of the non-GABA-ergic Purkinje cells contain motilin or a motilin-like compound (motilin is a 22-amino acid polypeptide isolated from the porcine gut that influences gastric motility and emptying; see Brown et al. 1973). Interestingly, Chan-Palay et al. (1981) also reported that a remaining population of between 30% and 40% of the Purkinje cells contains neither GABA nor motilin and, hence, may well use another neuromediator. (c) From high-affinity uptake and transport studies with tritiated amino acids, Wilkin et al. (1981) concluded that in the cerebellar cortex two biochemically separate populations of Golgi neurons (Fig. 48) are present, one transporting glycine, the other GABA.

That morphologically homogeneous neuron populations may contain different combinations of neuromediators is well known. It is a consistent finding that neurons containing a classical neuromediator and a neuropeptide form a subset within the set of neurons containing only the classical neuromediator (Lundberg and Hökfelt 1983). Subsets, the elements of which contain the same classical neuromediator but different neuropeptides, have also been found in several centres, such as the medullary raphe nuclei (see Table 3).

11. Although it cannot be excluded that within a number of fibre tracts all of the constituent fibres contain one and the same neuromediator, many pathways appear to be composed of subpopulations of fibres, each containing a different neuromediator or combination of neuromediators. The neuromediator profiles of fibre systems may vary considerably.

In Table 4 the total number of the neuromediators and the neuromediator profiles of a number of representative fibre systems are listed. It should be emphasized that we know little of the neuromediators present in a number of long pathways of classical neuroanatomy. This may be illustrated by the following examples.

The large cells present in the most superficial zone of the spinal dorsal horn are provided with long axons which pass via the spinothalamic tract to the nucleus ventralis posterolateralis thalami. Because many of these cells are known to contain enkephalin (Glazer and Basbaum 1981), it seems likely that a certain proportion of the fibres in the spinothalamic tract contain this peptide. However, we have no idea which neuromediators are present in the other components of this composite fibre system.

Concerning the medial lemniscus, all we know at present is that this large projection contains a limited number of substance P-immunoreactive fibres (Cuello and Kanazawa 1978). Within the dorsal column nuclei CCK-, DYN- and GABA-containing neurons have been observed (Kiyama et al. 1983; Khachaturian et al. 1983b; Rustioni et al. 1984). However, which of these cell populations represent projection neurons remains to be determined.

The nature of the neuromediators present in such massive fibre systems as the brachium conjunctivum and the pallidothalamic projection is entirely unknown.

In the visual geniculocortical projection, the presence of only one neuromediator, the peptide CCK, has been determined so far (Fallon and Seroogy 1984).

The neuromediators present in the medial longitudinal fasciculus (MLF), a conspicuous fibre system extending throughout the length of the brain stem and consisting of several well-myelinated fibre components (interstitiospinal, vestibulomesencephalic, vestibulospinal, reticulospinal), have not been determined as yet. All we know is that at mesencephalic levels some serotoninergic fibres penetrate the domain of this bundle. It seems possible that substance P-, enkephalin- and GABA-containing cells present in the medial vestibular nucleus contribute fibres to the vestibulospinal component of the MLF (Nomura et al. 1984), but there is no direct proof for this assumption.

Our obvious lack of knowledge concerning the nature of the neuromediators present in many of the fibre systems of classical neu-

Table 4. Neuromediator profiles and total numbers of neuromediators present in a number of fibre systems

	acetylcholine	dopamine	noradrenaline	adrenaline	serotonin	histamine	GABA	glutam. + aspart.	glycine	substance P	VIP	CCK	neurotensin	CRF	LHRH	somatostatin	TRH	vasopr. + oxytoc.	ACTH	αMSH	γMSH	endorphin	enkephalin	dynorphin	angiotensin II	total number
tr. olfactorius	+	+	+		+			**+**							+											6
tr. corticospinalis								**+**																		1
stria terminalis			+		+	+				+	**+**	+	+	+	**+**	**+**		+			+		+			13
fibr. amygdalofugales ventr.	+	+	**+**		**+**	+					+			+	**+**	+			+	+			+			12
band. diagonalis Brocae			+		+	+						+		+				+			+		+			8
fornix	+	+	+		+		+	+				+				+				+						9
stria medullaris	+		+		+		+								+	+		+								7
tr. geniculocorticalis (visual)												+														1
tr. habenulo-interpedunc.	+	+	+		+					+					+	+										7
f. telencephalicus med.	+	**+**	**+**	+	**+**	+				+	+	**+**	**+**	**+**	+	**+**			+				+			15
f. longitudinalis dors.		+	**+**	+	+	+				+				**+**	+	+		+	**+**				+			12
tr. strionigralis							**+**			**+**														+	+	4
tr. nigrothalamicus							**+**																			1
tr. nigrotectalis							**+**																			1
tr. supraopticohypophyseos												+					**+**						+	+	+	5
tr. infundibularis		+				+							+		+				+							5
nervus cochlearis	+							+															+			3
lemniscus lateralis						+				+		**+**				+										4
brach. colliculi inf.																+										1
tr. olivocerebellaris								**+**																		1
f. longitudinalis med.					+																					1
fibrae raphespinales					**+**					+		+					+									4
radix dorsalis spinalis								**+**		**+**	+	+				+								+	+	7
radix ventrales spinalis	**+**																									1

+, present; **+**, present in a large number of fibres

roanatomy may have to do with the fact that these systems use classical neuromediators which are produced locally in the terminals and, hence, cannot be demonstrated in the axons. Techniques aimed at visualizing not a neuromediator itself, but rather its synthesizing enzyme may yield negative results in the axons if the concentration of that neuromediator in the terminals is to a large extent maintained by reuptake after its release, rather than by continuous synthesis; again this mechanism is known to play an important role in neurons containing classical neuromediators. It may also be worthy of note that no specific enzyme markers of glutamate- and aspartate-containing neurons are available at present. The substances just mentioned are important excitatory neurotransmitters which may well be used in several of the classical pathways enumerated above. Finally, it should be appreciated that even the axonal levels of neuromediators produced in the soma, such as the neuropeptides, may be too low to be detected with immunohistochemical techniques.

The multiplicity of neuromediators present in many of the fibre bundles listed in Table 4 is doubtless related to the fact that most bundles in the central nervous system are composite, i.e. consisting of components arising from several, or even many different sources. For instance, the fasciculus telencephalicus medialis (or medial forebrain bundle) is known to be composed of at least 50 different shorter and longer, ascending and descending components (Nieuwenhuys et al. 1982).

In fibre systems in which classical neuromediators as well as neuropeptides are present it may well be that these substances co-exist within the same axons. It stands to reason that this situation may be expected in fibre systems arising from centres which contain projection neurons of 'multiple-neuromediator' types. Thus, certain periolivary neurons, which are known to contribute to the cochlear nerve via the olivocochlear bundle, contain acetylcholine and enkephalin, and in raphespinal fibres serotonin may co-exist with substance P and TRH, as well as with other neuropeptides (see Table 3 for references). Co-localization of two different peptides most likely occurs in a certain proportion of dorsal root fibres (substance P+CCK: Dalsgaard et al. 1982a) and in fibres of the supraoptico- and paraventriculo-hypophyseal tracts (vasopressin+dynorphin and vasopressin+angiotensin II: for references see Table 3).

12. The relations between the pathways of classical hodology and the various populations of neuromediator-specified fibres are often complex and require further exploration.

First of all, it should be stated that the conventional idea that the white matter of the central nervous system is composed mainly of tight, discrete and well-myelinated fibre systems is incorrect. In fact most pathways present in a given area of white matter show a considerable overlap. Moreover, the white matter contains enormous numbers of thin and ultrathin, unmyelinated fibres, which have escaped the attention of previous investigators, mainly because of the limitations of the techniques they employed. To give a single example: the pyramidal tract of the rat contains at medullary levels a population of ultrathin, unmyelinated axons (mean diameter 0.2 μm) which by far outnumbers the population of myelinated fibres present within the bundle (Leenen et al. 1982). In the light of the frequency of occurrence of small-diameter fibres, it is not surprising that many of the axons visualized with immunohistochemical techniques are described as thin and unmyelinated. This holds in general for monoaminergic axons (Lindvall et al. 1974b; Moore and Bloom 1978; Steinbusch 1981; Hayashi et al. 1984), but also for fibres containing neuropeptides (e.g. neurotensin: DiFiglia et al. 1984; CRF: Swanson et al. 1983; enkephalin: DiFiglia and Aronin 1984b). However, the total picture we have at present of the neuromediator-specified axon populations may nevertheless be biased towards the thin fibres, because of the difficulties with visualizing the neuromediators used by the coarser, myelinated fibres which form the core of most of the pathways of classical neuroanatomy. These difficulties have already been indicated under item 11 and need not be repeated here.

The data assembled in Table 4 give an impression of the neuromediators known to be present within a number of fibre systems. However, it should be appreciated that many of the fibre populations indicated as present within a certain bundle are by no means confined to that bundle. Moreover, several of the fibre systems included in the Table, as for instance the ventral amygdalofugal pathway and the medial forebrain bundle, are in fact loosely textured fibre assemblies.

The complexity of the relations between neuromediator-specified fibre populations and classical hodology may perhaps be best illustrated by the behaviour of monoaminergic, particularly noradrenergic and serotoninergic fibres. Throughout most of the brain these fibres constitute diffuse networks, which bear no relation whatsoever to conventional pathways. However, they do enter a great many such pathways for what has been called 'epiphytic guidance' to more remote targets (Azmitia and Segal 1978). Finally, in certain regions of the brain stem and spinal cord which are not, or not clearly related to

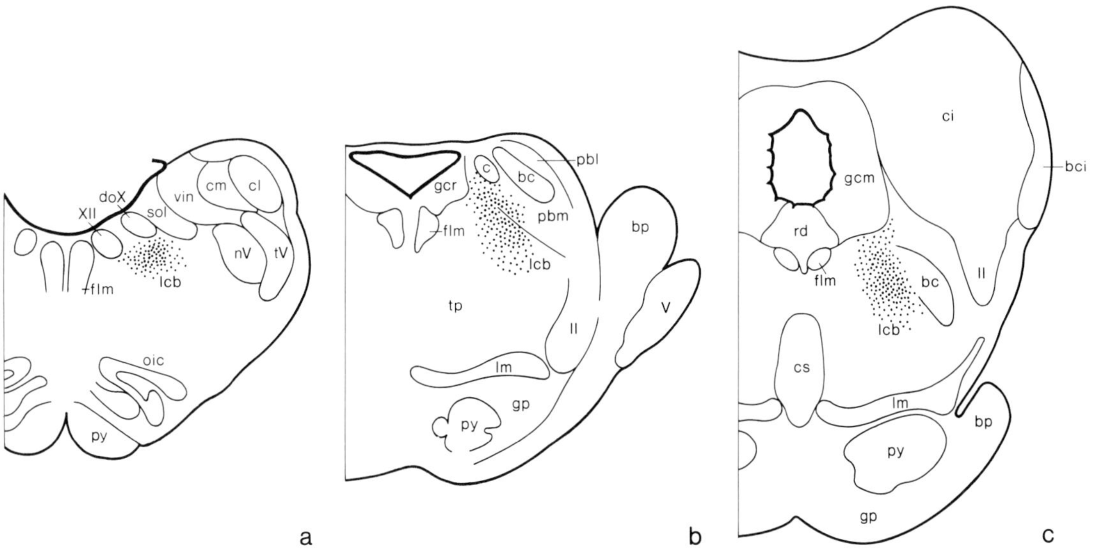

bc	Brachium conjunctivum	*ll*	Lemniscus lateralis
bci	Brachium colliculi inferioris	*lm*	Lemniscus medialis
bp	Brachium pontis	*nV*	Nucleus spinalis nervi trigemini
c	Locus coeruleus	*oic*	Oliva inferior complex
ci	Colliculus inferior	*pbl*	Nucleus parabrachialis lateralis
cl	Nucleus cuneatus lateralis	*pbm*	Nucleus parabrachialis medialis
cm	Nucleus cuneatus medialis	*py*	Tractus pyramidalis
cs	Nucleus centralis superior	*rd*	Nucleus raphes dorsalis
doX	Nucleus dorsalis nervi vagi	*sol*	Nucleus solitarius
flm	Fasciculus longitudinalis medialis	*tp*	Tegmentum pontis
gcm	Griseum centrale mesencephali	*tV*	Tractus spinalis nervi trigemini
gcr	Griseum centrale rhombencephali	*vin*	Nucleus vestibularis inferior
gp	Griseum pontis	*V*	Nervus trigeminus
lcb	Longitudinal catecholamine bundle	*XII*	Nucleus nervi hypoglossi

Fig. 34 a–c. Diagrammatic frontal sections through the right half of the brain stem of the cat to illustrate the position of the longitudinal catecholamine bundle. (Based on Fig. 6 of Jones and Friedman [1983]) **a** caudal rhombencephalon; **b** rostral rhombencephalon; **c** caudal mesencephalon

the trajectories of classical pathways, these fibres may attain such a high local concentration that they have been designated as special, transmitter-specific bundles. The large, 'longitudinal catecholamine bundle' described by Jones and Friedman (1983; see Fig. 34) represents a striking case in point.

There is one additional aspect concerning neuromediator-specified axons which requires discussion, and that is that these fibres are often described not only as thin and unmyelinated, but also as varicose or beaded (e.g. by Moore and Bloom 1978; Steinbusch 1981; Swanson et al. 1983; Jones and Friedman 1983; Hayashi et al. 1984; DiFiglia and Aronin 1984a). Systematic electron-microscopical studies on non-terminal segments of neuromediator-specified axons have, to my knowledge, not been reported. However, it is known that the terminal and preterminal ramifications of such axons show similar varicosities which at the ultrastructural level appear to contain accumulations of synaptic vesicles. These terminal and preterminal varicosities may be engaged in typical synaptic relationships, but such specialized contacts may also be absent. The latter condition has been repeatedly described for the swellings observed in monoaminergic axonal ramifications (Pickel et al. 1976; Chan-Palay 1978; Léger and Descarries 1978; Beaudet and Descarries 1978). It is assumed that the vesicles present in these non-synaptic varicosities release their content in the extracellular space, after which it is free to act upon receptor sites located within the range of dispersal. Conceivably, non-terminal segments of unmyelinated, varicose axons are also able to influence other neuronal elements in their vicinity by either synaptic or non-synaptic chemical transmission. Bundles consisting mainly or entirely of such fibres could constitute 'open' pathways, i.e. pathways which have the potential for influencing other neurons throughout their extent, rather than only in their area of final termination. Systematic immunohodological studies might well reveal that in the brain there occur, apart from the relatively discrete, myelinated pathways of classical neuroanatomy, a number of *'open multineuromediator channels'*, i.e. channels composed of several contingents of thin, beaded fibres, each contingent containing a different neuromediator. The dorsal longitudinal bundle of Schütz and the greater part of the medial forebrain bundle may well represent such channels. It also seems possible that the longitudinal catecholamine bundle of Jones and Friedman (1983) is not an entity in itself, but rather forms part of an open channel in which fibre contingents using other neuromediators participate. Fibres containing CRF (Fig. 21), vasopressin and oxytocin (Fig. 25), ACTH (Fig. 27) and angiotensin II (Fig. 33) occupy positions in the brain stem which closely correspond to that of the catecholamine bundle.

13. The number of central projection systems in which the neuromediators present in each of the consecutive neuronal links are known is surprisingly small.

The reasons for this lack of coherence in our present day knowledge of chemical neuroanatomy are threefold: (a) Most immunohistochemical studies are neuromediator directed or area directed rather than system directed. (b) Certain links in the neuronal chain may use a known neuromediator, the presence of which is, however, hard to demonstrate. (c) Certain links may involve a neuromediator of an as yet unknown nature.

As paradigms of central projection systems in which at least one of the neuromediators present in each of the consecutive elements forming their neuronal circuits are known, I will briefly discuss the viscerosensory projection originating from the site of entrance

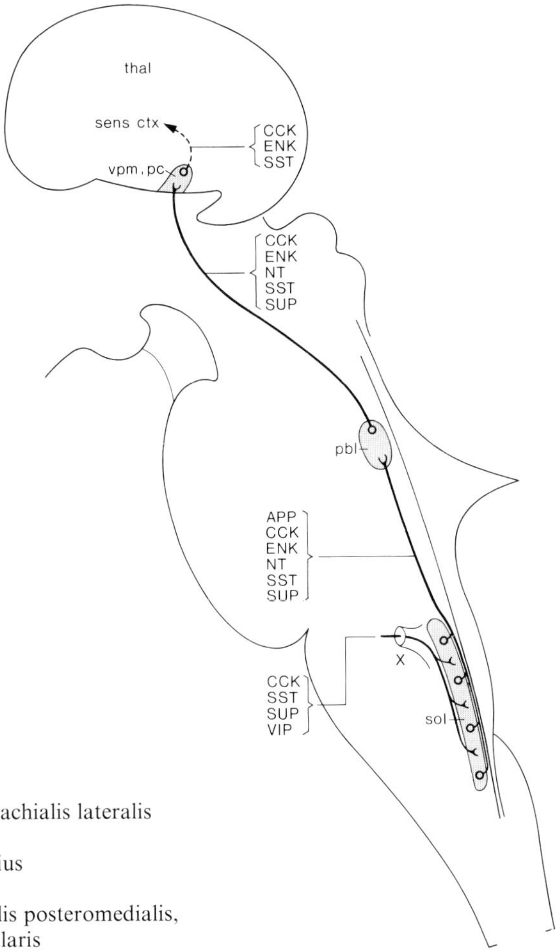

pbl	Nucleus parabrachialis lateralis
sens ctx	Sensory cortex
sol	Nucleus solitarius
thal	Thalamus
vpm, pc	Nucleus ventralis posteromedialis, pars parvocellularis

Fig. 35. Diagram showing the 'viscero-cortical' projection originating from the site of entrance of the vagus nerve and its relay centres. The neuromediators identified in the consecutive sets of neurons of this projection are indicated. (Based on Mantyh and Hunt [1984])

of the vagus nerve, the extrapyramidal system, and the descending pain-control system.

The *viscerosensory projection* originating from the site of entrance of the vagus nerve includes, from the periphery to the cortex, four consecutive sets of projection neurons (Fig. 35). The first of these is formed by primary afferent elements, the perikarya of which are situated in the nodose ganglion. Their axons, which carry information from the respiratory, cardiovascular, gastrointestinal and gustatory systems, terminate in the nucleus solitarius (Beckstead and Norgren 1979; Kalia and Mesulam 1980a, b; Kalia and Sullivan 1982). The solitary nucleus is known to project to the parabrachial nuclei (mainly the lateral one) located in the dorsolateral part of the pontine tegmentum (Ricardo and Koh 1978; Norgren 1978; King 1980). These nuclei project in turn to the most medial, parvocellular division of the nucleus ventralis posteromedialis of the thalamus (Saper and Loewy 1980; Voshart and Van der Kooy 1981). The fourth and final link in this viscerosensory projection is constituted by fibres which pass from the thalamic centre mentioned to the visceral and taste areas of the sensory cortex (Burton and Benjamin 1971; Saper 1982). Using combined immunohistochemical and retrograde tracing techniques, Mantyh and Hunt (1984b) have now demonstrated that the various components constituting the vagocortical projection all contain several neuropeptides. In the nodose ganglion many small neurons were observed to be immunoreactive to antibodies against one of the following four neuropeptides: CCK, SST, substance P and VIP. In the solitary nucleus many neuronal cell bodies having a projection to the parabrachial nuclei have been shown as immunopositive for one of six neuropeptides including avian pancreatic peptide, CCK, enkephalin, neurotensin, SST and substance P. In the parabrachial nucleus numerous elements projecting to the ventral posteromedial thalamic nucleus appeared to be immunopositive for one of the following five neuropeptides: CCK, enkephalin, neurotensin, SST and substance P. In the ventral posteromedial nucleus of the thalamus, finally, several neuronal perikarya were shown to be immunopositive for one of three neuropeptides (CCK, enkephalin, SST) and to project to the visceral and taste sensory cortex. The functional significance of these neurochemical findings remains to be clarified. However, it is remarkable that one of the two neuropeptides present in a certain proportion of all four sets of projection neurons which make up this viscerosensory projection is the gut-brain peptide CCK. As pointed out by Mantyh and Hunt (1984b), this peptide is capable of eliciting the same response (satiety) when administered either centrally or peripherally. They suggest that if other projection channels were to show a comparable 'repetitive' neuropeptide organization, it would be possible that activation at any level (either peripherally or centrally) could produce a similar response.

The basal ganglia and the other structures known as *extrapyramidal centres*, such as the subthalamic nucleus and the substantia nigra, constitute, with their emerging fibre systems, a number of interrelated loops or circuits from which output systems emerge at several points. Contrary to the classical view developed at the beginning of this century, the quantitatively most important outflow of the striatum converges via the globus pallidus and the thalamus upon the precentral cortex (areae 4 and 6), i.e. the cortical region from which the motor part of the pyramidal tract originates. Thus, it is primarily via this tract that the so-called extrapyramidal system influences motor activity. The very intricate circuitry of the basal ganglia and related structures is the subject of several recent reviews (e.g. Graybiel and Ragsdale 1979; Mehler 1981; Nauta and Domesick 1984; Nieuwenhuys 1984; Nieuwenhuys and Cools 1984), and need not to be detailed here. Within the frame of the present discussion on neuromediator-specified projection systems I will focus on a true extrapyramidal chain of connections along which the oculomotor and the

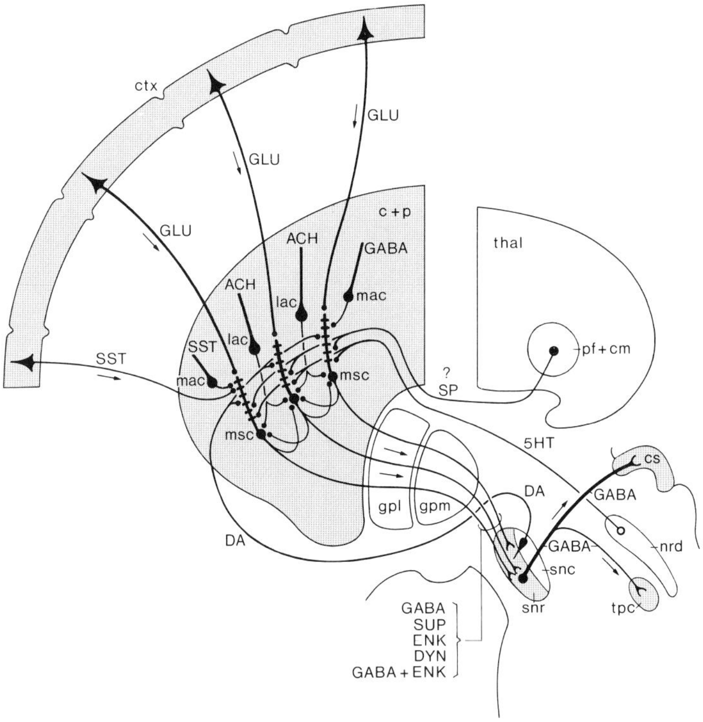

c Nucleus caudatus
cm Nucleus centromedianus
cs Colliculus superior
ctx Cortex cerebri
gpl Globus pallidus, pars lateralis
gpm Globus pallidus, pars medialis
lac Large aspiny cell
mac Medium-sized aspiny cell
msc Medium-sized densely spiny cell
nrd Nucleus raphes dorsalis
p Putamen
pf Nucleus parafascicularis
snc Substantia nigra, pars compacta
snr Substantia nigra, pars reticulata
thal Thalamus
tpc Nucleus tegmenti pedunculopontinus, pars compacta

Fig. 36. Diagram showing some connections in the extrapyramidal system. The neuromediators identified in the various neurons are indicated; for further explanation see text

bulbospinal motor apparatuses are affected. This chain comprises three consecutive links, formed by corticostriatal, striatonigral and nigrotectal and -tegmental projections (Fig. 36). The neurochemical data on the sets of neurons included in these and related projections are mainly derived from the preceding chapter and from the following reviews: Graybiel and Ragsdale 1983; Graybiel 1984; Bolam 1984; McGeer et al. 1984c, to which the reader is referred for details.

As regards the first link in the projection system to be discussed, it is known that the whole of the neocortex sends fibres to the nucleus caudatus and the putamen, and that all parts of these two cell masses receive fibres from the neocortex. Although this projection originates from different classes of cells located in different cortical layers, all of its fibres use one and the same neurotransmitter, namely the excitatory amino acid glutamate.

Efferent fibres from the caudate and putamen converge toward the globus pallidus and pass radially through both segments of that structure. During their transit through the globus pallidus these fibres emit numerous collaterals that synapse with pallidal neurons. They then leave the globus pallidus, attenuate considerably, and descend to the substantia nigra, to terminate in the pars reticulata of that structure. Retrograde labelling experiments have shown that most of the striatonigral fibres originate from a characteristic and very common cell type known as the medium-sized, densely spiny cell. In elements of this type the following neuromediators have been shown to be present: GABA, substance P, enkephalin and dynorphin, and the presence of all of these four substances has also been demonstrated in the pars reticulata of the substantia nigra. It is unclear at present whether different subpopulations of medium-sized, densely spiny cells contain different neuromediators. However, Aronin et al. (1984) have recently demonstrated that in numerous caudate neurons of medium size, glutamic acid decarboxylase (GAD), i.e. the GABA synthesizing enzyme, and enkephalin are co-localized. Quantitative analysis revealed that GAD and enkephalin co-exist in about one-half of the caudate cell populations containing each of these substances.

Apart from the direct striatonigral projection just discussed, there is evidence for the presence of an indirect striatonigral connection which is synaptically interrupted in the external pallidal segment. Retrograde labelling experiments have demonstrated the presence of pallidonigral neurons in the compartment last mentioned, and striatal projection neurons have been shown to make synaptic contacts with such elements (Bolam 1984). The pallidonigral neurons have been reported to contain GABA (McGeer et al. 1984). Whether the striatofugal fibres impinging on the pallidonigral neurons are axons of separate neurons, collaterals of striatonigral elements or a mixture of both (see Graybiel and Ragsdale 1983) remains to be determined. However, the presence of a very dense plexus of enkephalinergic fibres in the external pallidal segment (Fig. 29) suggests that striatofugal fibres containing this neuromediator may well influence the pallidonigral neurons.

The third link in the descending extrapyramidal projection under discussion is formed by GABA-ergic neurons situated in the pars reticulata of the substantia nigra, which send their axons to the superior colliculus and the midbrain tegmentum. The nigrotectal projection is massive and terminates in the deeper layers of the superior colliculus. This pathway has been shown to contribute to the control of eye movements (Hikosaka and Wurtz 1983a, b, c, d). Moreover, the deeper layers of the superior colliculus give rise to tectoreticular and tectospinal pathways, which affect the bulbospinal motor apparatus. The neuromediators used by these pathways are unknown.

The nigrotegmental fibres terminate mainly in the compact part of the nucleus tegmenti pedunculopontinus (TPc), a cell mass situated in the most caudal part of the midbrain tegmentum and embedded in the reticular formation. This nucleus also receives afferents from the motor cortex, the internal pallidal segment and the subthalamic nucleus. The main output of TPc feeds back into the extrapyramidal circuitry impinging upon, among other structures, the globus pallidus, the subthalamic nucleus and the pars compacta of the substantia nigra (Edley and Graybiel 1983), but this centre also gives rise to a limited number of descending fibres which terminate in the pontine and medullary reticular formation. The neuromediators present in TPc and its efferents are unknown, but according to Graybiel (1984), the area in which this centre is embedded contains numerous cholinergic and peptidergic elements.

From the foregoing it appears that the projection system discussed comprises three consecutive sets of neuromediator-specified neurons: first, the neurochemically homogeneous corticostriate neurons, using the excitatory amino acid glutamate; second, the probably neurochemically heterogeneous striatonigral neurons, containing the inhibitory amino acid GABA, or one of the neuropeptides substance P, enkephalin or dynorphin, or combinations of these substances; and third, again a set of neurochemically homogeneous, GABA-ergic nigrofugal neurons. As a matter of fact, this is a very simplified way of representing things. In reality, these consecutive sets of neurons are not arranged in a simple array. Rather, influences from many different sources converge on each element. This may be exemplified by briefly considering the various non-cortical afferents of the medium-sized densely spiny neurons, i.e. the elements which, according to Bolam (1984), provide the basic framework of neostriatal circuitry. The extrinsic afferents of these cells include: (a) fibres arising from the intralaminar thalamic cell groups, in particular the parafascicular nucleus and centrum medianum – the neuromediator of these fibres, which together form a massive projection, has not been identified as yet; however, most recently Sugimoto et al. (1984) have provided experimental evidence indicating that at least some of the thalamostriate projection neurons contain substance P; (b) fibres ascending from the dopaminergic neurons in the compact part of the substantia nigra – these fibres are known to form an extremely dense terminal network in the caudatoputamen. Dendrites of the dopaminergic compacta cells, directed ventrally into the pars reticulata of the substantia nigra, have been shown to be contacted directly by terminals of striatonigral fibres (Wassef et al. 1981). Thus, at least a certain proportion of the dopaminergic nigral projection neurons may close a feedback loop between the compact part of the substantia nigra and the caudatoputamen; (c) serotoninergic fibres originating from the nucleus raphes dorsalis; (d) somatostatin-containing fibres, which possibly arise from the frontal cortex, cingulate cortex, hippocampal formation and amygdala. In addition to the extrinsic afferents just discussed, the medium-sized, densely spiny cells have been shown to be the postsynaptic targets of several groups of intrinsic elements. Under this heading the medium-sized, densely spiny cells include large elements with aspiny dendrites containing acetylcholine, GABA-ergic medium-sized, aspiny cells, and neurons of the same morphological type containing somatostatin. Thus, it appears that fibres containing the following neuromediators converge upon the medium-sized, densely spiny striatal cells: acetylcholine, dopamine, serotonin, GABA, substance P, SST and probably enkephalin and dynorphin. It is worth noting that the striatum is also in receipt of extrinsic fibres containing noradrenalin, neurotensin, α-MSH and CCK. However, the axons containing these substances have not so far been demonstrated to synapse directly with medium-sized, densely spiny neurons.

The so-called *descending pain-control system* (Fig. 37) is extensively discussed in a recent review article (Basbaum and Fields 1984), to which the reader is referred for details. In the present survey only a few morphological aspects of this system will be highlighted (see also p. 96 and p. 101). Essentially, the descending pain-control system or 'endogenous

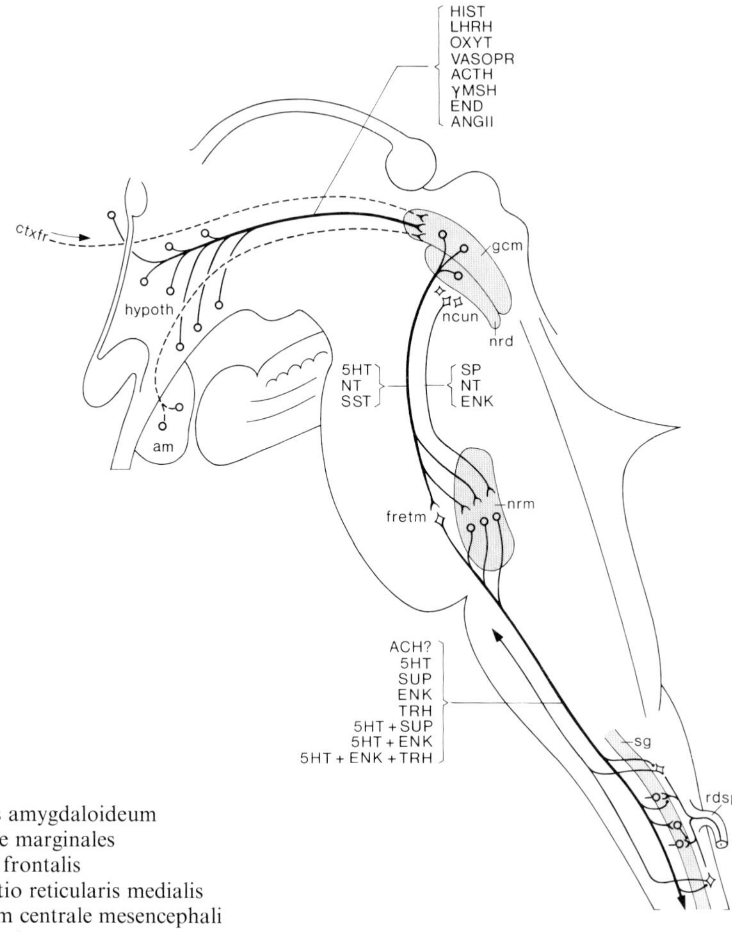

am	Corpus amygdaloideum
cm	Cellulae marginales
ctxfr	Cortex frontalis
fretm	Formatio reticularis medialis
gcm	Griseum centrale mesencephali
hypoth	Hypothalamus
n cun	Nucleus cuneiformis
nrd	Nucleus raphes dorsalis
nrm	Nucleus raphes magnus
rdsp	Radix dorsalis nervi spinalis
sg	Substantia gelatinosa

Fig. 37. Diagram showing some fibre connections related to the descending pain-control system. The neuromediators identified in the fibres of the various pathways are indicated

analgesia' system consists of a series of interrelated centres or areas situated at three different levels of the neuraxis: the mesencephalon, the rhombencephalon and the spinal dorsal horn. It is in the area last mentioned that the system inhibits noxiously evoked activity. The mesencephalic complex involved in pain control consists of the griseum centrale mesencephali or periaqueductal grey (PAG), the nucleus raphes dorsalis, which is partly embedded in the central grey, and the adjacent reticular formation, i.e. part of the cuneiform nucleus. Within this complex are found populations of neurons containing numerous different neuromediators including serotonin, GABA, substance P, CCK, neurotensin, enkephalin and dynorphin.

At the rhombencephalic level the nucleus raphes magnus and adjacent portions of the medial reticular formation constitute an important relay between the mesencephalic complex discussed above and the spinal cord. Like the mesencephalic complex, this way station may be characterized as a multineuromediator centre. Neuronal populations containing serotonin, substance P, CCK, TRH, enkephalin and dynorphin have been found within its confines. Moreover, in certain cells serotonin and substance P or serotonin and enkephalin co-exist, and in other cells even three different neuromediators – serotonin, substance P and TRH – are co-localized. It is important to note that neither the mesencephalic nor the rhombencephalic complex discussed above are exclusively involved in pain control.

The spinal portion of the endogenous analgesia system comprises, apart from the terminal parts of fibres descending from the rhombencephalon and possibly the mesencephalon (see below), local circuit neurons situated in the substantia gelatinosa. Again, in this part of the dorsal horn there are populations of cells containing many different neuromediators, including GABA, substance P, neurotensin, enkephalin and dynorphin. It seems likely that the enkephalinergic elements form part of the descending pain-control system, but to what extent the elements containing the other neuromediators mentioned are involved in this system remains to be established.

The mesencephalic central grey receives significant inputs from the cerebral cortex, particularly the frontal, somatosensory and cingular regions, the amygdala, the preoptic area and the hypothalamus (Hardy et al. 1981; Beitz 1982c; Bragin et al. 1984). The hypothalamic projection comprises fibre contingents each containing a different neuromediator, among which are histamine, LHRH, vasopressin, oxytocin, ACTH, γ-MSH, endorphin and angiotensin II (for references, see the pertinent preceding sections). Apart from the descending projections mentioned above, the PAG also receives afferents from numerous brain stem centres, including the nucleus cuneiformis, the ventral tegmental area, the locus coeruleus, the parabrachial area and the pontine and medullary reticular formation, as well as from the spinal cord (Beitz 1982c; Mehler 1969).

The PAG sends a large projection to the rhombencephalic relay centre of the descending pain-control system, i.e. the nucleus raphes magnus and adjacent parts of the reticular formation (Abols and Basbaum 1981; Carlton et al. 1983; Fardin et al. 1984), and the experimental work of Beitz (1982a) has shown that this projection originates from serotonin- and neurotensin-containing cells in the PAG. In addition, fibres originating from the PAG and the dorsal raphe nucleus project directly to the spinal cord (Mantyh and Peschanski 1982). These fibres, some of which originate from serotoninergic cells situated in the dorsal raphe nucleus, may well contribute to pain control by influencing neurons in the spinal dorsal horn. Another mesencephalic centre implicated in the endogenous control of pain, i.e. the nucleus cuneiformis, also distributes fibres to the nucleus raphes magnus. However, whereas the fibres arising from the PAG contain serotonin or neurotensin, those descending from the cuneiform nucleus have been demonstrated to contain enkephalin and substance P (Beitz 1982b).

The important question of which elements actually influence the neurons projecting from the mesencephalic to the rhombencephalic centres implicated in pain modulation has not been answered as yet. Given the facts that (a) the PAG contains enkephalinergic and dynorphinergic neurons, (b) it receives a substantial endorphinergic projection from the hypothalamus, (c) the vast majority of enkephalin-immunoreactive terminals in the PAG are presynaptic to dendrites, (d) opiate actions on target neurons are generally inhibitory, and (e) excitation of the PAG output neurons is required to initiate descending control, Basbaum and Fields (1984) proposed that endogenous opioid peptides (either enkephalin, dynorphin, endorphin, or all three) activate neurotensinergic output neurons in the PAG by inhibiting an inhibitory – presumably GABA-ergic – interneuron.

The rhombencephalic pain-control relay centre receives, as we have seen, major afferent connections from the PAG and the adjacent midbrain nucleus cuneiformis. Additional afferents to this centre arise from many different sources, including the frontal cortex, the dorsomedial hypothalamic nucleus, the zona incerta, the deep superior colliculus, the rhombencephalic reticular formation, the spinal trigeminal nucleus, the dorsal column nuclei and the spinal cord (Abols and Basbaum 1981; Carlton et al. 1983). The histochemical and immunohistochemical studies reviewed in the previous sections indicate that the neuromediators present in the cells of origin of a number of these afferents can be specified as follows: PAG – serotonin, somatostatin, neurotensin; cuneiform nucleus – substance P, enkephalin, neurotensin; rhombencephalic reticular formation – substance P, enkephalin. These studies also indicate that the nucleus raphes magnus receives non-coerulean noradrenergic fibres, ACTH-containing fibres from the mediobasal hypothalamus, neurotensin- and enkephalin-containing fibres from the parabrachial area, and substance P-, neurotensin- and enkephalin-containing fibres from the nucleus solitarius.

The efferents of the rhombencephalic pain-control relay centre project to the spinal cord via the dorsolateral funiculus, where their terminal fields include laminae I, II and V, i.e. the laminae known to contain neurons which are maximally responsive to noxious stimuli (Basbaum et al. 1978; Leichnetz et al. 1978; Martin et al. 1982). The cells in the rhombencephalic relay centre which give rise to spinal projections employ serotonin, enkephalin, substance P, TRH, and presumably acetylcholine as neuromediators. A large fraction of these cells contain serotonin coexisting with one or more neuropeptides (Bowker et al. 1983; Basbaum and Fields 1984). The internal circuitry within the rhombencephalic relay centre is unknown. However, Basbaum and Fields (1984) supposed that within this centre neurotensin-containing fibres descending from the PAG are in direct synaptic contact with bulbospinal projection neurons.

Turning now to the spinal cord, there is ample evidence that influences exerted on elements in the spinal dorsal horn by the bulbospinal projections discussed above play a crucial role in the endogenous pain-control system. Thus, focal electrical stimulation of the nucleus raphes magnus causes behavioural analgesia and inhibits activity of most dorsal horn nociceptive neurons (see Miletic et al. 1984). However, with regard to the neuronal targets or the sites of action of the bulbospinal axons there are still many uncertainties. The following list of possibilities, which is by no means exhaustive, is based mainly on the work of Basbaum and Fields (1984).

1. Serotonin-containing bulbospinal neurons exert a presynaptic inhibitory action on nociceptive primary afferent fibres. There is physiological evidence supporting the existence of this connection (see Miletic et al. 1984).
2. Serotoninergic fibres postsynaptically inhibit nociceptive projection neurons situated in lamina I of the dorsal horn. Miletic et al. (1984) have recently provided morphological as well as physiological evidence for the presence of this direct postsynaptic inhibition of nociceptive projection neurons.

3. Descending bulbospinal fibres releasing an excitatory neuromediator (e.g. substance P) synapse with endogenous opioid (enkephalin or dynorphin)-containing local circuit neurons, which in turn exert an inhibitory influence on nociceptive primary afferents. There is ample evidence that the axons of bulbospinal projection neurons synapse with spinal local circuit neurons. However, no morphological substrate has been found for presynaptic control of primary afferents by (endogenous opioid-containing) interneurons. Basbaum and Fields (1984) suggested that this control of primary afferents may be exerted via a 'nonsynaptic' action, in a manner similar to that occurring in the peripheral autonomic system.
4. Serotoninergic fibres inhibit opioid-containing interneurons, and these exert an inhibitory action on nociceptive projection neurons via a second inhibitory (presumably GABA-ergic) interneuron. Although GABA-containing neurons do occur in the superficial dorsal horn, in terms of circuitry this arrangement is purely hypothetical.

In the preceding paragraphs I have discussed the following three projection systems: (a) the viscerosensory projection originating from the site of entrance of the vagus nerve, (b) a chain of connections forming part of the extrapyramidal system, and (c) the descending pain-control system. The viscerosensory projection comprises consecutively, primary afferent vagus fibres, neurons situated in the solitary nucleus projecting to the parabrachial area, neurons projecting from the parabrachial area to the most medial, parvocellular part of the nucleus ventralis posteromedialis thalami (VPMpc), and finally neurons projecting from the VPMpc to the visceral and taste areas of the sensory cortex (Fig. 35). The extrapyramidal projection discussed consists of three consecutive links formed by corticostriatal, striatonigral, and nigrotectal and nigrotegmental projections (Fig. 36). The descending pain-control system appears to comprise: (a) neurons situated in the telencephalon, the hypothalamus and the brain stem, projecting to a mesencephalic complex, encompassing the periaqueductal grey and parts of the nucleus raphes dorsalis and the cuneiform nucleus; (b) neurons projecting from this mesencephalic complex to a rhombencephalic relay centre, consisting of the nucleus raphes magnus and certain adjacent parts of the reticular formation; (c) neurons projecting from the rhombencephalic relay centre to the spinal cord; and (d) certain sets of neurons present in the spinal dorsal horn (Fig. 37).
These three central circuits were selected because at least one of the neuromediators in each of the consecutive populations of projection neurons forming these circuits is known. Although this knowledge is a definite achievement of chemical neuroanatomy and allows cautious functional interpretations at certain points (see e.g. Mantyh and Hunt 1984b), it should be emphasized that even in such a simplified setting, where central circuitry is reduced to array of neurons, numerous unsolved problems become visible. In the first paradigm, i.e. the vagocortical projection, we do not know what synaptic relations are involved in any of the relay centres. Hence, the question as to which of the consecutive sets of neuromediator-specified projection neurons are actually in direct synaptic contact remains to be answered. With regard to the extrapyramidal circuit taken as our second paradigm, it appears that the medium-sized, densely spiny neurons which constitute the striatonigral projection receive a large direct glutamatergic input from cortical neurons, and that several other sets of neuromediator-specified neurons also impinge on these striatal elements. However, although the axonal endings of the medium-sized, densely spiny cells most probably synapse with GABA-ergic elements in the pars reticulata of the substantia nigra, direct and convincing evidence for the existence of such contacts is lacking at present. Finally, it seems that it is only at the lowest level of the three-tiered descending pain-control system that some insight into the local synaptic relations exists.

14. The terms 'terminal field' and 'terminal plexus', as appearing in the literature and in the following sections of this treatise, should be taken with some caution.

In immunohistochemistry it is customary to designate any accumulation of fibre- or 'dot-like' immunoreactivity within the grey matter of the central nervous system as a terminal field. However, in a number of cases it cannot be excluded that the structures observed represent the initial parts of axons whose parent cell bodies have remained unstained, rather than axonal terminal segments. Moreover, the term 'terminal field' implies that the structures observed release the neuromediator they contain within that field, a phenomenon which is difficult to demonstrate. The probability that the structures observed actually release their neuromediator increases if receptors for that neuromediator appear to be present within the same area. The only strong morphological evidence for the presence of a true terminal field is, of course, the demonstration of presynaptic differentiations containing the neuromediator by immunoelectron-microscopy.

15. The terminal plexuses present within most grisea have been demonstrated to be composed of subpopulations of axonal ramifications, each containing a different neuromediator. The neuromediator profiles of the terminal fields present within the various grisea may differ considerably.

The total numbers of known neuromediators and the neuromediator profiles of the fibres and terminals present within the confines of a number of representative grisea are listed in Table 5. It will be seen that in a number of cell masses, such as the central amygdaloid nucleus, the caudatus-putamen complex and the nucleus solitarius, the number of subpopulations of terminal ramifications each containing a different neuromediator is very large. It is also noteworthy that in the caudatus-putamen complex, apart from a number of neuropeptides, practically all 'classical' neuromediators are present. The various subpopulations of neuromediator-specified terminal ramifications within a given griseum may show considerable differences in density. In order to give some idea of these differences, two degrees of density have been indicated in Table 5. It must be borne in mind, however, that in many mapping studies these differences are expressed in a range of five or six levels.

16. Some subpopulations of neuromediator-specified fibres and terminals may conform exactly to cytoarchitectonic entities, but others may not be restricted to the boundaries of classical nuclear groups.

With regard to these phenomena I will confine myself to a few examples. In the spinal cord many subpopulations of fibres, each containing a different neuropeptide, show a striking congruency with the substantia gelatinosa, the nucleus intermediolateralis and the central grey (lamina X). The external segment of the globus pallidus contains a very dense plexus of enkephalinergic fibres, whereas in the internal pallidal segment nu-

Table 5. The total number of neuromediators and the neuromediator profiles of a number of cell masses. Only the neuromediators occurring in the fibres and terminals are listed

	acetylcholine	dopamine	noradrenaline	adrenaline	serotonin	histamine	GABA	glutam. + aspart.	glycine	substance P	VIP	CCK	neurotensin	CRF	LHRH	somatostatin	TRH	vasopr. + oxytoc.	ACTH	αMSH	γMSH	endorphin	enkephalin	dynorphin	angiotensin II	total number
n. centralis amygdalae		+	**+**							+	+	**+**	**+**			**+**			+		+	+		+	**+**	12
n. septi medialis			+		+	+							+	+	+											6
n. caudatus + putamen	+	**+**	+		+	+	**+**	**+**	+	+		**+**	+			+				+			**+**	+	+	16
globus pallidus, p. lateralis		+			+		**+**			+			+										**+**			6
globus pallidus, p. medialis		+			+		**+**			**+**													+	+		6
n. habenulae medialis	+		+		+	+	+								+	+			+				+			9
n. habenulae lateralis		**+**	+				+			**+**						+		+	+				+			8
corpus geniculatum mediale	+		**+**				+									+										4
n. subthalamicus	+	+								+																3
n. supraopticus		+	**+**			+				+														+		5
n. suprachiasmaticus					**+**	+				+					+									+		5
corpus mamillare					+	**+**		+							+	+										5
n. nervi oculomotorii																+	+									2
n. ruber							+																			1
substantia nigra, p. compacta		**+**			**+**			+	+				+			+							+			7
substantia nigra, p. reticulata					+		**+**	+		**+**						+							**+**	**+**		7
area tegmentalis ventralis	+	**+**			+		+			+		**+**	+		+	+			+		+	+				12
n. interpeduncularis	+		+		**+**					+		**+**			+	+							+			8
nn. lemnisci lateralis			+									**+**				+										3
n. raphes magnus			+		+								+			+						+	+			6
n. parabrachialis lateralis		+	+		+	+				+	+	+	+			+	+	+	**+**		+	+	**+**	+		16
nn. pontis			+			+		**+**		+						+										5
n. nervi facialis			+		+	+				+						+	+						+	+		8
n. solitarius		+	**+**	**+**	+	+				**+**	**+**	+	**+**			+	+	+	+		+	+	**+**	+		17
n. intermediolateralis		+	**+**	**+**	**+**							+		+		+		+							**+**	9

+, present; **+**, present in a large number of structures

merous dynorphin- and substance P-containing fibres are found. 'Non-congruency' is shown by the diffuse plexus of serotoninergic fibres which extends over virtually the entire central nervous system. Local concentrations of these fibres may or may not conform to nuclear boundaries (see Steinbusch 1981). According to a recent mapping study (Johansson et al. 1984), somatostatin-containing fibres behave similarly in many different parts of the neuraxis. Diffuse networks of α-MSH-containing fibres have been observed in the hippocampus and spinal cord (Köhler et al. 1984b), and Haber and Elde (1982) noted that in the mediobasal telencephalon of the monkey, striking and discrete zones of enkephalinergic fibres occur which do not conform to known nuclear boundaries. Finally, it should be mentioned that in the preoptico-hypothalamic continuum the terminal plexuses of many neuromediator-specified subpopulations of fibres often extend beyond the boundaries of the cell masses to encompass the adjacent non-cellular regions to which dendrites of the nuclei project. Comparable phenomena can be observed in other parts of the brain, as for instance the solitarius region in the lower rhombencephalon (Kalia et al. 1984b).

17. Neuromediator-specified subpopulations of fibres and terminals occurring within the confines of grisea often show a spatial segregation within these grisea. Such spatially segregated terminal fields may correspond to other neuromediator terminal fields, as well as to cytoarchitectonic, cytochemical or functional subdivisions.

I will illustrate these important phenomena with a number of examples taken from various parts of the brain: (a) the nucleus solitarius, (b) the parabrachial region, (c) the interpeduncular nucleus, (d) the substantia nigra, (e) the paraventricular nucleus, (f) the suprachiasmatic nucleus, (g) the preoptic region, (h) the caudatus-putamen complex and (i) the cerebral cortex.

The *nucleus solitarius* is responsible for integrating respiratory, cardiovascular (baroreceptive and chemoreceptive), and gastrointestinal functions, and within the confines of this nucleus a division of discrete subnuclear units of functional significance can be made. Respiratory subnuclei are found mainly in the ventrolateral part of the solitary nucleus, whereas the cardiovascular and gastrointestinal subnuclei occupy dorsal and medial positions respectively. In an attempt to correlate the cytoarchitecture and functions of the solitary nucleus with immunohistochemistry, Kalia et al. (1984b) prepared detailed maps of the distribution of a number of neuropeptides within that nucleus. Although a remarkable degree of co-representation of neuropeptides within the various subnuclei of the solitary nucleus was found, the following differences between the distribution patterns of individual neuropeptides could be detected. The subnuclei associated with respiratory control show somatostatin, enkephalin, and substance P immunoreactivity, with somatostatin being most prominent in these nuclei; the subnuclei involved in baroreflexes and chemoreflexes show mainly substance P and neurophysin II immunoreactivity, whereas the subnucleus associated with gastrointestinal function receives predominantly substance P-immunoreactive nerve terminals, although somatostatin, enkephalin and neurophysin II are also present in scattered amounts. In a second study, Kalia et al. (1984c) demonstrated that neurophysin-immunoreactive nerve terminals are selectively distributed to the dorsal and medial subnuclei and their adjacent dendritic regions. They pointed out that this selective distribution of neurophysin-immunoreactive nerve terminals in the cardiovascular and gastrointestinal subnuclei of the nucleus solitarius implicates the presence of a direct, descending, hypothalamic, oxytocin–neurophysin II-containing pathway influencing these functions of the nucleus solitarius.

It has already been mentioned under item 13 that the *parabrachial region* receives from the nucleus solitarius a substantial projection consisting of several subpopulations of fibres, each of which contains a different neuropeptide. Milner et al. (1984) studied the distribution of substance P, neurotensin and enkephalin in the parabrachial region and their association with efferents from the solitary nucleus in the rat, by combining the immunohistochemical localization of these neuropeptides with the autoradiographic labelling of tritiated amino acids anterogradely transported from the caudal portion of the medial part of the solitary nucleus (Fig. 38). Neurotensin- and substance P-containing fibres were located primarily in the ventrolateral quadrant of the parabrachial region, but they were differentially localized with respect to each other (Fig. 38b, c). However, both peptides were detected in an area of the parabrachial region containing dense autoradiographic labelling (Fig. 38e). In contrast, enkephalin-containing fibres were located primarily in the dorsolateral parabrachial quadrant (Fig. 38d), an area which appeared to contain only sparse autoradiographic label-

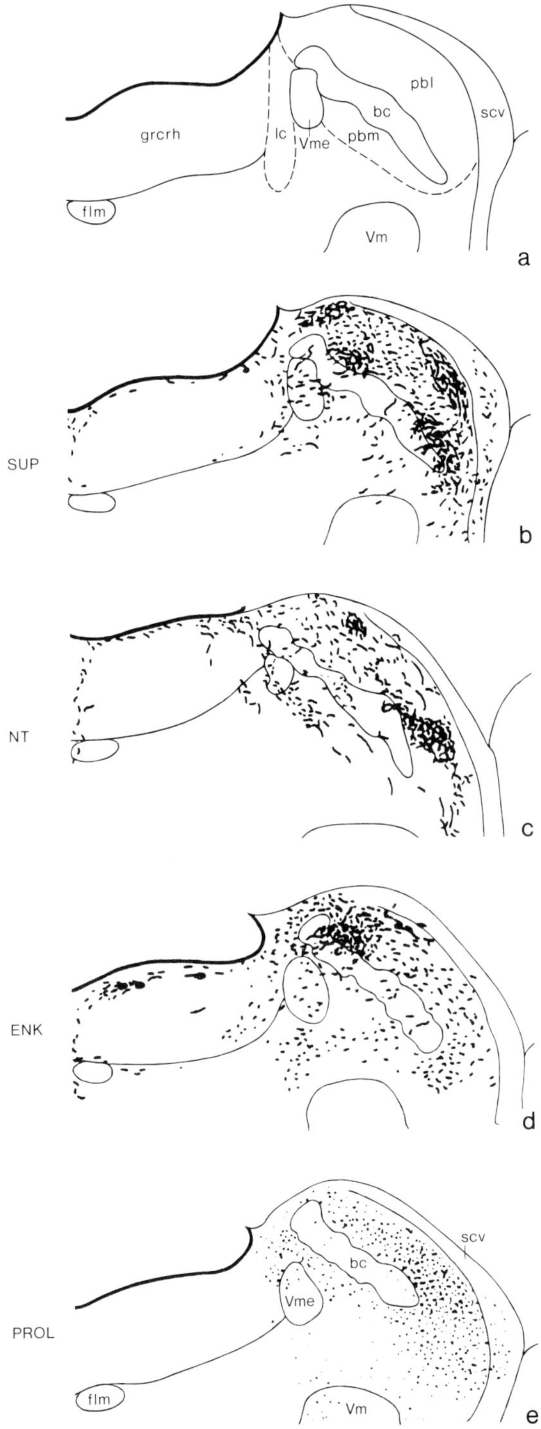

bc Brachium conjunctivum
flm Fasciculus longitudinalis medialis
grcrh Griseum centrale rhombencephali
lc Locus coeruleus
pbl Nucleus parabrachialis lateralis
pbm Nucleus parabrachialis medialis
scv Tractus spinocerebellaris ventralis
Vm Nucleus motorius nervi trigemini
Vme Nucleus mesencephalicus nervi trigemini

Fig. 38 a–e. Schematic drawings of frontal sections through the parabrachial region of the rat; **a** subdivisions of the region; **b, c, d** distribution of substance P-(*SUP*), neurotensin-(*NT*), and enkephalin-(*ENK*) immunoreactive fibres; **e** anterograde labelling after an injection of tritiated proline (*PROL*) into the caudomedial part of the nucleus solitarius. (Modified from Milner et al. 1984)

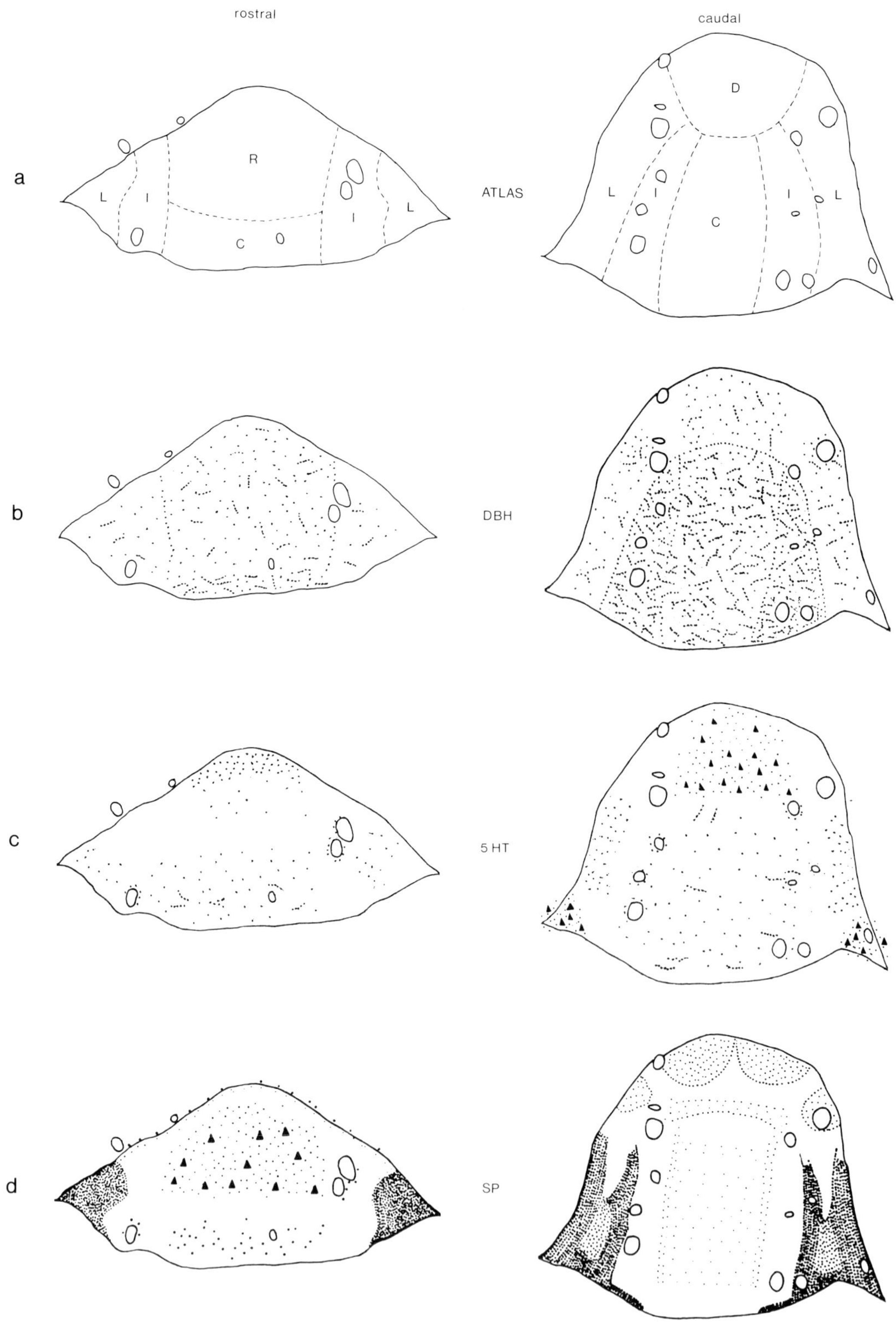

Fig. 39

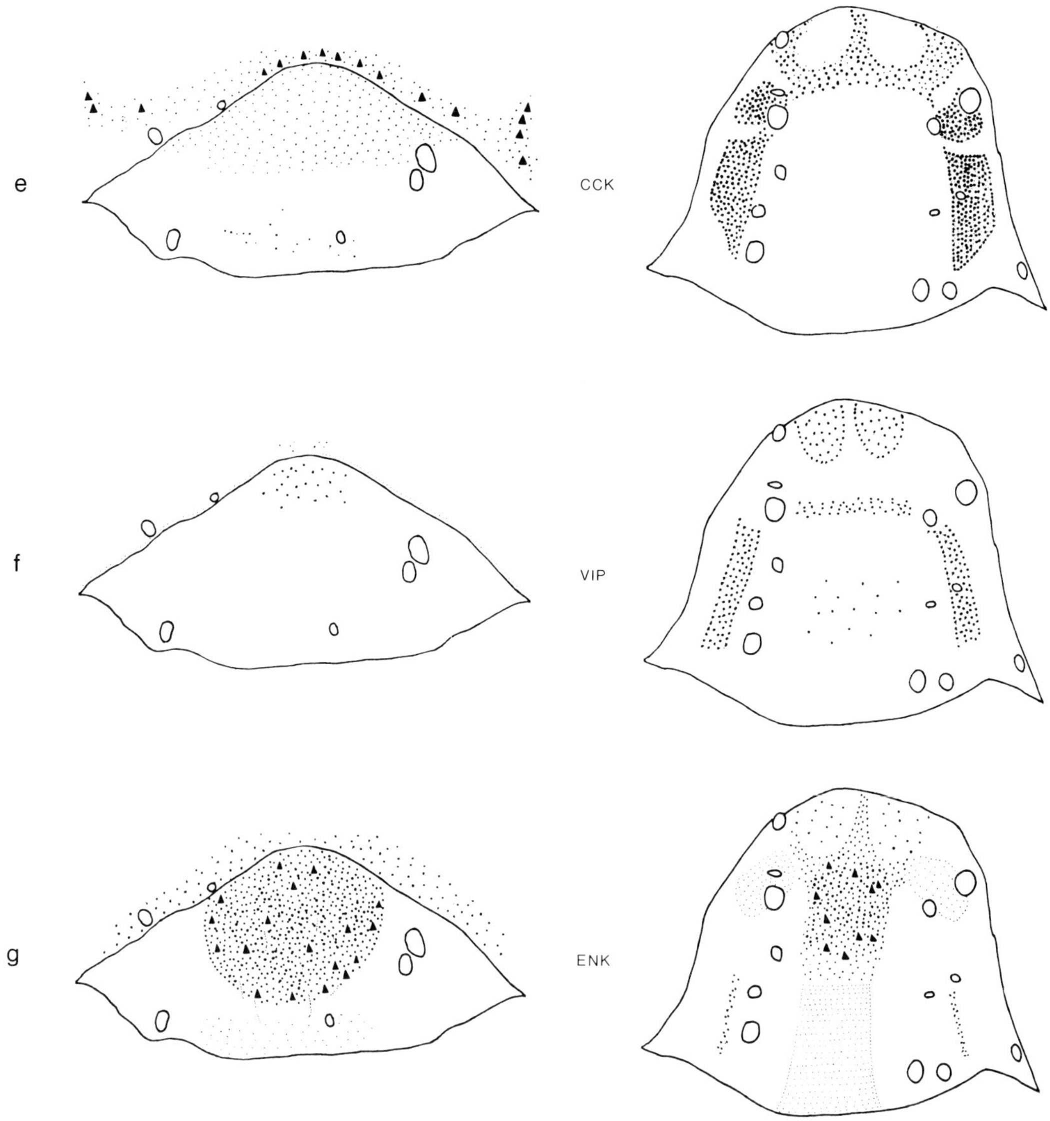

Fig. 39 a–g. Diagrammatic frontal sections through the rostral (*left*) and caudal (*right*) portions of the interpeduncular nucleus of the rat. **a** Subnuclear boundaries based on the *atlas* of Hamill and Lenn (1984); **b–g** identical levels of the interpeduncular nucleus illustrating the distribution of immunofluorescent cell bodies (*triangles*) and processes (*dots*) reactive for dopamine-β-hydroxylase (*DBH*), serotonin (*5HT*), substance P (*SP*), cholecystokinin (*CCK*), vasoactive intestinal polypeptide (*VIP*) and Leu-enkephalin (*ENK*) respectively. (Redrawn from Hamill et al. 1984)

Abbreviations: *R,D,C,I,L* rostral, dorsal, caudal, intermediate and lateral subnuclei of the interpeduncular nucleus

ling of transported amino acids (Fig. 38e). The authors concluded that in the parabrachial region of the rat, substance P and neurotensin are contained within two different populations of afferents which may originate from the caudal portion of the medial part of the nucleus solitarius, whereas enkephalin is more likely to be found in other afferents or possibly in intrinsic neurons.

The *interpeduncular nucleus* is an unpaired cell mass situated in the basal region of the mesencephalon. It receives a large afferent projection via the fasciculus retroflexus and may be considered part of the so-called limbic midbrain area (Nauta 1958). It is not a homogeneous nucleus, but is composed of several subnuclei. In a recent study, Hamill and Lenn (1984; see also Lenn and Hamill 1984) distinguished three single subnuclei (rostral, central, dorsal) and four paired subnuclei (interstitial, intermediate, lateral, dorsal lateral). The positions of five of these subnuclei are indicated in Fig. 39. Hamill et al. (1984) have mapped in detail the localization of structures containing dopamine-β-hydroxylase (i.e. the synthesizing enzyme of noradrenalin), serotonin, substance P, CCK, VIP, enkephalin and somatostatin in the interpeduncular nucleus of the rat. They demonstrated that perikarya and processes containing these substances are heterogeneously distributed over the nucleus and conform in their distribution partly, but not entirely to the subnuclear areas (Fig. 39). Thus, the rostral subnucleus contains numerous enkephalin-positive perikarya and co-extensive field of enkephalin-containing processes. Caudally, however, this complex of enkephalinergic structures extends through both the dorsal and central subnuclei. The dorsal subnucleus is characterized by the presence of numerous serotonin-containing perikarya. In this subnucleus a remarkable histochemical differentiation was observed. Substance P- and VIP-containing processes were concentrated in bilateral ovoid areas located in the dorsal portion of the subnucleus, whereas CCK- and enkephalin-containing processes surrounded these areas. Comparable complementary patterns in the distribution of these four peptides appeared to be present in the rostral subnucleus. The lateral subnucleus, finally, was found to contain a very dense plexus of substance P-positive fibres, a dense plexus of CCK-positive fibres, and a moderately dense plexus of VIP-positive fibres. However, all of these three plexuses occupied only a part of that subnucleus. Since few functional studies of the effects of neuroactive substances have been performed on the interpeduncular nucleus, Hamill and collaborators (1984) considered it premature to speculate on the physiological significance of the monoamines and peptides present in this nucleus. However, they suggested that the abundance of neuroactive substances contained in this region may well indicate that the interpeduncular nucleus is an important link in the limbic system. Similar suggestions are frequently encountered in the neuroimmunohistochemical literature. In my opinion, such an abundance of neuromediators has to do with the mode of information processing in a centre, rather than being indicative of the functional importance of that centre.

It has already been discussed that the *substantia nigra,* which is situated in the ventral part of the midbrain, consists of two histologically different portions, the pars compacta and the pars reticulata, and that both of these portions represent important nodal points in the extrapyramidal system (see item 13 and Fig. 36). Inagaki and Parent (1984) studied the distribution of substance P and enkephalin in the substantia nigra of the rat, the cat and the squirrel monkey by means of the indirect immunofluorescence technique. Their finding can be summarized as follows (Fig. 40): In the rat a dense network of fine substance P-containing fibres is uniformly distributed throughout the rostrocaudal extent of the substantia nigra pars reticulata (SNr) and in the most ventral part of the substantia nigra pars compacta (SNc). Enkephalin-containing fibres are scattered among SNc neurons, but also occur in the SNr, particularly in its caudolateral part. In the cat, numerous fine substance P-positive

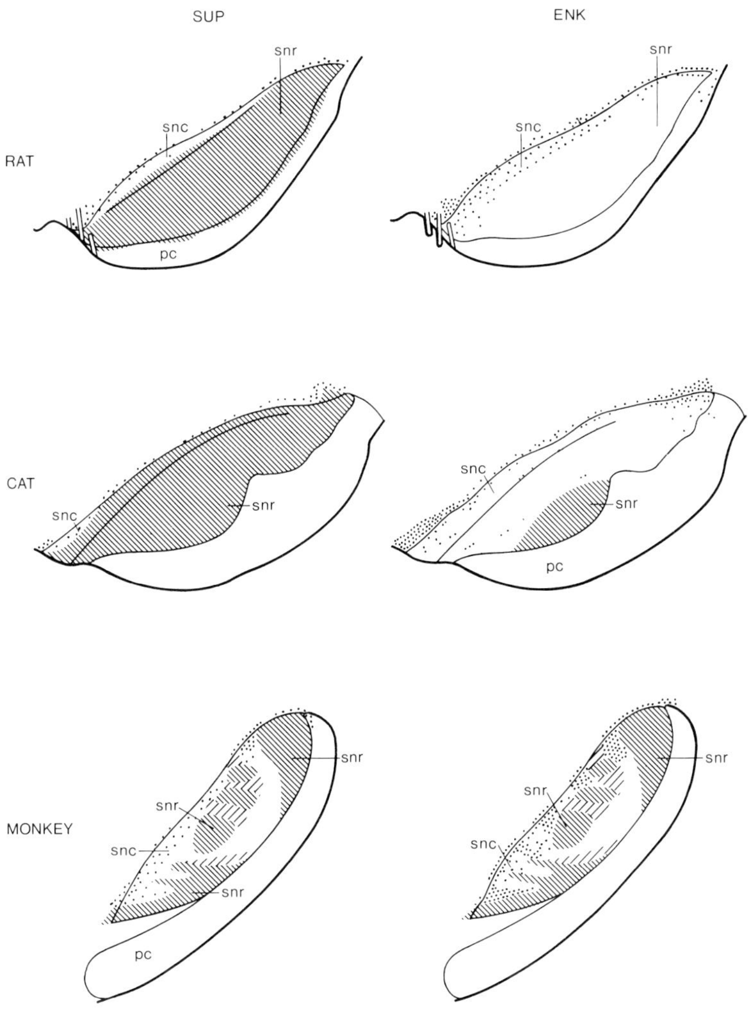

Fig. 40. Schematic drawings of frontal sections through the substantia nigra of the rat, the cat and the squirrel monkey to illustrate the distribution of substance P- (*SUP*) and enkephalin- (*ENK*) positive fibres. The *hatched areas* represent networks of fine immunoreactive fibres, whereas the *dots* illustrate the more scattered and coarse fibres. (Modified from Inagaki and Parent 1984)

Abbreviations: *pc*, pedunculus cerebri; *snc*, substantia nigra, pars compacta; *snr*, substantia nigra, pars reticulata

fibres are distributed in the SNr according to a pattern similar to that found in the rat. Fine enkephalin-containing fibres are densely packed in the ventromedial portion of the SNr, whereas coarse enkephalinergic fibres are scattered in both the SNr and the SNc. In the monkey, fine substance P- as well as enkephalin-containing fibres occur in very large numbers in the SNr. These two types of fibres are distributed according to a strikingly similar, patch-like pattern. In addition, coarse fibres displaying either substance P or enkephalin immunoreactivity are scattered among the SNc neurons in the monkey. Thus, it appears that substance P-containing fibres are present in large numbers in the substantia nigra of the rat, the cat and the monkey, and that these fibres are distributed according to a rather similar pattern in the three species. However, the number of enkephalin-containing fibres within the substantia nigra is much larger in the cat than in the rat and still larger in the monkey than in the cat. Moreover, in the monkey the fibres containing substance P and enkephalin are distributed according to complex patterns which display a remarkable congruency.

As has already been discussed in a previous section (see p. 86) the *hypothalamic paraventricular nucleus* consists of two basic divisions: magnocellular and parvocellular. A detailed analysis based on both cytoarchitectonic and connectional criteria revealed the presence of no less than eight subdivisions within the nucleus (Swanson and Kuypers 1980). Cells in the magnocellular division project primarily to the posterior lobe of the pituitary gland, and are clustered in separate anterior, medial and posterior parts. Cells in the parvocellular division project primarily to the external lamina of the median eminence and to autonomic centres in the lower brain stem and spinal cord; they can be divided into five distinct parts: anterior, medial, lateral, dorsal and periventricular. In Fig. 41, which shows frontal sections through the intermediate and caudal parts of the paraventricular nucleus, the topographical relationships of the posterior magnocellular part and of the medial, dorsal and periventricular parvocellular parts are indicated.

Swanson et al. (1981) studied the distribution of catecholaminergic fibres in the paraventricular nucleus with immunohistochemical techniques, using antisera to the catecholamine synthesizing enzymes DBH and PNMT. Structures stained with an antiserum to PNMT are most probably adrenergic, because the enzyme, which converts noradrenaline to adrenaline, is not found in noradrenergic neurons. The demonstration of noradrenergic structures is less direct with these techniques. However, if a terminal field is stained with anti-DBH but not with anti-PNMT it is presumably noradrenergic. Swanson et al. (1981) also determined the localization of the oxytocin- and vasopressin-containing cells within the magnocellular subdivisions of the nucleus, by using antibodies raised against these two nonapeptides. Some of their results may be summarized as follows (Fig. 41): Adrenergic PNMT-stained fibres were distributed to the entire parvocellular division of the paraventricular nucleus, particularly to the medial part of the division. Only sparsely scattered adrenergic fibres were found in the magnocellular division of the nucleus. As regards the distribution of DBH-stained fibres, it appeared that all parts of the parvocellular division were at least moderately innervated. The greatest density of labelled fibres and varicosities was found in the medial parvocellular part, with only slightly fewer fibres found in the dorsal parvocellular part. The periventricular part of the parvocellular division showed an intermediate density of DBH-stained fibres. Because this distribution of fibres was similar to that observed with anti-PNMT, it was not possible to determine with certainty whether most parts of the parvocellular division contained noradrenergic fibres. However, the periventricular zone appeared to be more densely innervated by DBH-stained fibres than by PNMT-stained fibres, which suggests that this zone, at least, may contain both adrenergic and noradrenergic fibres. The posterolateral aspect of the posterior magnocellu-

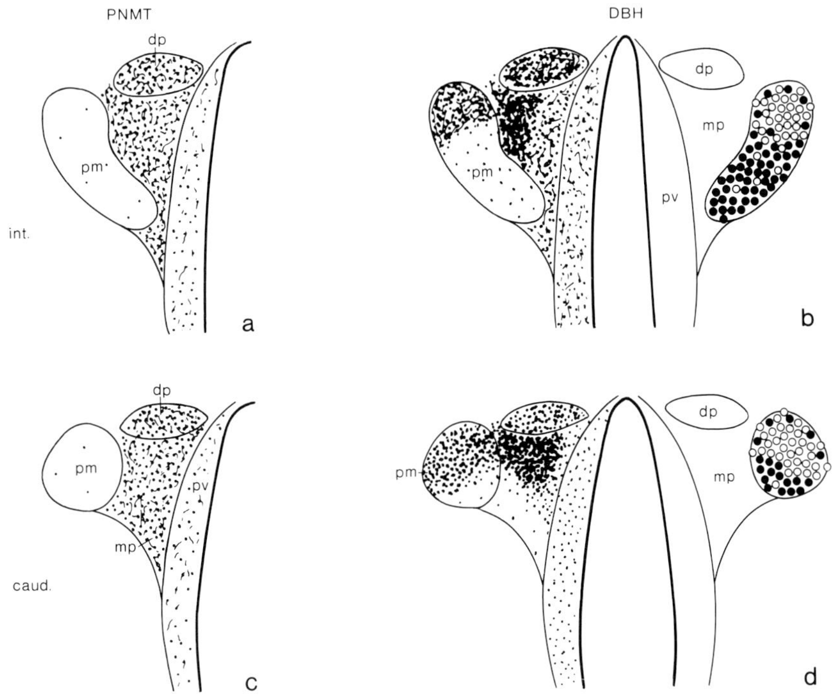

Fig. 41 a–d. Diagrammatic frontal sections through the intermediate (**a, b**) and caudal part (**c, d**) of the nucleus paraventricularis of the rat to illustrate the distribution of *PNMT*-stained (**a, c**) and *DBH*-stained fibres (**b, d**). The distribution of oxytocin-stained (●) and vasopressin-stained (○) cells in the posterior magnocellular part of the nucleus is shown on the *right side* of drawings **b** and **d**. (Modified from Swanson et al. 1981)

Abbreviations indicate the following parts of the paraventricular nucleus: *dp*, dorsal parvocellular; *mp*, medial parvocellular; *pm*, posterior magnocellular; *pv*, periventricular parvocellular

lar part of the paraventricular nucleus was densely innervated by DBH-stained fibres, while the anteromedial aspect of the same cell group contained only scattered fibres. Analysis of the localization of cells containing the posterior pituitary hormones revealed that vasopressin-stained cells were concentrated in the posterolateral part of the posterior magnocellular cell group, whereas oxytocin-stained cells were found primarily in the anteromedial part of that cell group. Swanson et al. (1981) concluded from these results that the adrenergic input to the paraventricular nucleus may influence cells that project to preganglionic autonomic centres in the medulla oblongata and spinal cord, and that the noradrenergic input to the paraventricular nucleus may influence primarily vasopressinergic cells that project to the posterior lobe of the pituitary, as well as cells in the periventricular part of the paraventricular nucleus that project to the median eminence.

The *suprachiasmatic nuclei* in the hypothalamus may be considered the endogenous clock of the brain. They play a critical role in the generation and entrainment of circadian rhythms. The retinohypothalamic tract, a visual pathway which is known to participate in entrainment of circadian rhythms, terminates in the suprachiasmatic nuclei. It has already been mentioned that the suprachiasmatic nucleus of the golden hamster, as that of the rat, can be divided on cytoarchitectonic grounds into separate ventrolateral and dorsomedial parts, and that the investigations of Card and Moore (1984) have shown that VIP-containing perikarya are localized within the ventrolateral subdivision, whereas vasopressin- and somatostatin-containing cells are localized within the dorsomedial subdivision (see p. 119 and Fig. 42). Card and Moore (1984) also studied the distribution of fibres and terminals, containing serotonin, GAD (i.e. the synthesizing enzyme of GABA), VIP, avian pancreatic polypeptide, somatostatin and vasopressin in the suprachiasmatic nucleus of the golden hamster. Serotoninergic axons appeared to form a plexus in the ventromedial portion of each nucleus. Dense plexuses of varicose axons positive for GAD and for VIP were found throughout the rostrocaudal extent of the nucleus. Avian pancreatic polypeptide-containing fibres appeared to arborize within the ventrolateral aspect of each nucleus. Axons containing vasopressin were found to form a dense plexus in the dorsomedial suprachiasmatic nuclei and in a vertical column at the lateral aspect of each nucleus. A largely similar pattern of distribution was shown by somatostatin-containing axons (Fig. 42). Card and Moore (1984) pointed out that the retinohypothalamic projection terminates exclusively in the ventrolateral subdivision of the suprachiasmatic nucleus and that the avian pancreatic polypeptide-containing fibres, which are also distributed to the ventrolateral subdivision, originate from the lateral geniculate body. Thus, the ventrolateral subdivision receives both primary and secondary optic fibres. The dorsomedial subdivision, on the other hand, does not receive any extrinsic optic projections. However, the VIP-containing neurons found in the ventrolateral subdivision most probably represent a population of local circuit neurons with a large portion of their axonal arbor spreading within the dorsomedial subdivision, thus subserving an integrative function within the suprachiasmatic nucleus. Card and Moore (1984) consider it likely that this intrinsic ventrolateral-to-dorsomedial connection is at least partly reciprocated by the axons of vasopressin- and somatostatin-containing local circuit neurons present in the dorsomedial subfield. They also held that the dense plexus of GAD-containing fibres found in the suprachiasmatic nucleus probably largely originates from intrinsic neurons. Lightly stained GAD-positive perikarya were found to be present throughout the nucleus. Taken together, the findings of Card and Moore (1984) provide evidence of four different populations of neuromediator-specified local circuit neurons in the suprachiasmatic nucleus of the golden hamster.

The *preoptic region* can be divided into periventricular, medial and lateral zones. In the preceding sections no distinction has been

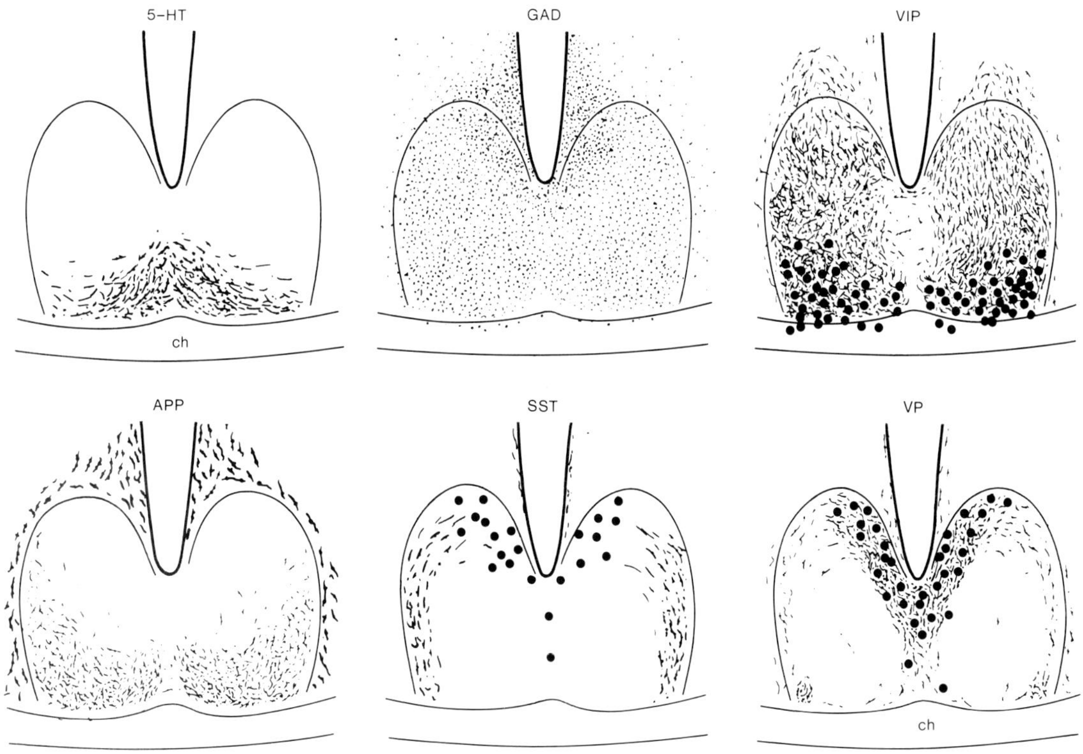

Fig. 42. Diagrammatic frontal sections through the nucleus suprachiasmaticus of the golden hamster to illustrate the distribution of serotonin (*5 HT*)-, glutamic acid decarboxylase (*GAD*)-, vasoactive intestinal polypeptide (*VIP*)-, avian pancreatic polypeptide (*APP*)-, somatostatin (*SST*)- and vasopressin (*VP*)-like immunoreactivity within that centre. (Modified from Card and Moore 1984)
Abbreviation: *ch*, chiasma opticum

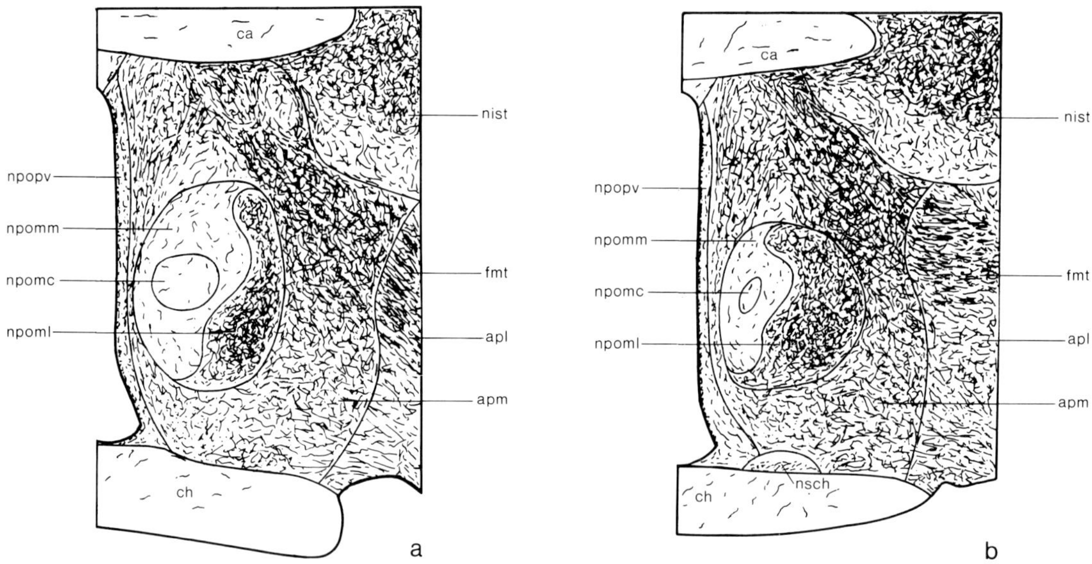

apl Area preoptica lateralis
apm Area preoptica medialis
ca Commissura anterior
ch Chiasma opticum
fmt Fasciculus medialis telencephali
nist Nucleus interstitialis striae terminalis
npomc Nucleus preopticus medialis, pars centralis
npoml Nucleus preopticus medialis, pars lateralis
npomm Nucleus preopticus medialis, pars medialis
npopv Nucleus preopticus periventricularis
nsch Nucleus suprachiasmaticus

Fig. 43 a, b. Drawings of frontal sections through the medial part of the preoptic region in the male (**a**) and female (**b**) rat, showing the boundaries of the various cell masses and the distribution of serotoninergic fibres in that region. (Redrawn from Simerly et al. 1984)

made between the medial preoptic nucleus and the medial preoptic zone. However, in a recent study of the preoptic region of the rat, Simerly et al. (1984) pointed out that the medial preoptic area represents an undifferentiated grey in which several cellular condensations, or nuclei are embedded. The largest of these is the centrally placed medial preoptic nucleus (npom), which extends nearly the length of the medial preoptic area. This nucleus can be subdivided into three parts: a medial cell-dense part (npomm), a lateral cell-sparse part (npoml), and a central very cell-dense part (npomc) that is embedded in the medial part (Fig. 43). The npomc corresponds to the sexually dimorphic nucleus of the preoptic region identified by Gorski et al. (1980). Simerly et al. (1984) carried out a volumetric analysis of the medial preoptic nucleus, which revealed a marked sexual dimorphism in the relative size of each of its parts. In the male, the volumes of the cell-dense npomm and npomc appear to be notably larger, whereas in the female more than half of the nucleus is occupied by the cell-sparse lateral part. The npomm as a whole appears to be slightly larger in the male (see Fig. 43). Simerly et al. (1984) also studied the distribution of serotoninergic fibres in the preoptic region of male and female rats, using the antiserum directed against serotonin prepared by Steinbusch (1981). It appeared that in both the male and the female the three subdivisions of the npom contain a characteristic density of serotoninergic fibres. The npoml was found to be filled with a dense plexus of varicose immunoreactive fibres. In contrast, the npomm contained only a low density of stained fibres, whereas the npomc appeared to be virtually devoid of serotoninergic fibres (Fig. 43). Thus, according to Simerley et al. (1984), in the sexually dimorphic medial preoptic nuclear complex of the rat there is a remarkable congruency between cytoarchitecture and the patterns of distribution of serotoninergic axons.

With regard to the distribution of fibre systems within its confines, the *caudate-putamen complex* shows a marked neurochemical heterogeneity at two different levels of organization. First, it is known that several neuromediator-specified fibre systems are not equally dispersed over this complex. Thus, a subpopulation of mesencephalic neurons which contain both dopamine and a CCK-like peptide projects only to a restricted caudomedial zone of the caudate nucleus (Hökfelt et al. 1980b), and a plexus of somatostatinergic fibres shows a ventromedial to dorsolateral gradient of decreasing density (Johannson et al. 1984; Fig. 44a). Second, staining for AChE has revealed that in the caudate-putamen complex, 300–600-μm-wide zones of low AChE activity stand out against an otherwise AChE-rich background. Graybiel and Ragsdale (1978, 1979), who first identified these zones, designated them as striatal bodies or *striosomes* (Fig. 44b). Graphical reconstructions have shown that in most places the striosomes form part of a complex three-dimensional labyrinth.

During the past few years it has become clear that the striosomes and the 'matrix' in which they are embedded represent throughout most of the caudate-putamen complex chemoarchitectonically distinct tissue compartments, which are related to the intrinsic structure of this complex as well as to the organization of its afferent and efferent connections. The following features should be mentioned in this context (for details and references the reader is referred to the review articles of Graybiel and Ragsdale 1979, 1983 and of Graybiel 1984, and to the recent studies of Gerfen 1984 and Herkenham et al. 1984):

1. Apart from a low AChE concentration, the striosomes show high enkephalin, substance P, GABA and neurotensin immunoreactivity.
2. The complementary matrix compartment shows, in addition to a high AChE concentration, a dense plexus of somatostatin-containing fibres.
3. The striosomes show a remarkably high concentration of opiate receptors.
4. Studies of the fetal development of the striatum in the cat with tritiated thymidine

autoradiography have shown that clusters of cells born contemporaneously correspond to the striosomes.

5. In the caudate nucleus of the adult rhesus monkey, cell clusters, some with capsules poor in cells, appear to match the striosomes (Goldman-Rakic 1982).

6. Cells containing somatostatin and the large cholinergic striatal interneurons tend to lie outside the striosomes (Graybiel 1984; however, see Gerfen 1984).

7. The striosomes appear to coincide with patches of high dopamine density, the so-called 'dopamine islands', which correspond to a distinct, early-developing contingent of nigrostriatal fibres.

8. Studies with anterograde tracers have revealed that not only the nigrostriatal pathway, but also the thalamostriate and corticostriate projections terminate in a patchy fashion. It has been demonstrated that the terminal patches of the thalamostriatal fibres avoid the striosomes. The massive corticostriate projection can be divided into two fibre contingents, one that avoids the striosomes and another that projects specifically to these structures. Thus, sensory and motor cortical areas preferentially project to the matrix compartment, but the efferents from the prelimbic cortex (i.e. a medial frontal cortical area) are distributed to the striosomes (Gerfen 1984).

9. After large injections of HRP into the globus pallidus there appears to be a marked heterogeneity in the pattern of distribution of the retrogradely labelled cells in the caudate-putamen complex. Patches of poor cell labelling alternate with zones of pronounced cell labelling, and the former have been shown to correspond with the AChE-poor striosomes. This could mean that the striosomes represent regions with a high density

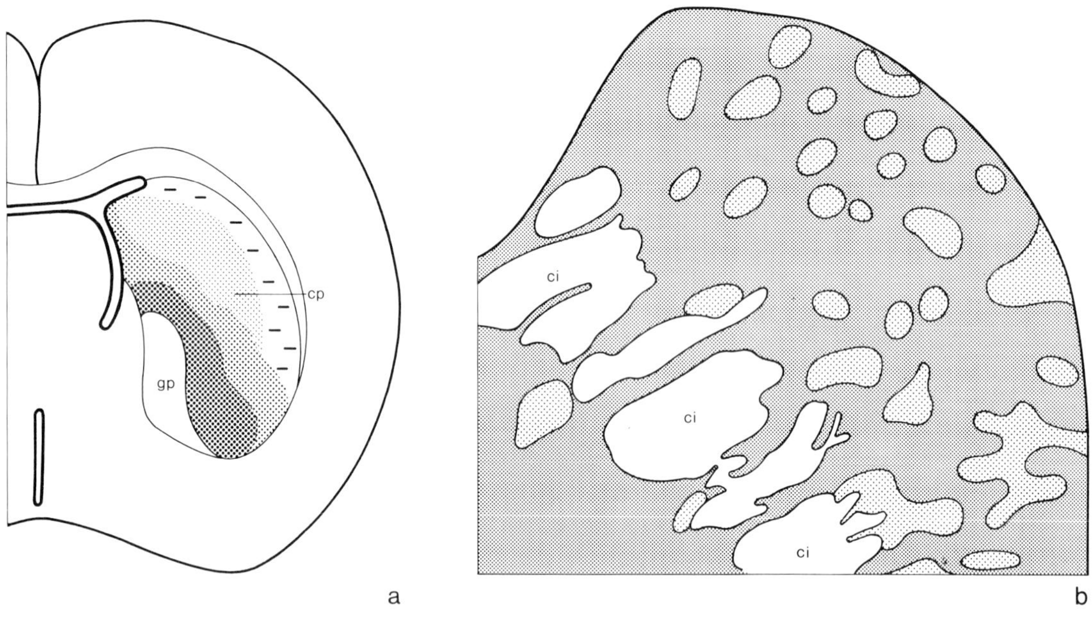

Fig. 44 a, b. Aspects of the histochemistry of the striatum: **a** distribution of somatostatin-immunoreactive fibres in the basal ganglia of the rat (based on Fig. 5C in Johannson et al. 1984); **b** diagrammatic representation of a transverse section through the caudate nucleus of the rhesus monkey, stained for AChE; striosomes poor in AChE embedded in an AChE-rich matrix (based on Fig. 11 in Graybiel and Ragsdale 1983)

Abbreviations: *cp*, caudatoputamen complex; *ci*, capsula interna; *gp*, globus pallidus

of local circuit neurons, whereas the neurons with extrinsic connections are found mainly in the striatal matrix. However, Gerfen (1984) recently demonstrated that following fast-blue injections in the compact part of the substantia nigra retrogradely labelled cells are distributed preferentially in the striosomes, whereas after injections centred in the pars reticulata of the substantia nigra labelled neurons appeared to be most densely distributed in the matrix compartment. Moreover, it has recently been shown that the striosomes contain clusters of substance P- and dynorphin-B-positive perikarya. According to Graybiel (1984), who reported these findings, this could suggest that some elements of the substance P- and dynorphin-containing striopallidal and strionigral projection systems may have preferential origins within the striosomes.

From the foregoing synopsis it appears that the caudate-putamen complex displays an intriguing mosaic-like chemical heterogeneity, and that several structural and connectional features fit into this mosaic. The fact that the input-output connections of the caudate-putamen complex are to a large extent broken up into spatially segregated and chemically specified compartments raises, according to Graybiel (1984), the possibility that these compartments represent units in which groups of neurons can be modulated in a coordinated way. In her opinion, such group modulation could occur through a high density of conventional synaptic contacts mediating synchronized inputs, or by local nonsynaptic extracellular diffusion of neuromediators. In relation to the latter possibility, Graybiel considered to be of importance the finding that the striatal compartments are at least partly rimmed by septa (Graybiel and Ragsdale 1978, 1983).

On the basis of his findings on the organization of corticostriatal and striatonigral connections, Gerfen (1984) suggested that the striosomes (the 'patches' in his terminology) and the matrix represent functionally different compartments. The striosomes receive input from the prelimbic cortex and project to the substantia nigra pars compacta, whereas the matrix receives inputs from sensory and motor cortical areas and projects to the substantia nigra pars reticulata. He emphasized that the prelimbic cortex receives direct limbic inputs from the amygdala and the hippocampus. A few years ago, Kelley et al. (1982) reported that in the rat a voluminous amygdalostriatal projection is present, which is distributed to all parts of the caudatoputamen except its anterodorsolateral quadrant (see also Russchen and Price 1984). They pointed out that this projection widely overlaps the striatal projections from the hippocampus, the cingulate cortex, the ventral tegmental area and the mesencephalic raphe nuclei. Like the amygdalostriatal system, all of these striatal afferents avoid the anterodorsolateral striatal sector. Kelley and his colleagues further established that the striatal sector just mentioned is the main destination of the corticostriatal projection from the sensorimotor cortex. In view of these findings they interpreted the large striatal region receiving a direct projection from the amygdala as the 'limbic' and the remainder as the 'non-limbic' striatal compartment. Thus, Kelley et al. (1982), as well as Gerfen (1984), arrived at the conclusion that in the striatum, limbic and non-limbic (somatosensory) parts can be distinguished. However, whereas according to the former these two parts constitute separate major striatal sectors, according to the latter they interdigitate as striosomal and matrix compartments. Gerfen (1984) also provided some evidence suggesting that somatostatin-containing intrinsic neurons provide a link between these two striatal compartments.

Before leaving the chemoarchitecture and the related compartmentalization of the striatum, it is worth mentioning that, according to a recent study of Herkenham et al. (1984), the striosomes present in the nucleus accumbens (i.e. a ventromedial extension of the caudatoputamen) differ in some respects from those found in the remainder of the striatum. In this cell mass the striosomes stand out as dense cell clusters composed of a specialized form of medium-sized spiny striatal neurons. These clusters exclude thalamic and dopaminergic inputs but show, as in the remainder of the striatum, a low AChE activity and a very high density of opiate receptors. Herkenham and colleagues (1984) considered it likely that the clusters preferentially receive input from nearby striatal enkephalinergic neurons, and they suggested that these structures represented way stations devoted to intrinsic information processing. This concept of the possible functional significance of the striosomes differs considerably from those enunciated by Graybiel (1984) and Gerfen (1984).

With regard to the disposition of terminal fields in the *cerebral cortex* I will confine myself mainly to a few examples pertaining to the monoaminergic innervation of that part of the brain in the rat and in primates.

Using an antiserum directed against DBH, Morrison et al. (1978, 1980) studied the noradrenergic (NA) innervation of the cerebral cortex in the rat. Their principal findings can be summarized as follows: A rich network of NA fibres is present throughout all layers and regions of the dorsal and lateral cortex. The pattern of NA innervation is not diffuse, but is characterized by a geometric orderliness that is uniform throughout the neocortex, with only minor variations. Dense terminal fields are found in layers IV and V, whereas layers II and III are characterized by straight radial fibres, and layers I and VI by the presence of tangentially oriented fibres. In the prelimbic cortex the branching pattern and density of the fibres are similar to those in the dorsal and lateral cortex, but in the anterior cingulate cortex a low density of NA fibres was found in the superficial layers. The use of the immunohistochemical stain for DBH in conjunction with glyoxylic acid-induced histofluorescence (Lewis et al. 1979), as well as application of the same histofluorescence technique following surgical interruption of the NA innervation of the cortex (Morrison et al. 1980), revealed that in the anterior cingulate cortex the NA and dopaminergic (DA) innervation are complementary in that the DA fibres terminate in layers I–III, while the NA fibres, as already mentioned, are largely confined to the deep layers (IV and V).

An immunohistochemical analysis of the serotoninergic innervation of the cerebral cortex in the rat was performed by Lidov et al. (1980), employing the antiserum against serotonin prepared by Steinbusch (1981). They found that throughout the dorsal and lateral cortex, as well as in the anterior cingulate cortex, serotoninergic fibres form a densely arborizing plexus which is relatively uniform across all cortical layers. However, a distinctive and different pattern of serotonin innervation was found in the posterior cingulate cortex. In this area the serotoninergic axons appeared to be restricted largely to laminae I and III.

The organization of the NA innervation of the cerebral cortex of the squirrel monkey was studied by Morrison et al. (1982a; see also Morrison and Magistretti 1983), again using an antibody directed against DBH. Their analysis revealed that although the general laminar and tangential features of NA axons are similar to those in the rat, the primate cortex exhibits far greater regional specificity, in that many diverse neocortical areas possess specific patterns of laminar distribution and density of NA fibres. The authors presented a detailed description of the NA innervation in the dorsolateral prefrontal cortex (Brodmann's areae 9 and 10), the primary somatosensory cortex (areae 3, 1, 2) and the primary visual cortex (area 17). The NA innervation of these three regions is similar, in that fibres are present in all six layers. However, the primary somatosensory cortex is much more densely innervated than the prefrontal and primary visual cortices. Moreover, layer IV, which shows a dense and terminal-like innervation in the prefrontal and somatosensory cortices, contains only very few fibres in the primary visual cortex (Fig. 45 a–c). In another study Morrison et al. (1982b) compared the noradrenergic and serotoninergic innervation of the primary visual cortex in the same primate species. They found that these innervations exhibit a high degree of laminar complementarity. As shown in Fig. 45c and d, layers V and VI receive a dense NA projection and a very sparse serotoninergic projection, whereas layer IV receives a very dense serotoninergic projection and is largely devoid of NA fibres.

Morrison and colleagues (Morrison et al. 1982b; Morrison and Magistretti 1983) have pointed out that the complementary laminar patterns of monoamine terminal fields, as found in the primate visual cortex and also in the anterior cingulate cortex of the rat, support the notion that in these areas monoaminergic fibres are directed at specific postsynaptic targets, and that different monoamines may affect different stages of cortical information processing. Thus, in the primate visual cortex the serotoninergic projection may preferentially innervate spiny stellate cells in laminae IVa and IVc. These intrinsic neurons are present in great abundance in the fourth cortical layer; they are known to receive a direct thalamocortical input and to synapse on pyramidal cells. The NA fibres, on the other hand, presumably directly engage pyramidal elements in the third and fifth layers.

A detailed analysis of the catecholamine innervation of the cerebral cortex of the rhesus monkey was recently carried out by Levitt et al. (1984), using the glyoxylic acid histofluorescence technique. This study provided further support for the existence of a considerable regional variation in cortical monoaminergic innervation in primates (Fig. 46). The patterns of innervation found could be clearly related to the cytoarchitecture of the areas studied. Remarkably, the particular laminar distributions of catecholaminergic afferents in the cortex of the squirrel monkey, as described by Morrison et al. (1982a), appear to differ in some respects from the fluorescence data presented by Levitt et al. (1984) for the rhesus monkey. For example, according to Morrison et al. (1982a), the catecholaminergic innervation is dense and terminal-

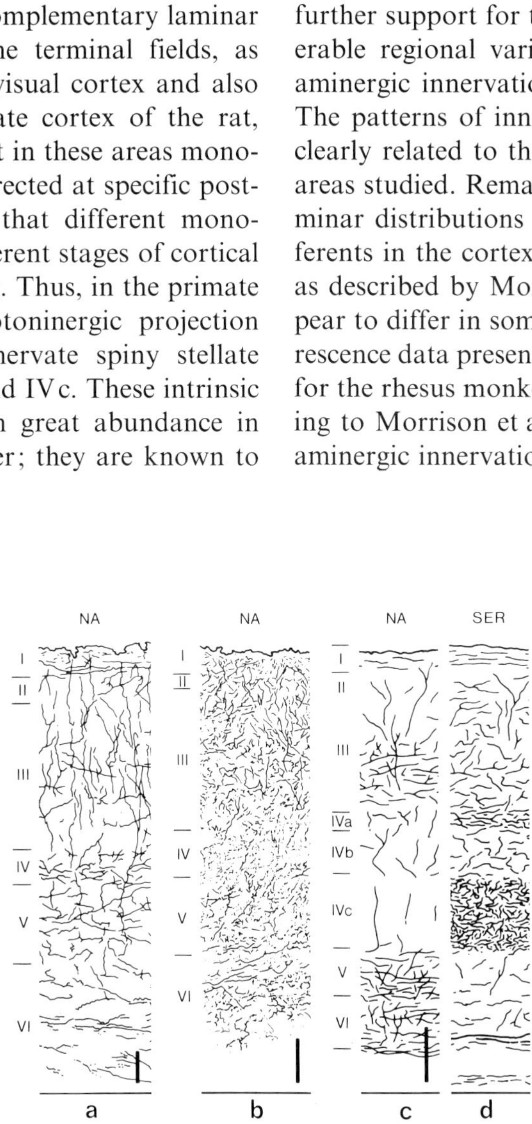

Fig. 45 a–d. Monoaminergic innervation of three different regions of the cerebral cortex of the squirrel monkey. **a, b, c** Noradrenergic (*NA*) innervation of the dorsolateral prefrontal cortex (areae 9, 10), the primary somatosensory cortex (areae 3, 1, 2) and the primary visual cortex (area 17) respectively; **d** serotoninergic (*SER*) innervation of the primary visual cortex. *Bars* represent 200 μm (Slightly modified from Morrison and Magistretti 1983)

like in lamina IV in both the prefrontal and the somatosensory cortex of the squirrel monkey, whereas in the corresponding areas of the macaque, catecholaminergic fibres tend to be less dense in lamina IV relative to other layers (cf. Fig. 45 with Fig. 46). Finally, it may be mentioned that a recent study of Mesulam et al. (1984b) has revealed that in the primate cerebral cortex not only the monoaminergic innervation, but also the AChE-containing fibres display marked regional variations in laminar distribution and density, variations which respect cytoarchitectonic boundaries as well as patterns of cortical specialization (Fig. 47). Mesulam et al. (1984b) presented evidence suggesting that most of these fibres represent the terminal segments of cholinergic projections originating from the nucleus basalis of Meynert (see Fig. 2).

In the preceding pages I have reviewed studies on the patterns of neuromediator-specified terminal fields in nine different centres of the brain, ranging from the solitary complex in the lower rhombencephalon to the cerebral cortex. The phenomena described by the various authors can be categorized as follows:

1. Congruity of different neuromediator-specified terminal fields
2. Complementarity of different neuromediator-specified terminal fields
3. Congruity of neuromediator-specified terminal fields with cytoarchitectonic entities (areas, subfields)
4. Congruity of neuromediator-specified terminal fields with subsets of neuromediator-specified perikarya

Phenomenon 1 (i.e. congruity of different neuromediator-specified terminal fields) has been observed in, among other structures, the interpeduncular nucleus (substance P- and VIP-containing fibres: Fig. 39), the primate substantia nigra (substance P- and enkephalin-containing fibres: Fig. 40) and the striosomal and matrix compartments in the caudate-putamen complex, the former containing high enkephalin, substance P, GABA and neurotensin immunoreactivity, the latter being characterized by a high AChE concentration and a dense plexus of somatostatin-containing fibres.

Fig. 46. Catecholaminergic innervation of six representative areas of the cerebral cortex of the rhesus monkey. *Heavier lines* represent noradrenergic fibres, *thinner lines* represent dopaminergic fibres. (Slightly modified from Levitt et al. 1984)

Phenomenon 2 (i.e. complementarity of different neuromediator-specified terminal fields) appears to occur in many of the centres selected: the parabrachial nuclei (substance P versus neurotensin: Fig. 38), the interpeduncular nucleus (substance P plus VIP versus CCK plus enkephalin: Fig. 39), the cingulate cortex of the rat (dopamine-containing fibres terminating in the superficial layers versus noradrenalin-containing fibres terminating in the deeper layers), the primate visual cortex (layer IV receiving a very dense serotoninergic innervation versus layers V and VI receiving a dense NA innervation: Fig. 45) and, most strikingly, the striosomal and matrix compartments of the caudate-putamen complex, showing the differences already indicated.

Phenomenon 3 (i.e. congruity of neuromediator-specified terminal fields with cytoarchitectonic entities) was also frequently reported. Thus, Milner et al. (1984) stated correctly that the lateral parabrachial nucleus is subdivisible into cytoarchitectonic units which correspond to the fields of termination of substance P- and neurotensin-containing fibres on the one hand, and to those of enkephalin-containing fibres on the other hand (Fig. 38), and we have seen that in the nucleus accumbens of the rat and in the caudate nucleus of the rhesus monkey, patches of neuromediator-specified terminals correspond to cell clusters. However, I am under the impression that in a number of the publications reviewed, congruity of transmitter-specified terminal fields and cytoarchitectonic units was claimed somewhat too lightly, a tendency which might be designated as *'conformism'*. Inspection of the pictures in the admirably detailed studies of Kalia et al. (1984b, c) on transmitter-specified terminal fields in the solitary complex reveals that one of their main conclusions, namely that the four neuropeptides investigated display distinct patterns of terminations in the subnuclei of that complex, does not hold true for all of these peptides at all of the five levels examined. A similar criticism can be levelled against the study of Hamill et al. (1984) on the distribution of transmitter-specified fibres and terminals over the interpeduncular nucleus. Finally, it may be mentioned that in the preoptic region of the rat the correlation between the cytoarchitectonic divisions

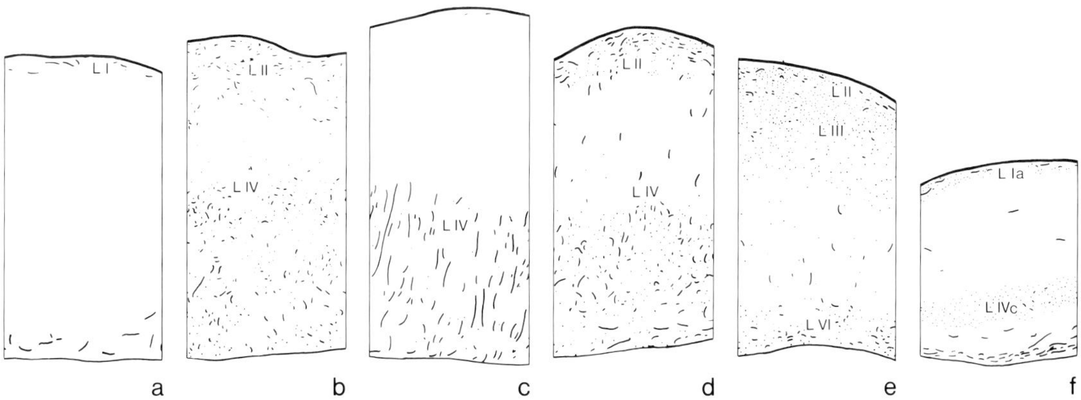

Fig. 47 a–f. Distribution of AChE-positive fibres within six representative cortical areas of the rhesus monkey; **a** rostral orbitofrontal; **b** caudal orbitofrontal; **c** motor; **d** insular; **e** entorhinal; **f** primary visual. (Slightly modified from Mesulam et al. 1984)

and the patterns of termination of serotoninergic fibres is presumably less striking than was indicated by Simerly et al. (1984). Geeraedts (1985), who made a meticulous cytoarchitectonic analysis of the region in question, remained unable to confirm the delineation of the medial preoptic nucleus as presented by these authors.

Phenomenon 4 (i.e. congruity of neuromediator-specified terminal fields with subsets of neuromediator-specified perikarya) appeared to occur in, among other structures, the suprachiasmatic nucleus (APP-containing fibres with VIP-positive cell bodies: Fig. 42), in the striosomes (enkephalin-, GABA- and neurotensin-immunoreactive fibres with substance P- and dynorphin B-immunoreactive cell bodies) and, most strikingly, in the posterior magnocellular subnucleus of the paraventricular nucleus, where NA fibres specifically address the vasopressinergic neurons located in the posterolateral part of that subnucleus (Fig. 41).

In several of the studies on the patterns of neuromediator-specified terminal fields, reviewed earlier in the present section, it was attempted to interpret the results obtained in terms of a neurochemcial specification of the internal circuitry in the centres studied. Thus, Card and Moore (1984) pointed out that their findings suggest the presence of four different types of neuromediator-specified local circuit neurons in the suprachiasmatic nucleus, and Morrison et al. (1982b) surmised that in the primate visual cortex serotoninergic fibres preferentially innervate interneurons, whereas NA fibres synapse mainly with projection neurons. It should be emphasized that our knowledge of the morphological organization of practically all of the centres in the mammalian neuraxis is by far too fragmentary to render possible a detailed neurochemical elucidation of their internal circuitry. At present, a coherent picture of the intrinsic organization is available for only a relatively small number of centres which, owing to the simple geometric arrangement of some of their constituent neuronal elements, are particularly accessible to morphological analysis. The lack of any trace of a simple geometric organization in a given centre is in itself a formidable obstacle for the clarification of its internal circuitry. This is, for instance, apparently the case in the caudate-putamen complex, and that is why the problem of how the striosomes are embedded in the microcircuitry of this centre is so hard to tackle.

18. To my knowledge there are no centres in the mammalian central nervous system for which the identity of the neuromediators present in all of the categories of their constituent elements has been clarified.

In order to elucidate this point I will briefly discuss the cerebellar cortex and the olfactory bulb; both the microcircuitry and the chemical signature of the constituent elements of these two structures are relatively very well known.

The *mammalian cerebellum* is a convoluted structure divided by transverse fissures of different depth into lobes and lobules. Its surface is formed by a cortex that presents a uniform, three-layered histological structure throughout its extent. These layers are, passing from superficial to deep, the molecular layer with few cells, the Purkinje cell layer and the granular layer (Fig. 48a). The following survey of the internal circuitry of the cerebellar cortex (Fig. 48b, c) is based on the works of Cajal (1911), Eccles et al. (1967), Palay and Chan-Palay (1974) and Ito (1984). The data on the neuromediators present in the various neuronal elements are, unless otherwise stated, derived from the reviews of Schulman (1983) and Ito (1984).

The large Purkinje cells, the somata of which form a single layer, constitute the output neurons of the cerebellar cortex. Their richly

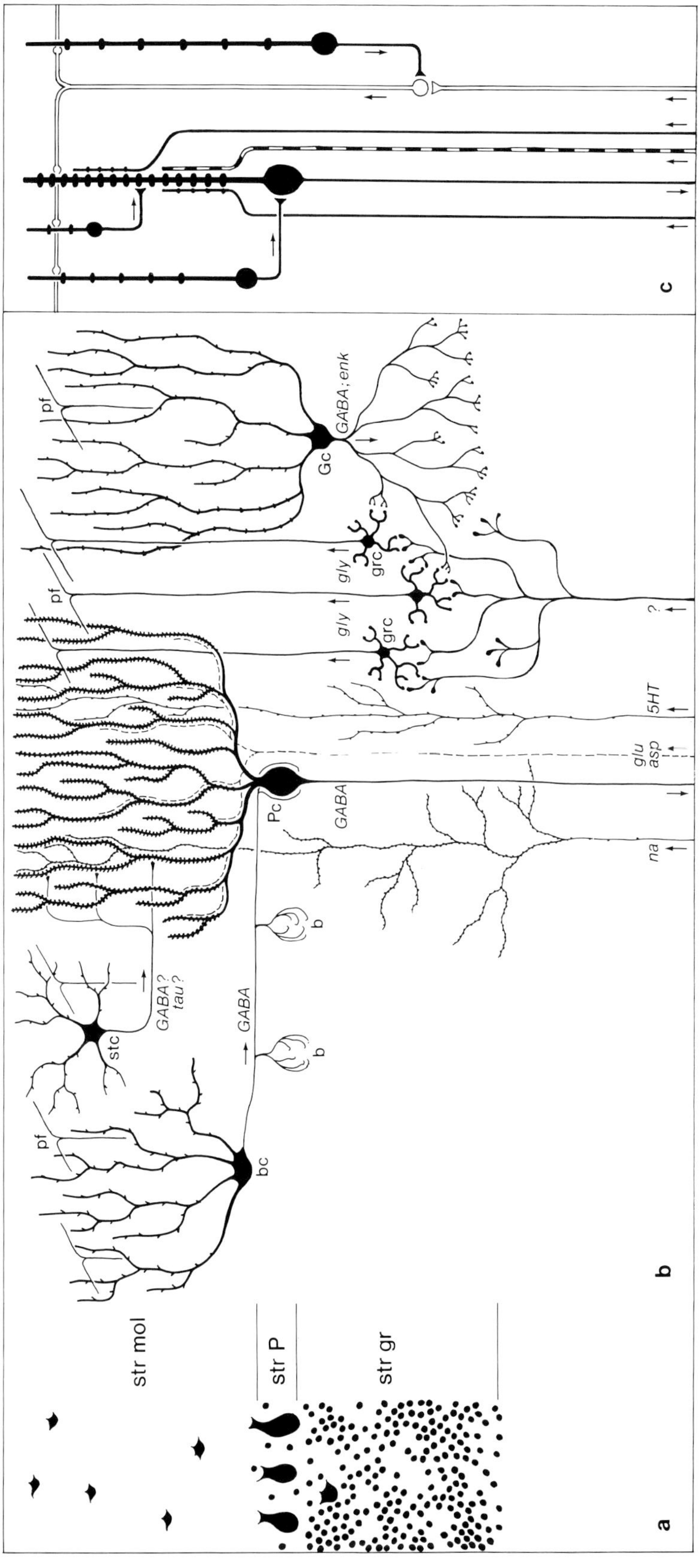

Fig. 48 a–c. The cerebellar cortex. **a** Cytoarchitecture; **b** neuronal elements and fibres as seen in Golgi or (noradrenergic and serotoninergic fibres) histofluorescence preparations; **c** circuit diagram. Inhibitory neurons and synapses are in *black* and excitatory ones have been left *unfilled*.

Abbreviations: *b*, basket; *bc*, basket cell; *cf*, climbing fibre; *Gc*, Golgi cell; *grc*, granule cell; *mf*, mossy fibre; *Pc*, Purkinje cell; *pf*, parallel fibre; *stc*, stellate cell; *str gr*, stratum granulare; *str mol*, stratum moleculare; *str P*, stratum Purkinje

ramifying dendritic trees, which extend into the molecular layer, are flattened and spread out at right angles to the longitudinal axis of the cerebellar lobules. Their smaller dendritic branches are densely covered with spines. The axons of the Purkinje cells descend through the granular layer and the cerebellar white matter to terminate in the central cerebellar and lateral vestibular nuclei. During their passage through the granular layer the Purkinje axons emit numerous recurrent collaterals (not shown in Fig. 48), which contact other Purkinje cells, as well as Golgi and basket cells.

Purkinje cells are known to exert an inhibitory influence on their target neurons, and until recently GABA was generally held to be the universal neurotransmitter released by all of these elements. However, as mentioned before, Chan-Palay et al. (1981) have provided immunohistochemical evidence suggesting that a certain proportion of the Purkinje cells contain the peptide motilin, and that some contain GABA as well as motilin. In some still more recent studies Chan-Palay and collaborators (Chan-Palay et al. 1982a, b; Chan-Palay 1984) claimed that Purkinje cells may also contain the amino acid taurine, or taurine plus GABA, or taurine plus motilin. They also reported that at least 30% of all Purkinje cells have neither taurine, GAD (i.e. the synthesizing enzyme of GABA), nor motilin immunoreactivity, and hence may well use another, as yet unknown, neuromediator. GABA, motilin and taurine all produce inhibitory effects in the cerebellum and in the lateral vestibular nucleus (Chan-Palay 1984).

The input to the cerebellar cortex is conveyed mainly by two types of axons, the climbing fibres and the mossy fibres. The climbing fibres originate from the inferior olivary complex, although it is still not established that this complex is the exclusive source of these fibres (Palay 1982). After having traversed the cerebellar granular layer, the climbing fibres reach the base of the dendritic tree of a Purkinje cell and, dividing and redividing, ascend along its branches. The climbing fibre arborization forms a large number of synaptic junctions on the proximal dendritic branches of a Purkinje cell, by which they are able to exert an extremely powerful excitatory action on that element. The amino acid neurotransmitters glutamate and aspartate have been suggested as possible mediators of this excitatory action, but the lack of a histochemical technique to demonstrate the presence of these substances precludes any but the most tentative conclusion (Schulman 1983).

The mossy fibres, the other main group of afferents, arise from various sources throughout the rhombencephalon and the spinal cord, including the vestibular nuclei, the pontine nuclei, certain parts of the reticular formation and the nucleus thoracicus, or column of Clarke. After having passed through the cerebellar white matter, these fibres ramify and terminate in the granular layer by forming characteristic large swellings. These 'mossy' terminals constitute the core of synaptic complexes, which are enclosed by glial membranes. They synapse with dendrite terminals of granule cells, upon which they exert an excitatory influence. As regards the neuromediators involved in this influence, Schulman (1983) and Ito (1984) agree that ACh may account for small populations of mossy fibres, but that candidate substances for the vast majority of these fibres are still unknown.

The small, densely packed granule cells give rise to some short dendrites, each of which terminates in a small, claw-like expansion. The axons of the granule cells ascend through the granular and Purkinje layers into the molecular layer, where they divide in a T-shaped fashion to give rise to two branches that run parallel to the surface and at right angles to the plane of branching of the Purkinje cell dendrites. These so-called parallel fibres in the cat have a length of 5–7 mm (Mugnaini 1965). They traverse the dendritic trees of a number of Purkinje cells and form excitatory synapses with spines from small dendrites (spiny branchlets). Each Purkinje cell receives about 80000 parallel fibre synapses. There is good evidence that the amino acid glutamate is the neurotransmitter of the granule cells.

Passing through the molecular layer, the parallel fibres exert their excitatory influence not only on the Purkinje cells, but also on the dendritic trees of three types of local circuit neurons, the Golgi, stellate and basket cells. The Golgi cells are relatively large neurons, the perikarya of which are scattered throughout the superficial zone of the granular layer. Their dendritic trees, which, unlike those of the Purkinje cells, are not confined to a single plane, ascend and branch in the molecular layer. The axons of the Golgi cells ramify profusely in the granular layer. The varicose terminal branches of these axons enter the glomeruli to form inhibitory synapses with granule cell dendrites. Because the Golgi cells are excited by parallel fibres, i.e. the axonal branches of granule cells, their inhibitory synapses on the dendrites of the same category of cells complete a simple negative feedback loop.

There is strong evidence that Golgi cells utilize GABA as a transmitter although, as mentioned before, Wilkin et al. (1981) adduced evidence suggesting that in the cerebellar cortex two biochemically separate populations of Golgi neurons are present, one containing glycine, the other GABA. It is also worth mentioning that enkephalin immunoreactivity has been reported in Golgi neurons. Whether the enkephalin-positive cells represent a population different from the GABA-ergic elements, or whether the Golgi cells may represent another case of coexistence of a neuropeptide and a classical neuromediator remains to be determined (Schulman 1983).

The stellate cells are the intrinsic neurons of the molecular layer. Their dendrites, like those of the Purkinje cells, radiate transversely in the cerebellar lobules. Their axons are also transversely orientated and establish inhibitory synaptic contacts with Purkinje cell dendrites. The nature of the neuromediator of the stellate cells has not been fully established. Immunohistochemical studies (e.g. Ribak et al. 1978; Ottersen and Storm-Mathisen 1984) indicate GABA as a serious candidate, but on the other hand, biochemical (Nadi et al. 1977) and neurophysiological (Okamoto et al. 1983) evidence is available suggesting that the inhibitory amino acid taurine might be the stellate cell neuromediator.

The basket cells are to be regarded as deep stellate cells. Their axons arise from one side of the cell body and travel, again transversely orientated, just above the Purkinje cell layer, about 0.6 mm in either direction. During their course the axons emit descending collaterals which ramify around the somata of Purkinje cells, on which they end as inhibitory synapses. Each Purkinje cell receives axon collaterals from several different basket cells, and each basket cell participates in the formation of about ten pericellular 'baskets'. It will be clear that if a narrow beam of activated parallel fibres excites a number of Purkinje cells and basket cells in its path, the basket cells, by means of their transversely orientated axons, will exert their inhibitory action on series of Purkinje-cell somata flanking the beam, a typical example of lateral inhibition. Immunohistochemical, pharmacological, and electrophysiological evidence favour the identification of GABA as the basket cell neuromediator.

Histofluorescence studies have shown that, apart from the long-known climbing and mossy fibres discussed above, there is a third morphological category of cerebellar afferents. This comprises NA fibres originating from the locus coeruleus (Fig. 6) and serotoninergic fibres which arise from several raphe nuclei (Fig. 9).

The NA fibres enter the granular layers where they ramify and form a plexus. Branches of this plexus ascend to the molecular layer, where they either give rise to a few radial branches or divide in a T-fashion, just like parallel fibres (Mugnaini 1965). Electrophysiological studies have shown that the NA coeruleocerebellar projection inhibits the firing of Purkinje cells. However, there is evidence suggesting that noradrenalin does not act directly upon the ionic permeability of

the postsynaptic membrane, but enhances the generation of cyclic AMP, which in turn affects the membrane potential as a second messenger.
Histofluorescence (Hökfelt and Fuxe 1969) and immunohistochemical studies (Takeuchi et al. 1982) have shown that the serotoninergic cerebellar afferents, much like the NA ones, ramify in both the granular and molecular layers. The role of the serotoninergic fibres in the cerebellar cortex remains to be established.
The NA and serotoninergic cerebellar afferent fibres are both thin and varicose. Several authors (Palay and Chan-Palay 1974; Palay 1982; Beaudet and Sotelo 1970) have suggested that these fibres rarely form synaptic contacts, and apparently release their neuromediators en passant into the extracellular space without respect to precise and localized postsynaptic structures.
The data concerning the functional activity of the various cerebellar cortical neurons and the neuromediators involved, discussed above, are summarized in Fig. 48c. It may be concluded that:

1. The number of neuromediators in the cerebellar cortex is limited.
2. The identity of the neuromediator(s) present in the mossy fibres, i.e. one of the two main categories of cerebellar afferents, is unknown.
3. Amino acid transmitters figure largely, but neuropeptides play presumably only a very modest role in the cerebellar cortex.
4. Remarkably, all neurons with cell bodies in the cerebellar cortex are inhibitory, except for the granule cells.

As a final comment, it may be stated that the dominance of inhibition just referred to is of great importance for the tempo and mode of operation of the cerebellar cortex. Since all inputs are transmuted into inhibitory actions within maximally two synaptic relays, prolonged reverberation of signals – which might occur in chains of excitatory neurons, some of which possess recurrent collaterals – is impossible in the cerebellar cortex. As Eccles (1973) has put it: an area of the cerebellar cortex, after carrying out some 'computation', is 'clean' within 0.1 s and ready for the next 'computation'.

The *olfactory bulb* is a separate forward extension of the forebrain which represents the first relay station in the olfactory system. This structure is organized into seven layers, concentrically arranged around the bulbar ventricular cavity. The latter is patent in some species, but obliterated in others. These layers are, passing from superficial to deep (Fig. 49a):

1. A layer of olfactory nerve fibres, consisting of densely interwoven, extremely fine axons originating from the olfactory epithelium
2. The glomerular layer, which contains the conspicuous glomeruli – special regions of neuropil in which the terminal arborizations of the olfactory nerve fibres synapse with the dendrites of three types of secondary olfactory neurons, i.e., the mitral, tufted and periglomerular neurons. The two cell types first mentioned lie in deeper zones of the olfactory bulb; the elements last mentioned are small and granular and, as their name implies, they surround the glomeruli
3. The external plexiform layer constitutes a complex meshwork of interlacing axonal and dendritic branches. However, it also contains the perikarya of superficial interneurons and of the relatively large tufted cells
4. The mitral cell layer, consisting of densely packed granule cells, in which the very large somata of the mitral cells are embedded
5. The internal plexiform layer, which in some places cannot be clearly distinguished as a separate zone; it contains ascending dendrites of deep granule cells, axons of mitral and tufted cells, and axons of centrifugal fibres from other regions of the brain. Moreover, in this layer the perikarya of some granule cells and of somewhat large, intermediate short-axon cells (see below) are found
6. The granular layer, consisting of several concentric zones of densely packed granular

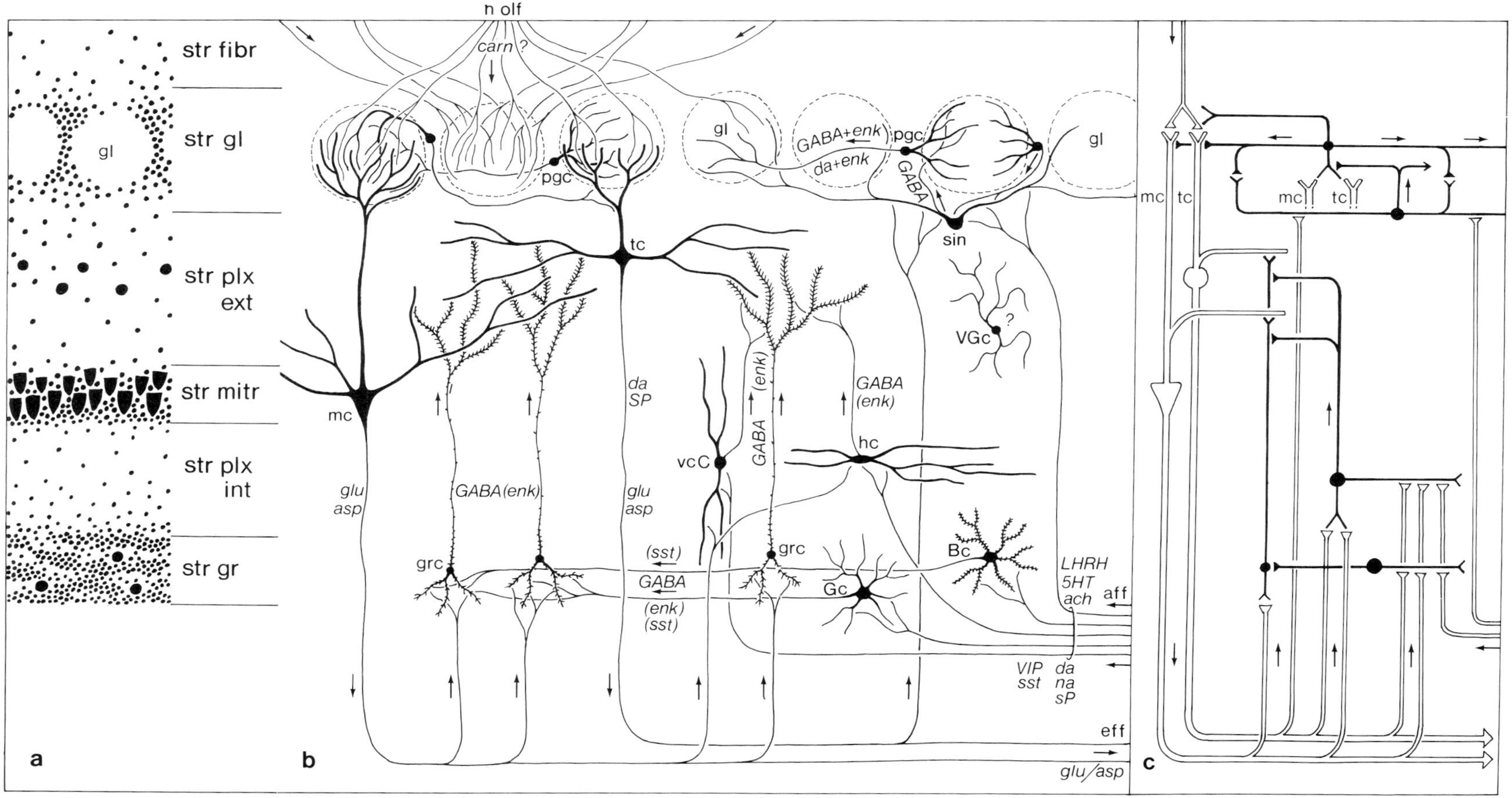

Fig. 49 a–c. The olfactory bulb. **a** Cytoarchitecture; **b** neuronal elements and fibres as observed in Golgi preparations (somewhat simplified); **c** circuit diagram. Inhibitory neurons are in *black* and excitatory ones have been left *unfilled*.

Abbreviations: *aff*, centrifugal afferents; *Bc*, Blanes cell; *eff*, bulbar efferents; *Gc*, Golgi cell; *gl*, glomeruli; *grc*, granule cell; *hc*, horizontal cell; *mc*, mitral cell; *n olf*, nervus olfactorius; *pgc*, periglomerular cells; *sin*, superficial interneuron; *str fibr*, stratum fibrosum; *str gl*, stratum glomerulosum; *str gr*, stratum granulare; *str mitr*, stratum mitrale; *str plx ext*, stratum plexiforme externum; *str plx int*, stratum plexiforme internum; *tc*, tufted cell; *VGc*, Van Gehuchten cell; *vcC*, vertical cell of Cajal

cells, separated from each other by bundles of nerve fibres; larger neurons are scattered in this layer
7. The periventricular zone, which is formed by a layer of ependymal cells or, in animals with obliterated olfactory ventricles, by the remnants of this layer

The microcircuitry of the olfactory bulb is much more intricate than that of the cerebellar cortex, and the number of neuromediators involved in the neuronal interactions is also much larger. Within the frame of this work only a brief survey can be presented of the structural and functional relations of the various bulbar elements and of their chemical signatures. In the preparation of this survey I have relied heavily on the recent review articles of Macrides and Davis (1983) and Halász and Shepherd (1983). The following aspects will be consecutively discussed: (a) the olfactory projection, (b) the roles of the periglomerular and granule cells, (c) the centrifugal fibres and their targets, and (d) the relationships of the deep and superficial interneurons (Fig. 49 b, c).

The *olfactory projection* is constituted by the primary olfactory elements and the second order olfactory projection neurons, i.e. the mitral and tufted cells. The receptive element of the olfactory apparatus is represented by slender bipolar cells situated in the mucosa of the nasal cavity. These elements give rise to extremely fine (0.2–0.4 μm), unmyelinated axonal processes, which carry the olfactory impulses directly to the brain. Upon entering the olfactory bulb the axons of the peripheral olfactory elements interlace in a most complex fashion and terminate by free arborizations in the glomeruli. In these spherical neuropil configurations the synaptic contacts between the terminals of the incoming fibres and the dendrites of the mitral and tufted cells are established. Electrophysiological studies have shown that the peripheral inputs to the olfactory bulb are excitatory. The primary olfactory elements and their processes have been found to contain high concentrations of the dipeptide carnosine (β-alanyl-L-histidine), as well as a specific olfactory marker protein (mol.wt. = 20000). However, the exact roles played by these substances remain to be elucidated.
The large mitral cells and the somewhat smaller tufted cells both have one main or primary dendrite entering a glomerulus and several secondary or accessory dendrites branching in the external plexiform layer. The main dendrites enter into synaptic contact with olfactory nerve fibres, as well as with axons and dendrites of periglomerular cells. The synaptic relations of the secondary dendrites will be dealt with below.
The axons of the mitral and tufted cells pass radially through the deeper layers of the olfactory bulb and then turn caudally to enter the telencephalon proper. During their caudal course through the bulb they emit numerous collaterals which contact granule cells, deep interneurons and, in the case of the tufted cells, also superficial interneurons. The main axons of the mitral and tufted cells gain myelin sheaths and become grouped together in bundles, which constitute the secondary olfactory projection. The fibres of this projection transmit olfactory signals to several telencephalic regions, including the anterior olfactory tubercle, the prepiriform cortex and certain parts of the amygdaloid complex (see e.g. Shipley and Adamek 1984). Several lines of evidence strongly suggest that glutamate and/or aspartate function as excitatory neurotransmitters in the efferents of the olfactory bulb to its principal target, i.e. the prepiriform cortex. As regards the tufted cells, glutamate/aspartate may well be the transmitter of only the more deeply situated elements of that category. Immunohistochemical studies have shown that the great majority of the external tufted cells and many of the middle tufted cells contain dopamine and that numerous external tufted cells contain substance P. Macrides and Davis (1983) found that the incidence of substance P-containing external tufted cells was less than the incidence of dopaminergic external tufted cells, but

they estimated that both are sufficiently high to indicate that in some of the external tufted cells dopamine and substance P co-exist. From the foregoing it appears that the tufted cells constitute a neurochemically heterogeneous population. The deep tufted cells presumably contain the amino acids glutamate or aspartate, whereas the more superficial ones may contain dopamine, or substance P, or both.

The *periglomerular cells* and the *granule cells* have several features in common. Both maintain reciprocal dendrodendritic synaptic contacts with mitral and tufted cells and both have these two cell types as their main targets. However, the periglomerular and granule cells also show marked differences, the most salient of which is that the former are regular short-axon cells, whereas the latter are amacrine, i.e. axonless elements.

Because of their position and small size, the periglomerular cells are often designated as superficial granule cells. Their dendrites enter glomeruli, where they receive impulses from olfactory nerve terminals and also enter into synaptic contact with the interglomerular ramifications of the main dendrites of mitral and tufted cells. The ultrastructural features of these dendrodendritic contacts suggest that mitral and tufted cell dendrites are excitatory to the periglomerular dendrites, and that these latter dendrites are inhibitory to mitral and tufted cell dendrites. The axons of the periglomerular cells pass to nearby glomeruli and provide for inhibitory intraction between glomeruli, analogous to the 'surround' inhibition in the retina. Other axonal branches of periglomerular cells terminate on the primary dendritic shafts of mitral and tufted cells, on which they form inhibitory synapses. Thus, the periglomerular cells exert an inhibitory influence on the mitral and tufted cells in two different places and in two different ways, namely via interglomerular dendrodendritic synapses and via subglomerular axodendritic synapses. The population of periglomerular cells is neurochemically heterogeneous. Immunohistochemical evidence indicated that many of these elements contain GABA, that many contain enkephalin, and that a smaller number contain dopamine. Two-colour immunocytochemistry has shown that the GABA- and dopamine-containing periglomerular cells constitute separate subpopulations (Mugnaini et al. 1984a, b). Because the incidence of enkephalin-immunoreactive cells is extremely high, Macrides and Davis (1983) considered it likely that enkephalin co-exists with GABA or dopamine in some periglomerular cells.

The amacrine (deep) granule cells have several short basal dendrites and a long, peripherally coursing dendrite which ramifies in the external plexiform layer among the secondary dendrites of mitral and tufted cells. The branching distal portions of these long granule cell dendrites are densely studded with conspicuous gemmules (Fig. 49b). The granule cells receive axodendritic synapses from recurrent collaterals of mitral and tufted cells, and there is also a heavy input to the granule cells from the telencephalon proper (see below).

The mitral/tufted cell secondary dendrites and the gemmules of the peripheral granule cell dendrites are richly interconnected by dendrodendritic synapses, organized in reciprocally oriented pairs. There is ultrastructural as well as electrophysiological evidence indicating that the mitral/tufted dendrite-to-gemmule synapses are excitatory, whereas the adjacent gemmule-to-mitral/tufted dendrite synapses are inhibitory. These peculiar synapse pairs constitute extremely short inhibitory pathways from mitral cell to mitral cell or from tufted cell to tufted cell. The loops formed by the collaterals of mitral and tufted cells, and by granule cells provide for surround inhibition of the elements first mentioned.

Extensive electrophysiological, immunohistochemical and biochemical evidence strongly suggests that GABA is an important neurotransmitter of the amacrine granule cells. Enkephalin immunoreactivity has been found in a small percentage of these elements. Co-existence of GABA and enkephalin in granule cells has not been reported so far.

The olfactory bulb receives a very strong *centrifugal projection,* which originates from a variety of sources. According to the data compiled by Macrides and Davis (1983), this projection includes the following neuromediator-specified components:

(a) Enkephalinergic fibres presumably originating from the anterior olfactory nucleus and the precommissural hippocampus.
(b) VIP-containing fibres arising from the anterior olfactory nucleus, the piriform cortex and some parts of the amygdaloid complex.
(c) LHRH-containing fibres originating from the anterior olfactory nucleus, the olfactory tubercle, the precommissural hippocampus and the medial septum-diagonal band complex. A certain proportion of these fibres pass to the most superficial zone of the olfactory bulb, then leave the brain, cross the cribriform plate and extend into the nasal epithelium. These fibres have been shown to form part of the nervus terminalis.
(d) SST-containing fibres originating from the piriform cortex and presumably also from the nuclei of the diagonal band.
(e) Substance P-containing fibres arising from the mesencephalic raphe nuclei.
(f) Cholinergic fibres, which originate mainly from the ventral nucleus of the diagonal band.
(g) Serotoninergic fibres arising from the mesencephalic raphe nuclei. Because in a small percentage of the neurons in these nuclei serotonin and substance P are co-localized, it might well be that some bulbofugal fibres also contain both of these neuromediators.
(h) Noradrenergic fibres originating from the locus coeruleus also contribute to the bulbofugal projections.

The various neuromediator-specified bulbopetal fibre contingents show marked differences in their laminar distribution within the olfactory bulb. Thus, the substance P, somatostatin and noradrenergic fibres are concentrated in the granular layer, LHRH-containing fibres attain their greatest density in the granular and glomerular layers, cholinergic fibres are found predominantly in the internal plexiform and glomerular layers, and serotoninergic fibres are found throughout the deep layers but are prevalent in the glomerular layer. Electron-microscopical studies have shown that the centrifugal bulbar afferents terminate predominantly on interneurons. The centrifugal fibres which are distributed to the glomerular zone confine themselves to the periglomerular regions. The nature of the action of most of the contingents of the neuromediator-specified bulbopetal fibres is unknown. However, the cholinergic and noradrenergic fibres are considered to exert an excitatory influence on their target neurons.
It has already been discussed that the activity of the bulbar output neurons (i.e. the mitral and tufted cells) is regulated by two types of small inhibitory interneurons, the interglomerular or *superficial granule cells* and the *deep amacrine granule cells* (Fig. 49 b, c). In addition to these small interneurons, the olfactory bulb contains several types of larger interneurons, all of which together form a link between the centrifugal bulbar afferents and the small interneurons. As indicated by Macrides and Davis (1983), these larger interneurons can be categorized into deep granule cells and periglomerular cells.
The larger deep interneurons include the Blanes cells, the Golgi cells, the vertical cells of Cajal and the horizontal cells. The Blanes and Golgi cells are multipolar neurons, the dendrites and axons of which are confined mainly to the granular layer. Their main difference is that the dendritic trees of the former, unlike those of the latter, are densely covererd with spines. The vertical cells of Cajal and the horizontal cells are located most commonly in the internal plexiform layer, their names referring to the orientation of their dendritic trees (Fig. 49 b). The axons of these elements are thought to form axodendritic synapses with the distal parts of the long, peripherally directed dendrites of the granule cells.

The larger deep interneurons as a group are thought to receive excitatory impulses from centrifugal fibres and from mitral and tufted cell axon collaterals, and to inhibit granule cell activities, thus ultimately disinhibiting the projection neurons, i.e. the mitral and tufted cells.

As regards the neuromediators present in the deep interneurons, Halász and Shepherd (1983) and Macrides and Davis (1983) cautiously suggested that GABA may act as a transmitter in a few of these elements. However, Mugnaini et al. (1984a, b) recently presented immunohistochemical evidence indicating that most if not all of the deep interneurons are GABA-ergic. Macrides and Davis reported that some deep interneurons show enkephalin or somatostatin immunoreactivity.

The larger superficial interneurons are, according to the description of Macrides and Davis (1983), located in the subglomerular zone and in the adjacent superficial zone of the external granular layer. Their dendrites branch among and around glomeruli, while their axons ramify predominantly in the periglomerular regions of the glomerular layer. Electron-microscopical studies indicate that they receive asymmetrical (i.e. 'excitatory-type') axodendritic and axosomatic synapses from tufted cell collaterals and centrifugal fibres as well as symmetrical (i.e. 'inhibitory-type') axodendritic and axosomatic synapses from periglomerular cells, and that they in turn form symmetrical axosomatic and axodendritic synapses with periglomerular cells. On the basis of these ultrastructural features it is reasonable to assume that the superficial interneurons, by inhibiting inhibitory interglomerular elements, exert a disinhibitory influence on mitral and tufted cells. Mugnaini et al. (1984a, b) suggested that the superficial interneurons may contain GABA.

Finally, brief mention may be made of the presence of the so-called Van Gehuchten cells in the olfactory bulb. These elements, which have diffusely branching dendrites, are situated in the internal plexiform layer (Fig. 49b). Their role in the circuitry of the bulb is entirely obscure, as is the nature of their neuromediator(s).

The microcircuitry of the olfactory bulb is summarized in Fig. 49c. Leaving details aside, it may be stated that the olfactory projection is constituted by two consecutive sets of excitatory projection neurons, namely the peculiar neurosensory cells, the somata of which are situated in the nasal mucosa, and the mitral and tufted cells, which represent the bulbar output neurons. The activity of these bulbar output elements is regulated by two categories of small, inhibitory interneurons, the periglomerular cells and the amacrine (deep) granule cells. These elements provide both for local and surround inhibition. The output of the olfactory bulb is also strongly influenced by large numbers of excitatory centrofugal afferent fibres. These fibres terminate on various types of medium-sized, inhibitory short-axon cells, which can be categorized into deep and superficial groups. The deep short-axon cells inhibit the inhibitory granule cells, whereas the superficial short-axon cells inhibit the inhibitory periglomerular cells. Through the mediation of these sets of inhibitory interneurons the excitatory bulbopetal fibres exert a disinhibitory influence on the bulbar output neurons.

If we compare now the cerebellar cortex (Fig. 48) with the olfactory bulb (Fig. 49), the following similarities and differences present themselves:

1. The number of cell types and the number of neuromediators present in the olfactory bulb are much larger than in the cerebellar cortex.
2. In both structures the identity of the neuromediator(s) present in some of its neuronal elements is unknown. In the olfactory bulb this holds for the primary olfactory afferents and for the enigmatic Van Gehuchten cells.
3. In both structures amino acid neurotransmitters figure largely, but the number of neuropeptides found in the olfactory bulb far exceeds that found in the cerebellar cortex. Most of the neuropeptides in the olfactory bulb are 'imported' by centrifugal fibres, but

at least a certain proportion of some of the bulbar types, for instance the tufted cells and the periglomerular cells, are peptidergic.
4. Processes of the all-pervading noradrenergic and serotoninergic cell groups in the brain stem penetrate the cerebellar cortex as well as the olfactory bulb.
5. Monoaminergic neuronal cell bodies do not occur in the cerebellar cortex, but in the olfactory bulb, subpopulations of the periglomerular and tufted cells are dopaminergic.
6. Axonless amacrine cells are lacking in the cerebellar cortex, but abundant in the olfactory bulb. The same holds true for dendrodendritic synapse complexes. In the olfactory bulb these remarkable structures form pairs in which an excitatory glutamatergic (or aspartatergic) action is immediately reciprocated by an inhibitory GABA-ergic action.

19. The elucidation of the question of where and to what extent non-synaptic chemical transmission plays a role in the central nervous system is of paramount importance for our understanding of the functioning of that organ.

Until recently it was generally assumed that chemical interneuronal communication in the central nervous system occurs mainly, if not exclusively, via synapses, i.e. morphologically specialized, intimate contacts between two neuronal elements. The presynaptic component of the synaptic junction is usually formed by a synaptic knob, a terminal swelling of an axonal branch (Fig. 50a). These so-called bouton terminaux are characterized by the presence of accumulations of small vesicular organelles which contain the neuromediator substance. The presynaptic and postsynaptic membranes, which are separated by an approximately 20-nm-wide synaptic cleft, show both morphological differentiations. On the presynaptic side, diffusely outlined dense patches protruding from the membrane into the cytoplasm of the terminal are observed. These local thickenings are arranged in a configuration known as the presynaptic vesicular grid. The postsynaptic membrane is characterized either by a continuous local membrane thickening or by the presence of small thickened patches. Freeze-etching studies have revealed that the external surface of the postsynaptic membrane is studded with tiny protrusions. These may well represent the exposed ends of postsynaptic receptor sites. Synaptic impulse transmission essentially involves the following sequence of steps: Upon the arrival of an impulse the synaptic vesicles move toward the surface, guided by the presynaptic vesicular grid. Their membrane fuses with the axolemma, after which they discharge their content into the synaptic cleft, a process which is called 'exocytosis' (Fig. 50b). The neuromediator molecules diffuse via the extracellular fluid in the synaptic cleft toward the postsynaptic membrane, where they bind with specific receptor sites. This interaction with the receptors elicits the opening of particular ionic channels and thereby the development of a local membrane depolarization or hyperpolarization. The synaptic cleft between the presynaptic and postsynaptic membranes contains a material of intermediate density which impedes the diffusion of neuromediator molecules away from the functional area into the general extracellular space.
It is important to note that not only the classical neuromediators (acetylcholine, monoamines, amino acids), but also neuropeptides may be involved in the typical synaptic transmission or neurocrine secretion just outlined. This may be inferred from the fact that many of these substances, including CCK (Conrath-Verrier et al. 1984; Takagi et al. 1984; Baali-Cherif et al. 1984), VIP (Gray et al. 1984), neurotensin (Ibata et al. 1984), SST

(DiFiglia and Aronin 1984a), LHRH (Silverman 1984), vasopressin (Buijs and Swaab 1979), and enkephalin (Somogyi et al. 1982; Moss and Basbaum 1983; DiFiglia and Aronin 1984b; Armstrong et al. 1984) have been demonstrated by means of immunoelectron microscopy to be present in terminals forming typical synaptic junctions.

During the past decade the concept has developed that in the central nervous system, apart from the classical synaptic transmission alluded to above, another form of chemical interneuronal communication occurs, a communication which is non-synaptic and involves a paracrine secretion process. Before discussing the observations which argue in favour of or against this concept, as well as the theoretical considerations to which it has led, it may be well to present an outline of its essential features.

It is known that the axons of many monoaminergic and peptidergic neurons give rise to profusely branching, thin, unmyelinated axons, and that these axons exhibit varicosities not only in their terminal areas, but throughout much of their extent (Fig. 50c). Ultrastructural observations indicated that these varicosities, although they contain large numbers of synaptic vesicles, frequently do not participate in the formation of synaptic junctional complexes. These observations led to the supposition that these non-synaptic varicosities release the neuromediator contained in their vesicular organelles by exocytosis freely into the extracellular space (Descarries et al. 1975, 1977; Beaudet and Descarries 1978; Chan-Palay 1978). After their release the neuromediator molecules are considered to travel by way of the intercellular fluid, to act on more distant as well as close target neurons that are provided with the appropriate receptors (Fig. 50d). By this non-synaptic mode of action particular sets of monoaminergic or peptidergic elements would be able to exert a diffuse tonic influence on vast neuronal assemblies, rather than

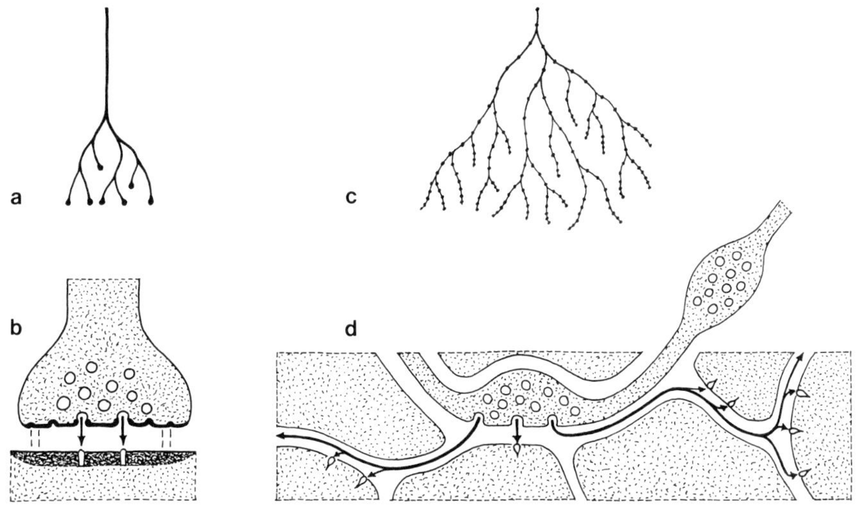

Fig. 50 a–d. Diagrammatic representation of: **a** an axonal branch with terminal knobs; **b** a classical synapse in which a presynaptic element influences a single postsynaptic element by means of a neurocrine secretory process; **c** an axonal branch showing numerous varicosities, both terminal and non-terminal; **d** an axonal varicosity which, by means of nonsynaptic, paracrine secretion of a neuromediator, influences several other neuronal elements. The width of the extracellular space has been exaggerated. The *arrows*, which indicate the flow of neuromediator molecules, point to membrane receptors. Further explanation in the text

controlling individual targets on a one-to-one basis. Moreover, the time course during which these influences were exerted could be considerably longer than the few milliseconds of classical synaptic transmission.

The evidence adduced in favour of the concept just discussed, as well as the arguments against it, may be summarized as follows:

1. Beaudet and Descarries (1978) pointed out that the fact that Hökfelt (1968) and Tennyson et al. (1974) remained unable to demonstrate the presence of synaptic junctions on the dopamine-containing boutons in the striatum, is in harmony with findings on noradrenergic and serotoninergic endings (see below), and seems to indicate that a relative paucity of synaptic junctions might be a feature common to all types of monoaminergic terminals. It should be emphasized, however, that, contrary to the initial findings of Hökfelt (1968) and Tennyson et al. (1974), several later authors, among them Groves (1980), Pickel et al. (1981) and Arluison et al. (1984), have demonstrated that dopaminergic terminals in the striatum do form typical synaptic contacts, although Arluison et al. reported that many of these synapses completely lack a post-synaptic density. It is also relevant in this context that typical dopaminergic synapses have been demonstrated in several other brain regions, including the lateral septum (Onteniente et al. 1984) and the supraoptic and paraventricular nuclei (Buijs et al. 1984).

2. Descarries et al. (1975) labelled serotoninergic axons and terminals in the superficial part of the cortex of the rat by superfusion of the cortical surface with tritiated serotonin followed by fixation with glutaraldehyde and electron-microscopic autoradiography. They found that in the frontoparietal cortex only a very small fraction of the serotonin varicosities exhibit the membrane differentiations of typical synaptic terminals. Extensive sampling in serial thin sections revealed junctional complexes in only 5% of labelled boutons, as opposed to 50% of unlabelled nerve endings in the surrounding neuropil. A corresponding study on the morphology of the noradrenergic terminals in the cortex yielded similar results. Using tritiated noradrenalin as a marker, Descarries et al. (1977) observed that likewise, only a very low proportion (less than 5%) of the noradrenergic varicosities in the frontoparietal cortex are engaged in genuine synaptic relationships.

The results of Descarries et al. (1975, 1977) just reviewed have been challenged by Molliver and colleagues (1982). These authors studied the ultrastructural features of noradrenergic and serotoninergic axons by means of DBH and serotonin immunohistochemistry and found that many (about 50%) of the DBH-positive and serotonin-positive terminals in the cerebral cortex of the rat clearly form conventional synaptic complexes. They considered it likely that in the varicosities which lacked synaptic contacts these membrane specializations were simply out of the section plane. Using the false transmitter 5-hydroxydopamine, Molliver et al. (1982) also demonstrated that in the cortex of newborn and very young rats monoaminergic fibres form numerous synapses with specialized appositions. Taking their findings together, Molliver and colleagues arrived at the conclusion that in the cortex monoaminergic axons communicate with target neurons via typical synapses.

3. Chan-Palay (1978) studied the localization of serotonin and noradrenaline in the paratrigeminal nucleus (a small cell mass sandwiched between the fibres of the restiform body and the descending tract of the trigeminal nerve) at the electron-microscopic level by autoradiography following administration of tritiated serotonin and noradrenaline. She reported that of the monoaminergic population of labelled axons, 60% are synaptic and less than 40% are non-synaptic, and that this proportion is the same for serotonin as for noradrenaline. Both the serotoninergic and the noradrenergic axons were found to form heterogeneous populations. On the basis of the shape, structure and content of their varicosities, three different types of non-synaptic serotoninergic fibres and two types of non-

synaptic noradrenergic axons were described. Chan-Palay (1978) emphasized that monoaminergic fibres in several other parts of the brain are also provided with non-synaptic varicosities and quoted previous publications in which she has reported the presence of such structures in the cerebellar cortex, the cerebellar and vestibular nuclei, the inferior olive, the raphe nuclei and the nucleus paragigantocellularis lateralis.

Apart from the observations summarized above, the following pieces of indirect and circumstantial evidence have been advanced in support of the concept that non-synaptic exocytotic release of neuromediators occurs in central nervous tissue:

1. Serotoninergic axons emanating from the nucleus raphes dorsalis penetrate the ependymal lining of the ventricle to form a vast supra-ependymal plexus (Fig. 9). The fibres of this plexus contain numerous non-synaptic varicosities. Chan-Palay (1977, 1982) has presented experimental evidence suggesting that serotonin released by these non-synaptic varicosities penetrates into the brain via certain ependymal cells, thus gaining access to the intercellular space of the subependymal nervous tissue. She pointed out that the supra-ependymal serotoninergic axons are well placed to respond to physical and chemical changes in the CSF environment such as changes in pressure, ciliary CSF flow, ionic balance and concentration of amine metabolites or their precursors, and conjectured that the response of these serotonin-containing axons may affect such global brain functions as sleep-wake cycles, diurnal rhythms, and transport of various substances and metabolism.
2. Release of neuromediators from axon terminals devoid of synaptic membrane specializations has long been known to be the rule in the peripheral autonomic nervous system (Dismukes 1977; Beaudet and Descarries 1978; Moore and Bloom 1979). Interestingly, Jan and Jan (1983) recently demonstrated that in the peripheral system LHRH can diffuse for tens of micrometers before activating receptors on sympathetic neurons.
3. From a theoretical point of view, the concept is plausible and attractive for the following reasons (Roubos and Buma 1984; Buma and Roubos 1985): (a) It suggests a very economical use of the available space in the central nervous system, there being no need for voluminous fibre paths and synaptic structures for this mode of interneuronal communication. (b) It provides a satisfactory explanation for the remarkably large number of neuromediators demonstrated in the central nervous system. (c) Although the various neuromediators all travel in the same extracellular space, a certain degree of specificity or 'privacy' is maintained, since all of these messengers address only 'their own' specific receptors.

However plausible and attractive the idea of interneuronal communication by the non-synaptic release of neuromediators may be, it should be appreciated that the papers of Descarries et al. (1975, 1977), Beaudet and Descarries (1978) and Chan-Palay (1978) reviewed above do not contain any direct and convincing evidence for this concept. No analyses of complete serial sections have been reported showing that the varicosities studied, at the moment of fixation, indeed did not show any trace of synaptic membrane specializations. Rather, the only study of this type, i.e. that of Groves (1980) on dopaminergic terminals in the striatum, revealed that all of the structures analyzed made a specialized synaptic contact. Convincing evidence is to be obtained only by a technique with which the process of exocytosis can be directly visualized. Fortunately, such a technique has recently become available. Buma et al. (1984) have shown that during tissue incubation in Ringer's solution containing tannic acid, exocytosis proceeds, but that the exteriorized contents of the secretory vesicles are immediately fixed by the tannic acid and do not diffuse away into the extracellular space. After conventional fixation the secretion products, the electron density of which is in addition considerably enhanced by the tannic acid, can be directly visualized. Preliminary studies with this so-called TARI (Tannic Acid Ringer Incubation) technique (Buma et al. 1984; Roubos and Buma 1984; Buma and Roubos 1985) have revealed that non-synaptic release sites are present in various parts of the central nervous system of invertebrates as well as in the area postrema of the rat. It is to be hoped that with the aid of this technique the questions of where and to what extent non-synaptic chemical transmission plays a role in the central nervous system can be tackled. As Schmitt (1984) recently stated, the solution of this problem may have a wide-ranging impact on basic neurobiology and its clinical applications. It is to be expected that the variety of neuropeptides present in the central nervous system play an important role in non-synaptic communication. However, one should not lose sight of the well-established fact that, as reviewed earlier in this section, terminals of many types of peptidergic neurons do form conventional synapses.

20. Degeneration of certain subsets of neuromediator-specified neurons has been reported for several neurological disorders.

A full discussion of this highly important subject is beyond the scope of the present book. Hence, I confine myself to an elementary survey of some major points.

1. In senile dementia of the Alzheimer's type (SDAT) there is a considerable loss of the large cholinergic neurons in the basal telencephalon, particularly in the nucleus basalis of Meynert (Whitehouse et al. 1981, 1982; McGeer 1984; McGeer et al. 1984b; Figs. 2, 51, 52). Whether this cell loss is due to a primary lesion of the perikarya or rather represents a retrograde phenomenon, following degeneration of the widespread cortical pro-

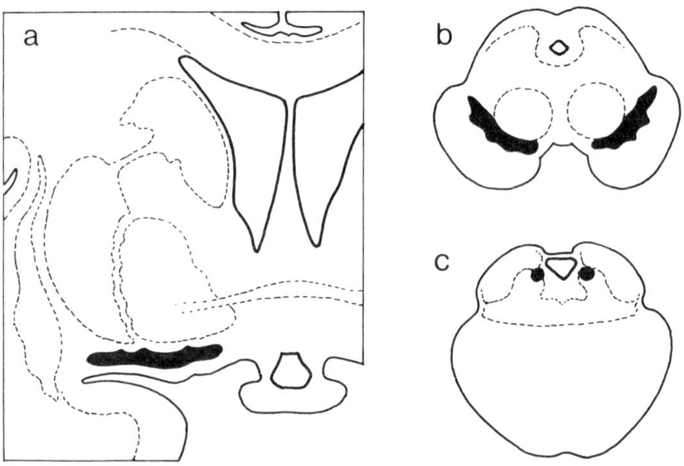

Fig. 51 a–c. Key diagrams indicating the positions of the areas in the human brain shown in Figs. 52, 53 and 54; the relevant areas are shown in *black*. **a** Frontal section through the basomedial telencephalon, illustrating the region containing the nucleus basalis of Meynert; **b** frontal section through the mesencephalon, showing the substantia nigra; **c** frontal section through the most rostral part of the rhombencephalon, demonstrating the position of the locus coeruleus

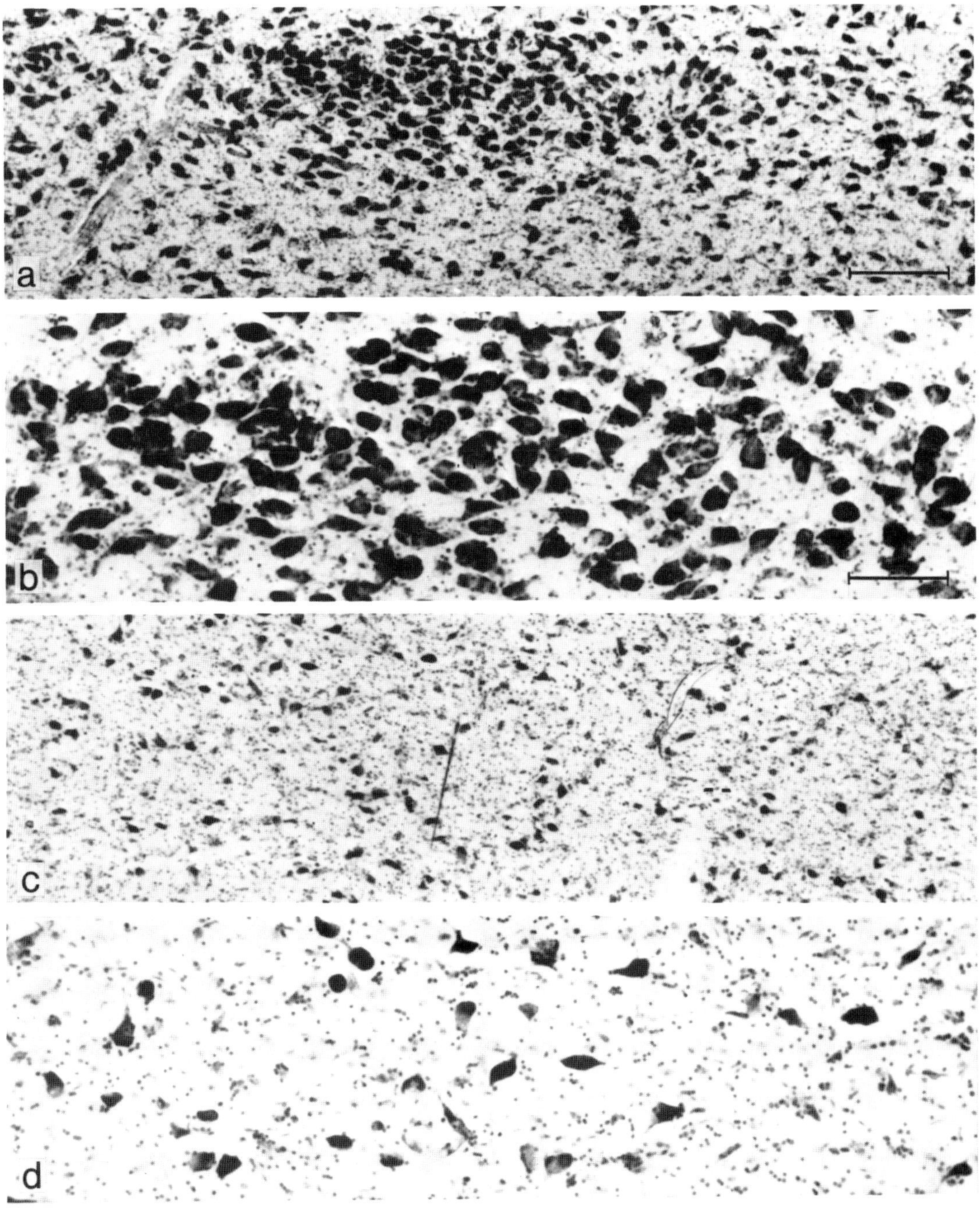

Fig. 52 a–d. Photomicrographs from the central part of the nucleus basalis of Meynert (nbM), of an age-matched control (**a** and **b**) and a patient with Alzheimer's disease (**c** and **d**). **a** and **b** show the normal density of nbM neurons in the control at low and high magnification respectively. **c** and **d** demonstrate the profound loss of neurons in the patient, again at low (**c**) and high (**d**) magnification. *Scale bar* is 200 μm in **a** and **c**, and 100 μm in **b** and **d**. (From Whitehouse et al. 1982, by kind permission of the authors and the publishers of Science)

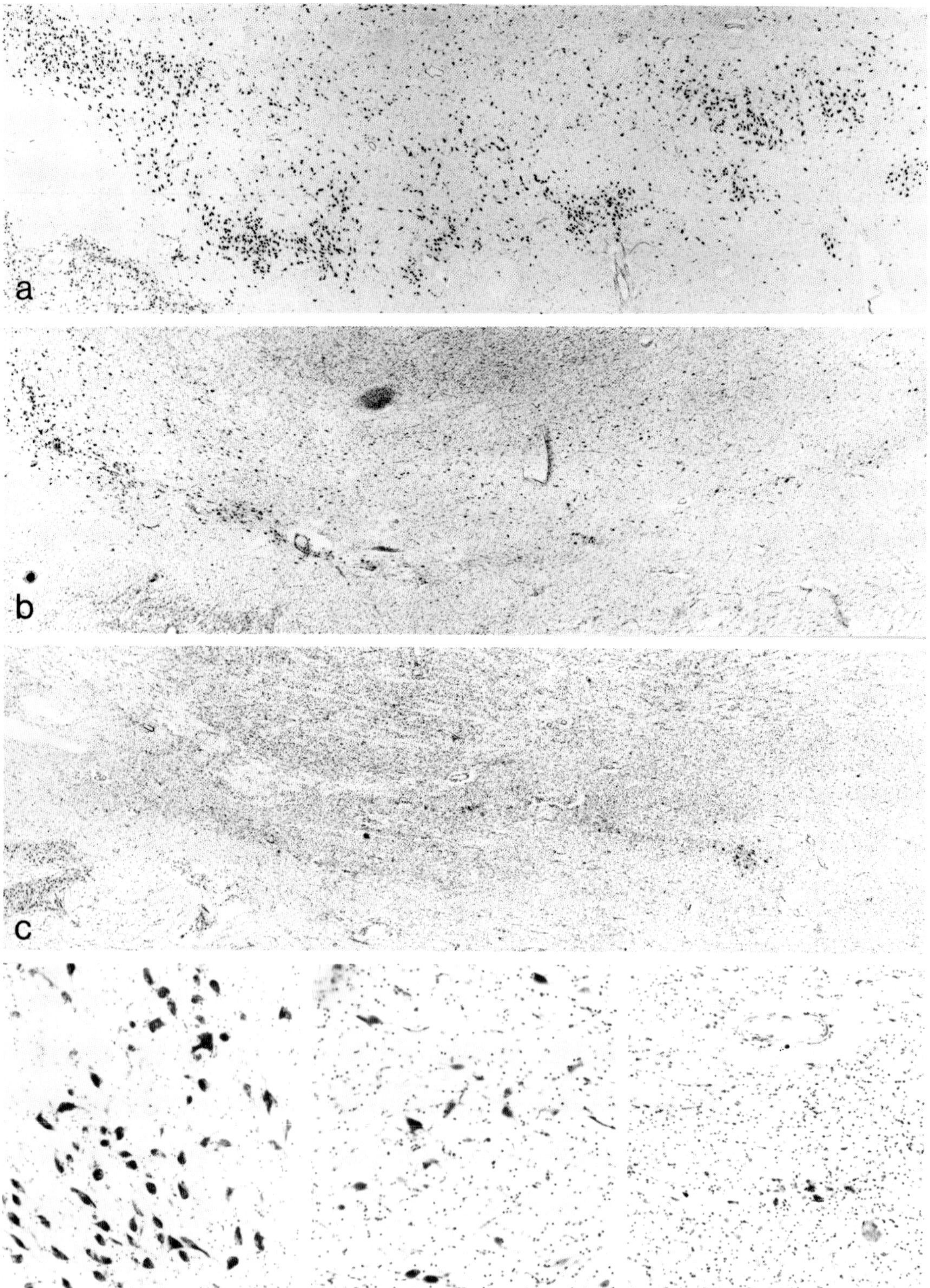
a
b
c
d
e
f

jections and terminals of the nucleus basalis neurons, is unknown at present. The latter possibility should not be ruled out. Price et al. (1982) have made the intriguing suggestion that the characteristic senile plaques found in the cortex of SDAT patients incorporate degenerating cholinergic endings, and Perry et al. (1982) presented evidence indicating that in SDAT the degeneration of nerve endings my be greater than that of the cells.

It is worthy of note that in SDAT several other centres, including the noradrenergic locus coeruleus (see e.g. Mann et al. 1984), are generally also affected, while at the same time such cholinergic cell groups as the large striatal local circuit neurons (Fig. 36) and the somatic and visceral motoneurons in the brain stem and spinal cord are spared.

2. Parkinson's disease is characterized by a progressive loss of dopaminergic neurons in the compact part of the substantia nigra (Figs. 51, 53), with consequent degeneration of the nigrostriatal projection. This degeneration is believed to be responsible for the akinesia and rigidity associated with this disorder (Hornykiewicz 1978). The dopaminergic elements in the adjacent area tegmentalis ventralis have also been shown to be affected in Parkinson's disease (Bogerts et al. 1983).

In the brains of many patients dying with Parkinson's disease a significant loss of neurons has also been found in two non-dopaminergic structures, namely the noradrenergic locus coeruleus (Figs. 51, 54) and the cholinergic nucleus basalis of Meynert. It has been suggested that the degeneration in the locus coeruleus may be related to depression, which is a major symptom in Parkinson's disease (Cash et al. 1984). Loss of neurons in the nucleus basalis is not surprising, in view of the high incidence of dementia in parkinsonian patients (Lieberman et al. 1979; Mayeux and Stern 1983). McGeer (1984) has shown that, contrary to many other centres of the brain, a striking loss of neurons occurs with age in the cholinergic nucleus basalis and in the dopaminergic substantia nigra pars compacta. He pointed out that these are the centres which are struck by SDAT and Parkinson's disease, affections that are sharply age dependent, and suggested that these diseases become overt when the number of cells in the nucleus basalis or the substantia nigra drops below a certain critical level. McGeer also expressed the opinion that the pathological processes underlying the two disorders must be somehow related; he pointed to the high coincidence of dementia and Parkinson's disease, and particularly to the remarkable fact that these two diseases frequently manifest themselves as one single complex among the autochthonous population of the island of Guam.

It is worthy of note that in the substantia nigra of patients dying with Parkinson's disease not only a loss of dopaminergic neurons, but also a reduction in the concentrations of substance P (Tenovuo et al. 1984) and M-enkephalin (Llorens-Cortes et al. 1984) has been found. Both of these peptides occur in fibres constituting the striatonigral projection (Fig. 36). It remains to be determined whether these reductions are primary or secondary to the degeneration of the dopaminergic nigrostriatal projection.

3. Chronic tardive dyskinesia, a disturbance occurring after prolonged (months to years) treatment with neuroleptics, may be the re-

◀ **Fig. 53 a–f.** The substantia nigra in normal and diseased conditions, as shown in frontal sections. **a** Substantia nigra of a healthy person, age 24; **b** the same region in a 63-year-old patient dying with paralysis agitans (6 years' duration). **c** The same region in a 54-year-old patient dying with postencephalitic parkinsonism (16 years' duration). **d, e, f** Details at higher magnification of **a, b** and **c** respectively. In **f** only some pigmented remnants are visible. Magnifications of **a, b** and **c**: ×18; **d, e** and **f**: ×80. (Kindly provided by Professor A. Hopf)

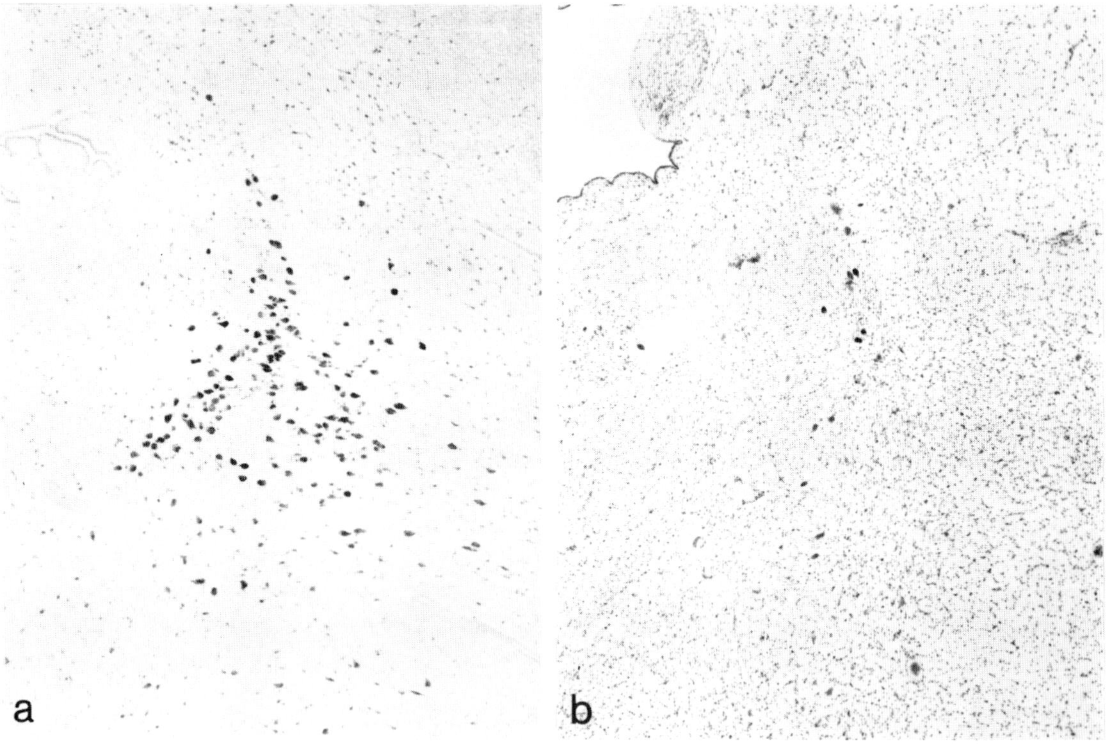

Fig. 54 a, b. The locus coeruleus in normal and diseased conditions, as shown in frontal sections. **a** Locus coeruleus of a healthy person, age 30 years; **b** the same region in a 46-year-old patient dying with postencephalitic parkinsonism (4 years' duration). Magnification of **a** and **b**: ×35. (Kindly provided by Professor A. Hopf)

sult of drug-induced damage to a subpopulation of striatal GABA-containing neurons (Fibiger and Lloyd 1984).

4. Huntington's chorea is a rare, autosomal, dominantly inherited degenerative disease of middle age, characterized by involuntary choreatic movements and dementia. The degenerative changes in this disorder involve neuron loss in the cerebral cortex and a marked degeneration and loss of neurons in the nucleus caudatus and the putamen. The nigrostriatal dopaminergic projection remains intact in Huntington's chorea, but there is evidence indicating that within the striatum there is a loss of neurons containing GABA, acetylcholine, substance P and L-enkephalin (Marshall et al. 1983; Hornykiewicz 1984). Although patients with Huntington's chorea generally develop severe dementia, there is no reduction of cortical ChAT activity and no significant loss of neurons from the nucleus basalis (Arendt et al. 1983; Clark et al. 1983).

CHAPTER 5

'New' Entities in the Central Nervous System: The [Paracrine?] Core and Its Adjuncts

In the preceding chapter I have attempted to characterize the present state of chemical neuroanatomy and its relations to classical neuroanatomy in twenty statements, and I have added shorter or longer comments to each of these. In this chapter I will develop a new view on the organization of certain parts of the neuraxis. In doing so I will not hesitate to go clearly beyond well-established data at some points. Thus, the character of this chapter is speculative and differs considerably from that of the preceding ones. Nevertheless, the format of statements followed by comments will be maintained, and the numbering of the statements will also be continued. An important advantage of this mode of recording is that it facilitates cross-reference of the various statements.

21. The centres and pathways which contain a particularly large number of neuromediators together constitute a readily recognizable entity (Table 6, Fig. 55).

After having written Chap. 3, I collected all the data on the localization of neuromediators in centres and fibre tracts gathered from the literature in a large table. On the left side of that table the names of more than 150 different structures, beginning with the glomerular layer of the olfactory bulb and ending with the spinal anterior horn, were placed in a column. At the top of the table the names of the 25 neuromediators discussed were placed side by side, beginning with acetylcholine and ending with angiotensin II. In the matrix thus prepared the presence of a particular neuromediator in a particular structure was recorded, using separate symbols for neuromediator present in perikarya (o), neuromediator present in a considerable number of perikarya (●), neuromediator present in fibres and/or terminals (+), and neuromediator present in a large number of fibres and/or terminals (**+**). Needless to say, the choice between using the symbols o or ● and + or **+** in many instances could be made only in an arbitrary way. The table was not expected to present a complete overview of the neuromediators present in all of the structures listed (see caveats 1 and 2 at the beginning of Chap. 4). However, it was hoped that the table would reveal some trends, and so it did. It showed first of all that the total numbers of neuromediators and the neuromediator profiles of the various grisea and fibre tracts demonstrate considerable differences. (This finding has already been recorded in statements 4, 11 and 15 and in Tables 2, 4 and 5 respectively.) I now decided to bring together all centres and fibre tracts containing a particularly large number of neuromediators (≥ 10 and ≥ 7 respectively). The result is shown in Table 6. Even if we take into consideration that in our analysis nine 'classical' neuromediators and 16 neuropeptides were involved, it is nevertheless quite remarkable that practically all of the structures brought together contain a variety of neuropeptides. The 'neuromediator-rich'

Table 6. Neuromediator profiles and total numbers of neuromediators present in a number of "neuromediator-rich" centres and fibre systems. Presence of estradiol- and testosterone-concentrating cells within the centres listed is indicated at the right

	acetylcholine	dopamine	noradrenaline	adrenaline	serotonin	histamine	GABA	glutam. + aspart.	glycine	substance P	VIP	CCK	neurotensin	CRF	LHRH	somatostatin	TRH	vasopr. + oxytoc.	ACTH	αMSH	γMSH	endorphin	enkephalin	dynorphin	angiotensin II	total number	estradiol	testosterone
n. centralis amygdalae	○	+	**+**				○			+	⊕	**+**	✦	●		✦			**+**		+	+	●	⊕	**+**	16	○	○
n. bas. + lat. amygdalae	⊕	+	**+**		+		○				○	+	⊕			**⊕**			+		+	+	○			13	○	○
n. cort. + med. amygdalae					+	+	○			**⊕**	○	⊕	⊕		+	**⊕**		●			+	+	○		○	14	●	●
n. septi lateralis		**+**			+	+	●	+		⊕	⊕	○	⊕	○		⊕	+	✦	**+**		+	+	⊕	+		18	○	○
n. gyri diagonalis	●		+		+	+	●	+					⊕	+	⊕			+								10	○	○
n. interstit. striae term.		**+**	**+**			+		+		+	**⊕**	+	⊕	●	●	⊕	**+**	●	**+**		+	+	**⊕**	⊕		18	●	●
n. accumbens	⊕	**+**			+	+	○	+			**+**	**+**	**+**	○		**⊕**	**+**				+	+	✦	⊕	+	17		
tub. olfactorium	○	**+**	+				○			+	○	+	⊕		⊕	⊕							○			11	○	○
n. caudatus + putamen	⊕	**+**	+		+	+	●	**+**	⊕	⊕	○	**+**	⊕			⊕				+			✦	⊕	+	17		
gyrus dentatus	+	+	+		+		○	+			⊕	⊕				⊕				+			⊕	⊕		12		○
cornu ammonis	⊕	+	+		+		⊕	⊕			⊕	⊕		○		⊕		+		⊕			⊕	+		14		○
n. habenulae med.	⊕		+		+	+	+			○					+	+			+				+			10	○	○
nn. periventriculares thalami		+	**+**	+	+	+	○			+			+	⊕				+	+			+	⊕			13	○	○
n. preopticus med.			**+**		+					⊕	+	○	●	●	●				**+**		+	+	✦	+		13	●	●
n. anterior hypothalami		+	**+**		+					⊕	+		●			✦			**+**				○	+		10	○	○
n. paraventricularis magn.		○	**+**	**+**							+	●	●	●	○	●	**+**	●	**+**		+	+				14	○	●
n. paraventricularis parv.		+	**+**	**+**		+					+	●	●	○				●	**+**		+	+	○	✦	○	15	○	●
n. supraopticus		+	**+**			+				+		●		○	○			●					○	✦	○	11	○	○
n. suprachiasmaticus					**+**	+				+	●				+	●	○	○					○	✦		10	○	○
lob. posterior hypophyseos		+				+				+		**+**	+		+	+		**+**	+				+	**+**	+	12		
n. dorsomedialis		+	**+**	+	+	○				⊕	+	**⊕**					**⊕**	⊕	**+**		+		○	+	+	15	○	○
n. ventromedialis					**+**	**+**			+	○	+	**+**				+	⊕						○	+		10	●	●
n. infundibularis	○	⊕	**+**		+					⊕			●	○	○	○			✦		⊕	○	○	⊕	+	15	●	●
eminentia mediana		**+**	**+**			+				+	**+**	**+**	**+**	**+**	**+**	**+**	**+**	**+**	**+**			+	+	+	**+**	17		
a. preopt.-hypoth. lat.	⊕		+		+	○	○		+	⊕			○	○	○	+	○			●	+	+	○	✦		17	○	○
gris. centrale mesencephali			+	+	+	+	○			⊕	✦	●	⊕	○	+	⊕		+	**+**		+	+	⊕	⊕	+	19	●	○
n. parabrachialis lat.	○	+	+		+	+				+	+	⊕	⊕	○		+	+	+	**+**		+	+	✦	⊕		18	●	
locus coeruleus		+	✦	+	**+**					+			⊕	○				⊕	**+**			+	⊕		+	12	●	
n. spinalis n. trigemini	+		+		+			+		**+**		⊕	✦			⊕							✦	⊕	**+**	11	○	
n. solitarius	○	+	**+**	**⊕**	+	+				**⊕**	✦	⊕	✦	○		⊕	+	+	**⊕**		⊕	⊕	✦	⊕		19	●	
s. intermedia spinalis	○	+	+		**+**						+	+	⊕	○		⊕	+	+	+				+	+		14	○	
s. gelatinosa spinalis	⊕	+	+		**+**		○			**+**	**+**	**+**	✦			**+**		+					**⊕**	**⊕**	**+**	14	○	
stria terminalis			+		+	+				+	**+**	+	+	+	**+**	**+**		+			+		+			13		
f. amygdalofugales ventr.	+	+	**+**		**+**	+					+			+	**+**	+			+	+			+			12		
band. diagonalis Brocae			+		+	+						+		+				+			+		+			8		
fornix	+	+	+		+		+	+				+				+				+						9		
stria medullaris	+		+		+		+								+	+		+								7		
f. telencephalicus med.	+	**+**	**+**	+	**+**	+				+	+	**+**	**+**	**+**	+	**+**			+				+			15		
f. longitudinalis dors.		+	**+**	+	+	+				+				**+**	+	+		+	**+**				+			12		
tr. hab.-interped.	+	+	+		+					+					+	+										7		

○, present in perikarya; ●, present in a considerable number of perikarya; +, present in fibres and/or terminals; **+**, present in a large number of fibres and/or terminals

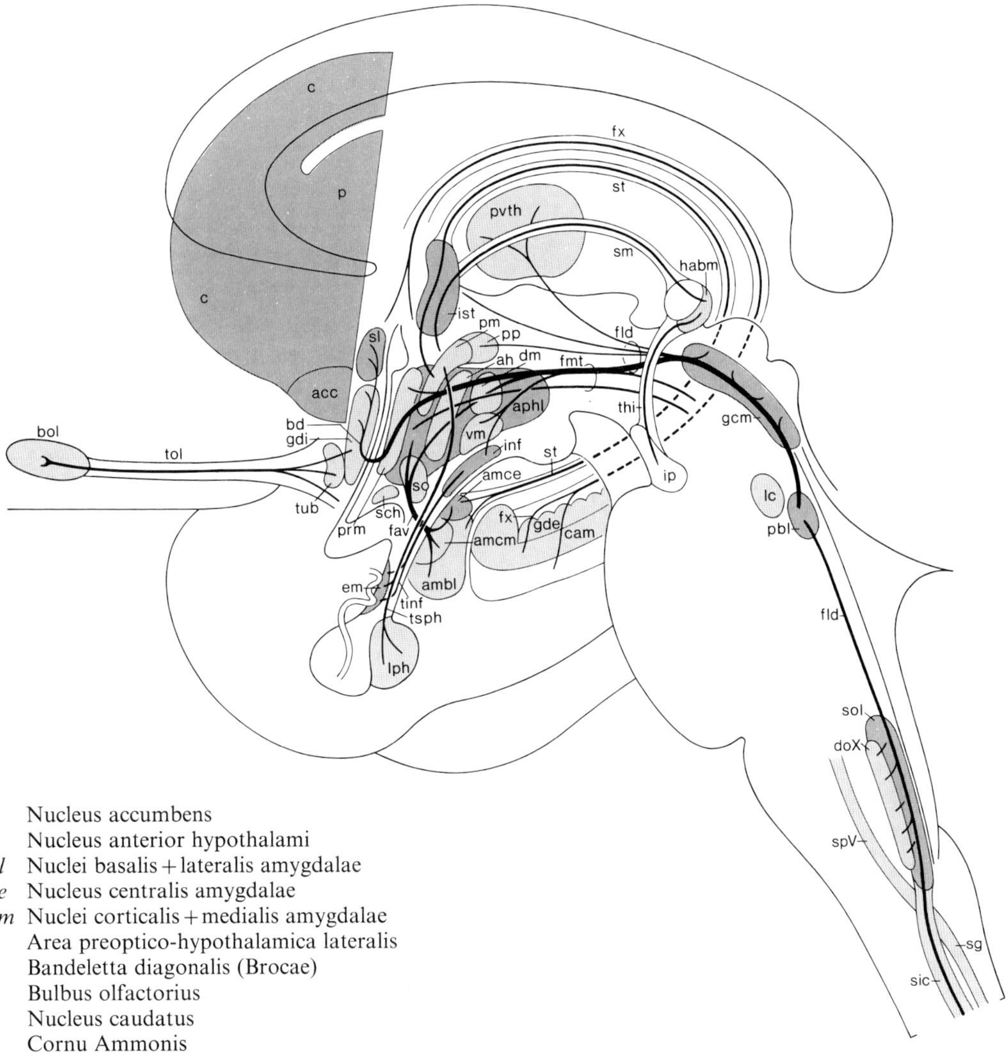

acc Nucleus accumbens
ah Nucleus anterior hypothalami
ambl Nuclei basalis + lateralis amygdalae
amce Nucleus centralis amygdalae
amcm Nuclei corticalis + medialis amygdalae
aphl Area preoptico-hypothalamica lateralis
bd Bandeletta diagonalis (Brocae)
bol Bulbus olfactorius
c Nucleus caudatus
cam Cornu Ammonis
dm Nucleus dorsomedialis
doX Nucleus dorsalis nervi vagi
em Eminentia mediana
fav Fibrae amygdalofugales ventrales
fld Fasciculus longitudinalis dorsalis
fmt Fasciculus medialis telencephali
fx Fornix
gcm Griseum centrale mesencephali
gde Gyrus dentatus
gdi Gyrus diagonalis
habm Nucleus habenulae medialis
inf Nucleus infundibularis
ip Nucleus interpeduncularis
ist Nucleus interstitialis striae terminalis
lc Locus coeruleus
lph Lobus posterior hypophyseos
p Putamen
pbl Nucleus parabrachialis lateralis
pm Nucleus paraventricularis, pars magnocellularis
pp Nucleus paraventricularis, pars parvocellularis
prm Nucleus preopticus medialis
pvth Nuclei periventriculares thalami
sch Nucleus suprachiasmaticus
sg Substantia gelatinosa
sic Substantia intermedia centralis
sl Nucleus septi lateralis
sm Stria medullaris
so Nucleus supraopticus
sol Nucleus solitarius
spV Nucleus spinalis nervi trigemini
st Stria terminalis
thi Tractus habenulointerpeduncularis
tinf Tractus infundibularis
tol Tractus olfactorius
tub Tuberculum olfactorium
tsph Tractus supraoptico-paraventriculo-hypophyseos
vm Nucleus ventromedialis

Fig. 55. The [paracrine?] core of the central nervous system. Explanation in text.

centres and pathways were subsequently transferred to the schematic diagram of the human central nervous system which forms the basis of numerous previous figures in the present treatise. The result is shown in Fig. 55. It will be seen that most of the structures assembled together constitute a readily recognizable entity. In the spinal cord and the brain stem the substantia intermedia centralis, the nucleus solitarius and the adjacent dorsal motor vagal nucleus, the nucleus parabrachialis lateralis and the periaqueductal grey constitute a chain of centres interconnected by the fasciculus longitudinalis dorsalis of Schütz. Rostrally this chain of centres is connected with the hypothalamus by way of the fasciculus longitudinalis dorsalis and the medial forebrain bundle. From the hypothalamus, fibre systems diverge into several directions: (a) the most dorsal fascicles of the fasciculus longitudinalis dorsalis fan out into the periventricular thalamic nuclei; (b) the stria medullaris-fasciculus retroflexus pathway projects to the nucleus medialis habenulae and the nucleus interpeduncularis; (c) the tractus infundibularis and the tractus supraoptico-paraventriculo-hypophyseos terminate in the median eminence and the posterior lobe of the pituitary respectively; (d) the stria terminalis and the so-called ventral amygdalofugal bundle connect the hypothalamus with the amygdaloid complex; (e) the diagonal band of Broca contains numerous fibres interconnecting the nuclei which bear its name and the septum with the hypothalamus; (f) the hippocampus and certain basomedial centres are reciprocally connected by the precommissural fornix; and (g) the precommissural fornix and particularly the stria terminalis contain numerous fibres which connect the bed nucleus of the stria terminalis with many of the telencephalic and diencephalic centres mentioned. On the basis of the data gathered in Chap. 4 (under statement 18), the olfactory bulb and olfactory tract were also included in Fig. 55. The tractus infundibularis and the tractus supraoptico-paraventriculo-hypophyseos are lacking in Table 6, because specific data concerning the neuromediators present in their constituent fibres are scant. However, the data available on the neuromediators present in their nuclei of origin and sites of termination justify their inclusion in Fig. 55. Some centres or complexes which, because of the large number of neuromediators present in them, have been included in Table 6 and Fig. 55 at first sight do not form part of the continuum constituted by the other grisea and fibre paths. These apparently separate structures include the substantia gelatinosa spinalis and its rostral continuation in the spinal trigeminal nucleus, the locus coeruleus and the caudatus-putamen-accumbens complex. A certain number of centres in the brain stem which, on the basis of their richness in neuromediators, should have been included in Table 6 and Fig. 55 have been left aside. These structures include the nucleus raphes magnus, the nucleus raphes dorsalis, the area tegmentalis ventralis, the nucleus cuneiformis and the rhombencephalic medial reticular formation. The reason for this exclusion is that, in my opinion, these centres belong to separate entities which will be discussed under statement 26. Finally, it should be mentioned that, although in the mesocortex and neocortex more than ten different neuromediators have been detected so far (the total number is 17 according to my records), these structures have nevertheless been left out of consideration. This has been done because I do not regard this vast histological continuum as a single centre comparable with the subcortical structures listed in Table 6.

22. The centres and pathways forming the continuum shown in Fig. 55 are not only particularly rich in neuromediators, but also show an extraordinarily high density of neuromediator-specified fibres and terminals.

The procedure which led to this finding was as follows: From the literature used in the preparation of Chap. 3 all adequately illustrated neuromediator mapping studies of the rat were selected. Standard outline drawings of six equidistant frontal sections through the brain and three sections through the spinal cord (cervical, thoracic, lumbar) were prepared. These drawings were modified from the atlas of Paxinos and Watson (1982).

Data concerning the localization and density of fibres and terminals containing each of the 25 different neuromediators dealt with in Chap. 3 were transferred from the illustrations in the pertinent mapping studies to separate sets of the standard outline drawings. With the aid of a copier all of the drawings – nine for each of the 25 neuromediators – were transferred to transparent sheets.

The 25 sheets representing the most rostral section through the central nervous system of the rat were stacked, matched and photographed with transmitted light, and the same procedure was repeated for the remaining eight levels. In this way I obtained pictures showing the total density of neuromediator-specified fibres and terminals at nine levels of the neuraxis of the rat.

These pictures, which will be published elsewhere (Nieuwenhuys, Veening and Van Domburg, in preparation), revealed that practically all of the structures brought together in Table 6 and Fig. 55, because of their wealth of different neuromediators, also show an extraordinarily high density in neuromediator-specified fibres and terminals.

The procedure described was repeated separately for the 16 neuropeptides included in Chap. 3, with practically identical results.

23. The set of 'neuromediator-rich' and 'neuromediator-dense' centres and pathways assembled in Table 6 and depicted in Fig. 55 coincides largely, but not entirely, with the limbic system.

The long and stirring history of the development of the concept of the 'limbic system' cannot be treated here in detail. However, I will attempt to highlight a few major aspects of it.

In 1878 Broca drew attention to the fact that a cortical zone situated along the medial margin of the mammalian cerebral hemisphere, which surrounds the area of fusion of brain stem and prosencephalon, constitutes a separate entity which he termed 'le grand lobe limbique'. He emphasized that he had introduced this term to denote an anatomical structure and not a functional unit. The cortical areas which Broca included in 'his' limbic lobe encompass the gyrus cinguli, the gyrus parahippocampalis and the adjoining cortex of the uncus region, and the hippocampal formation, i.e. the cornu Ammonis plus the gyrus dentatus.

During the last decades of the nineteenth century and the first decades of the twentieth century it was generally believed that most if not all of the structures included in Broca's limbic lobe are dominated by olfactory projections and thus form part of the rhinencephalon.

In 1937 Papez published a notable paper in which he claimed on theoretical grounds that a circuit, of which the hippocampal formation and the cingulate gyrus form important components, constitutes the neural substrate of emotional behaviour. This theory received some substantiation from the work of Klüver and Bucy (1937, 1939), who demonstrated that in monkeys, resections of the anterior portions of the temporal lobes (which included the hippocampal formation and the amygdaloid complex) have, among other effects, a profound influence on affective responses. Somewhat schematically it may be said that the impact of the publication of Papez and those of Klüver and Bucy was threefold: the idea that the rhinencephalon encompasses almost the entire limbic lobe fell into the background; a direct linkage between emotion and Broca's limbic lobe became established; and the amygdaloid complex, a subcortical structure, became incorporated into the limbic lobe.

MacLean (1952, 1958, 1962, 1970) drew attention to the fact that the various components of Broca's great limbic lobe are strongly and reciprocally connected with a number of subcortical structures, particularly the septum, the amygdala, the midline thalamic nuclei, the habenula and the hypothalamus. He suggested (see MacLean 1970, Fig. 4) that the cortical limbic ring is rostrally closed by two subcortical nodal points, i.e. the more dorsally situated septum and the more ventrolaterally located amygdaloid complex. MacLean cited clinical and experimental evidence suggesting that the lower part of the ring, fed by the amygdaloid complex, is primarily concerned with emotional feelings and behaviour that ensures self-preservation. In his wording, the circuits of this lower part of the ring are: "so to speak, kept busy with the selfish demands of feeding, fighting and self-protection" (MacLean 1970, p. 340). The structures associated with the septum in the upper part of the ring, on the other hand, would be involved in: "expressive and feeling states that are conducive to sociability and the procreation and preservation of the species" (MacLean 1970, p. 340). The data and aspects just epitomized led MacLean to the conclusion that the limbic cortex, together with the subcortical structures to which it is directly connected, comprises a functionally integrated system which he designated (in keeping with Broca's terminology) as the *limbic system* (for the first time in: MacLean 1952).

A notable extension to the limbic system concept was made by Nauta (1958, 1973; Nauta and Haymaker 1969). This investigator added to the telencephalic limbic 'arch' (within which he included the hippocampal formation and the amygdaloid complex, but not the gyrus cinguli and the gyrus parahippocampalis) a neural continuum which may be designated as the 'limbic axis'. This continuum includes, from rostral to caudal: the septal and preoptic regions, the hypothalamus, and a number of mesencephalic structures which occupy a paramedian position (Nauta's 'limbic midbrain area'). Nauta pointed out that these various entities are structurally heterogeneous, but that all of them are strongly interconnected by shorter and longer ascending and descending fibres. Taken together, these connections constitute, in Nauta's opinion, one large functional system which he designated as the 'limbic system-midbrain circuit'. He emphasized that the large telencephalic limbic structures, i.e. the hippocampus and the amygdaloid complex, are both reciprocally connected with the rostral pole of the limbic axis, and thereby with the limbic system-midbrain circuit.

Nauta's views on the functioning of the limbic system may be summarized as follows (see particularly Nauta 1973). The structures constituting the limbic axis are centrally involved in the regulation of endocrine and visceral effector mechanisms. The functional state of these structures is not affected solely by neural afferents from the hippocampus and the amygdala, but is also modulated by impulses travelling along visceral-sensory pathways ascending from the spinal cord and the medulla oblongata, as well as by various blood-borne agents. Moreover, the limbic system as a whole is implicated in the regulation of affective and motivated behaviours.

It was long thought that the limbic system and the extrapyramidal motor system represented strictly separated structural and func-

tional entities. However, as has already been discussed on p. 153, Kelley et al. (1982) and Gerfen (1984) have provided connectional evidence indicating that vast portions of the caudate-putamen complex may be adequately designated as limbic.
The usefulness of the concept of a limbic system has been questioned by several authors, among whom Brodal (1981) is the most prominent. After a brief discussion and evaluation of the terms 'limbic lobe' and 'limbic system', Brodal arrived at the conclusion that neither term has a clear meaning and advocated their total banishment. Brodal argued that, pari passu with our increase of knowledge, the limbic system 'appears to be on its way to including all brain regions and functions' (Brodal, p. 690). He took issue with an attempt of Isaacson (1975, p. 331) to characterize the limbic system as a system, 'only because each of its components has relatively direct connections with the hypothalamus', by stating: 'On this basis several nuclei of the brain stem, among them the raphe nuclei, the nucleus locus coeruleus, the nucleus of the solitary tract, and the dorsal motor vagus nucleus and further the intermediolateral column in the spinal cord, should be considered parts of the "limbic system", since they all have connections – *even direct ones* – with the hypothalamus' (Brodal 1981, p. 690).
It is most remarkable that the set of centres and pathways which I have brought together on the basis of their 'neuromediator richness' and 'neuromediator density' (Fig. 55) encompasses not only all structures which Nauta included in the limbic system, but also the centres added to this system by Kelley et al. (1982) and, ironically, most of the 'limbic candidates' proposed by Brodal!

24. The set of 'neuromediator-rich' and 'neuromediator-dense' centres assembled in Table 6 and Fig. 55 closely corresponds to the set of estrogen- and androgen-hormone-concentrating neuronal groups which Stumpf included in what he termed 'the periventricular brain'.

Stumpf and his associates (Stumpf 1970, 1975; Stumpf and Sar 1975, 1977, 1978; Stumpf et al. 1975; Keefer and Stumpf 1975; Sar and Stumpf 1975) studied the distribution of neurons concentrating sex steroids in the central nervous system of several mammals, including insectivores, rodents and primates, by administration of tritiated estradiol or testosterone followed by autoradiography. The essence of their findings is summarized in the two columns furthest on the right in Table 6 and in Fig. 56. It will be seen that most of the steroid hormone-concentrating neurons are located within the centres brought together on the basis of their 'neuromediator richness' and 'neuromediator density'.
Stumpf (1975) suggested that the various steroid hormone target neurons are involved in hormonal 'feedback' regulation, and that only a subpopulation of these cells is involved in the manufacture of hypophysiotropic releasing or inhibiting factors. He considered it likely that other groups of the labelled neurons may act primarily as threshold modulators for sensory input. Stumpf and Sar (1978) pointed out that in the lower rhombencephalon and spinal cord the estradiol- and testosterone-concentrating neurons show considerable differences in localization. Estrogen concentration appeared to prevail in sensory areas and cells that are known to modulate sensory perception, whereas testosterone prevailed in neurons that are associated with somatomotor functions. Stumpf (1975; see also Stumpf and Sar 1977) emphasized that most of the steroid hormone target neurons are situated close to the ventricular

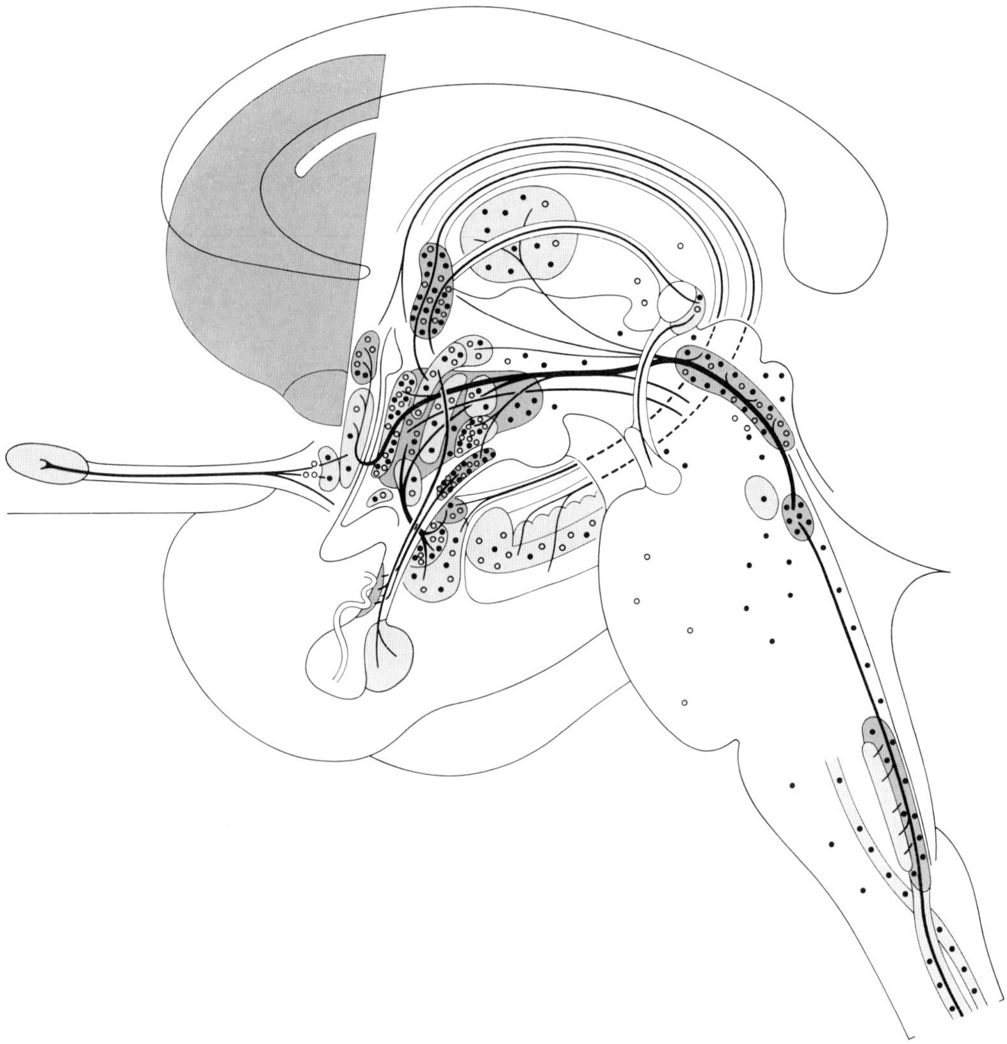

Fig. 56. The [paracrine?] core of the central nervous system; the location of estradiol (•)- and testosterone (○)-concentrating cells is indicated

system, i.e. in a zone which he regarded as a phylogenetically ancient part of the neuraxis. He suggested that this 'periventricular brain' fulfils a neuroendocrine function throughout. The richly vascularized circumventricular organs all contain accumulations of estradiol-concentrating cells and constitute, in Stumpf's opinion, nodal points in the periventricular brain.

25. The set of 'neuromediator-rich' and 'neuromediator-dense' centres and pathways in the mammalian central nervous system, which has many features in common with the limbic system as well as with the steroid hormone-concentrating periventricular brain, constitutes a unit which may be designated as the [paracrine?] core of the neuraxis.

The reason for selecting the term 'core' is threefold – positional, connectional and functional. As regards the position of the various centres, they extend throughout the length of the neuraxis from the olfactory bulb to the caudal spinal cord, and most though not all of them occupy a central, periventricular position. Figure 55 shows rather wide gaps between the dorsal vagal complex (i.e. the nucleus solitarius plus the dorsal motor vagal nucleus) and the lateral parabrachial nuclei, and between periaqueductal grey and the hypothalamus. It is known that these gaps are bridged by zones of relatively undifferentiated central grey, but the number of neuromediators present in these zones was not systematically recorded during our analysis of the literature.

It has been well-established by numerous experimental hodological studies that the various structures which make up the telencephalic and diencephalic portions of what is called here the [paracrine?] core of the neuraxis are strongly and reciprocally interconnected. The pertinent studies have been reviewed repeatedly (see e.g. Nieuwenhuys et al. 1981, 1982) and require no full documentation here. However, it may be well to briefly indicate the results of some recent experimental studies on the connections of the rostral pole of the 'paracrine core' with the remainder of that entity.

There is experimental neuroanatomical as well as physiological evidence indicating that the nucleus paraventricularis hypothalami projects directly to cardiovascular responsive sites in the upper thoracic spinal cord (Miura et al. 1983; Caverson et al. 1984).

Numerous more rostrally situated 'core' centres, including the bed nucleus of the stria terminalis, the medial and lateral preoptic areas, the nucleus centralis amygdalae, the paraventricular nucleus, the infundibular nucleus, the dorsomedial hypothalamic nucleus, the lateral hypothalamic area and the periaqueductal grey, project directly to the dorsal vagal complex (Van der Kooy et al. 1984; Schwanzel-Fukuda et al. 1984; Veening et al. 1984; Ter Horst et al. 1984).

Several of the connections just mentioned are reciprocated by projections ascending directly from the dorsal vagal complex to diencephalic and telencephalic 'core' centres. Such projections have been observed to terminate in the infundibular nucleus, the dorsomedial hypothalamic nucleus, the paraventricular nucleus, the medial preoptic area, the bed nucleus of the stria terminalis and, in addition, the periventricular thalamic nuclei (Ricardoh and Koh 1978).

The caudal (general visceroreceptive) part of the nucleus of the solitary tract has been shown to project to the lateral parabrachial nucleus (Ricardoh and Koh 1978; Milner et al. 1984; see Fig. 38e). The lateral parabrachial nucleus distributes fibres to the periaqueductal grey (Marchand and Hagino 1983) and to the lateral hypothalamic area, the ventromedial, dorsomedial and paraventricular hypothalamic nuclei, the median preoptic area, the bed nucleus of the stria terminalis and the nucleus centralis amygdalae (Fulwiller and Saper 1984; Zaborski et al. 1984; Lind and Swanson 1984). Conversely, the lateral parabrachial nucleus receives descending projections primarily from the paraventricular nucleus, but also from the preoptic region and from the anterior, lateral, dorsomedial and ventromedial hypothalamic nuclei (Takeuchi and Hopkins 1984).

It has already been indicated in the preceding chapter that the periaqueductal grey receives substantial projections from the amygdaloid complex and from various hypothalamic centres (see p. 135 and Fig. 37). For details of these descending projections I refer to the recent retrograde tracer studies of Mantyh (1982) and Marchand and Hagino (1983).

As regards the ascending efferent connections of the periaqueductal grey, the anterograde tracer study of Mantyh (1983) has shown that this griseum projects to several thalamic centres, including the intralaminar and midline nuclei, but that its heaviest ascending projections are directed to the preoptic region and to the anterior, dorsal, periventricular, ventromedial, lateral and posterior hypothalamic nuclei.

The studies just reviewed fully warrant the conclusion that the telencephalic and diencephalic 'core' centres are strongly and reciprocally connected with their more caudally situated counterparts. There is evidence that several of the projections discussed follow a periventricular route (see, e.g. Ricardoh and Koh 1978 and Mantyh 1982, 1983); however, because most of the studies cited are based on experiments with retrograde tracers, the actual courses of many of these connections remain to be elucidated.

Having indicated the positional and connectional grounds for selecting the designation 'core of the neuraxis' for the set of centres and pathways listed in Table 6 and depicted in Figs. 55 and 56, it now remains to unfold my functional arguments.

Current neurophysiological and neuroethological knowledge enables me to attach functional labels to practically all of the structures listed in Table 6. I am fully aware of the fact that such a catchword-like characterization of the functional significance of centres is hazardous and may be misleading. Many of the findings concerning the localization of functions derive from lesioning or stimulation experiments, and the results of experiments of these types should be judged with caution. As pointed out by Arnold (1969, p. 1045):

"Of course, it is obvious that a particular impairment after a lesion does not mean that the missing brain tissue had 'generated' or 'laborated' the behavior that is now defective. But it may mean that all lesions resulting in such impairment, no matter where located, have damaged structures or circuits that are required for normal responses. Similarly, electrical stimulation of some point in the brain does not mean that the behavior so produced is necessarily the normal behavior mediated by a 'center'. It may mean that the electrode has stimulated points in a circuit that normally mediate such behavior or that these connections have been preempted by the stimulating current and now cannot be used for the subject's normal behavior."

Moreover, even if the evidence that a given centre is implicated in a particular function is indeed conclusive, the question immediately presents itself: Does all, or only part of that centre subserve that function? Many grisea which are commonly designated as centres are in fact complex, and their structural non-homogeneity may well reflect involvement in a multiplicity of actions. Finally, Brodal (1981, p. 690) has remarked upon the fact that '... as research progresses it becomes increasingly difficult to separate functionally different regions of the brain. The borderlines between "functional systems" become more and more diffuse.' I concur with this statement, without drawing the conclusion that the ultimate result of all efforts in the realm of neurobiology will be a statement that the central nervous system is a single, totally integrated structural and functional unit, and just that.

Following these preliminaries (and the caveats included!), I now present the functional notes promised. (N.B. This matter will be discussed more in depth and fully documented in a forthcoming publication: Nieuwenhuys, Veening and Van Domburg, in preparation.)

Many of the centres included are of central importance in the regulation of *visceral effector mechanisms*. Thus, the dorsal vagal complex, the lateral parabrachial nucleus, the supraoptic, paraventricular and dorsomedial hypothalamic nuclei, the medial preoptic area, the bed nucleus of the stria terminalis and the nucleus centralis amygdalae have all been implicated in cardiovascular control. Modification of respiratory functions can be induced by stimulation of the dorsal vagal complex, the lateral parabrachial nucleus, the medial preoptic area, the bed nucleus of the stria terminalis and the nucleus centralis amygdalae. Finally, the motility and secretory activity of the digestive tract can be influenced by stimulating, among other structures, the dorsal vagal complex, the lateral parabrachial nucleus and the nucleus centralis amygdalae.

The substantia gelatinosa and its rostral continuation in the spinal trigeminal nucleus are recipient centres of *nociceptive stimuli*, whereas the periaqueductal grey plays a key role in the modulation of nociception.

The entire preoptico-hypothalamic continuum plays a particularly important role in the *maintenance of homeostasis, i.e. the regulation of the milieu intérieur*. This complex task is accomplished by (a) activation of visceral effector mechanisms, (b) modulation of the release of anterior and posterior pituitary hormones, and (c) initiation of the various behavioural patterns aimed at maintenance or restoration of homeostasis, i.e. the adaptive, goal-oriented foraging behaviours (see Swanson and Mogenson 1981).

Many of the centres included are involved in the regulation of *sexual and reproductive behaviour*. The lateral septal nucleus, the (sexually dimorphic!; see Fig. 43) medial preoptic nucleus, the bed nucleus of the stria terminalis, the amygdaloid complex, the nucleus suprachiasmaticus and the periaqueductal grey may be mentioned in this context. It is important to note (a) that the release of gonadotropic hormones from the pituitary gland is steered by regulation hormones manufactured by the hypothalamus, and (b) that numerous neurons in the central nervous system are targets of the male and female gonadal hormones which are produced under the control of the gonadotropic hormones. As pointed out by Stumpf (1975), a certain proportion of the target cells of the sex hormones are directly involved in the regulation of the production of hypothalamic regulating hormones (see Figs. 22 and 56), whereas others may serve as threshold modulators on both the sensory and motor sides. The latter may facilitate the fixed-action patterns related to reproductive behaviour.

Stimulation of many of the centres included has been reported to elicit *agonistic (i.e. combative) behaviour*. I confine myself here to mentioning the amygdaloid complex (all main divisions), the lateral septal nucleus, the nuclei of the diagonal band of Broca, the bed nucleus of the stria terminalis, the nucleus accumbens, the cornu Ammonis, several hypothalamic centres, among which is the lateral hypothalamic area, and the periaqueductal grey.

The *sense of smell* and its central apparatus plays an important role in both *feeding* and *reproductive behaviour*.

Taking together the data enumerated above, I venture the conclusion that the centres and pathways included in Table 6 constitute a functional complex which is involved most directly in processes aimed at the survival of the individual (organism) and of the species, and thus fully deserves the designation 'core of the neuraxis'.

I come now to an explanation of the denotation 'paracrine'. Let me state at once that the addition of this adjective to 'the core of the neuraxis' rests on a cascade of assumptions and, hence, is strongly speculative. First, however, a brief explanation of the word. The term 'paracrine secretion' is borrowed from endocrinology. It was introduced by Feyrter (1953) to denote that the specific secretion products of endocrine cells do not exclusively attain their target organs by way of the blood stream, but may also reach and influence groups of neighbouring cells via the interstitial tissue fluid.

My reasoning is now as follows:

1. There is a set of centres and pathways in the central nervous system which (a) together constitutes a single functional complex and (b) shows an extraordinary 'richness' and 'density' of neuromediators, particularly neuropeptides.
2. This accumulation of neuropeptides is not fortuitous; rather it reveals a trend.
3. It is to be expected that the 35-odd neuropeptides now known represent only a small fraction of the total number (Sternberger 1980; Snyder 1980).
4. It follows from 2 and 3 that in what I have called the core of the neuraxis some 200 different neuropeptides may well be operative.
5. The presence of such a multitude of specific messengers betrays the existence of a special language.
6. The pathways interconnecting and passing through the various centres constituting the core of the neuraxis consist largely of thin, unmyelinated fibres. I consider it likely that these pathways represent the 'open multi-neuromediator channels" already alluded to under statement 12 (p. 128). This implies that the constituent fibres of these bundles are able to influence numerous other neurons along their length, or during their passage through one or more interstitial nuclei, rather than only in their area of ultimate termination.
7. It is important for sets of fibres which contain and release a particular neuropeptide to contribute their specific signals to the neuropeptide language in the appropriate spatial and temporal context. As regards the spatial context a remarkable course of certain sets of peptidergic fibres, which I will denote as the *'detour phenomenon'*, deserves mention. From descriptions in the literature it appears that in the rat, LHRH- as well as somatostatin-containing fibres originating from the medial preoptic area pass laterally, join the medial forebrain bundle to follow that 'open channel' over a certain trajectory, then leave the bundle and pass medially again, to terminate in the median eminence. I consider it likely that this peculiar detour enables these fibres to make their specific contribution to the complex neuropeptide language by which the medial forebrain bundle communicates with the cells scattered within and around it, i.e. the lateral preoptico-hypothalamic continuum.
Given the presence of a multitude of specific messengers (the various neuropeptides), and given the potential presence of specific receptors for all of these messengers, it is very plausible that in areas where numerous thin, unmyelinated fibres containing many different neuropeptides occur, non-synaptic *paracrine* interneuronal communication prevails. As already mentioned under statement 19 (p. 171), this mode of communication implies a very economical use of the available space in the central nervous system. Moreover, although the various neuropeptides all travel in the same extracellular space (which opens up the possibility for as yet entirely unexplored interactions between neuropeptides), a certain degree of specificity is maintained, since all of these neuropeptides may well address only 'their own' specific receptors.
8. Finally, it may be that for various reasons including those outlined above, the multipeptide paracrine mode of interneuronal communication fits the specific functional demands of the core of the neuraxis particularly well with regard to both tempo and mode. However, this is not to imply that the other modes of communication, i.e. endocrine secretion and the strictly synaptic neurocrine secretion, do not occur in core regions. It is well known that endocrine secretion plays a prominent role in at least two such regions, the median eminence and the posterior pituitary. Moreover, peptidergic axons have been shown to participate in the formation of conventional synapses (see the enumeration under statement 19 on p. 168) and many such synapses have been observed in typical core regions.

So much for the explanation of the term *'the [paracrine?] core of the neuraxis'*.

26. The [paracrine?] core of the neuraxis has two adjuncts, termed here the 'median paracore' and the '(bilateral) lateral paracore' (Fig. 57).

Having already entered the realm of speculation, a few additions to the concept of a paracrine core in the neuraxis may be permissible.

I consider it likely that the series of raphe nuclei extending throughout the brain stem constitute an adjunct to the [paracrine?] core, which may be tentatively designated as the *median paracore*. In most places the raphe nuclei are directly adjacent to the core region, and in some they even penetrate into it (Fig. 57a). Moreover, as will be detailed below, fibres of the core region project heavily towards certain raphe nuclei.

A second adjunct to the core region is probably formed by a series of grisea which in most though not all levels of the brain stem extend from this region ventrolaterally into the tegmentum. At the mesencephalic level this series includes the continuum formed by the area tegmentalis lateralis, the pars compacta of the substantia nigra and the area tegmentalis ventralis (Fig. 57a). In the rhombencephalon it is the locus coeruleus, the subcoeruleus area and the cytoarchitectonically ill-defined cell groups A1, A2, A5, A7, C1 and C2 which form part of this series (Fig. 57b, c). Several other centres, such as the nucleus tegmentalis pedunculopontinus, pars compacta (TPc), the medial parabrachial nucleus and the so-called Kölliker-Fuse nucleus also fit into it. I summarize this longitudinally arranged chain of nuclei under the name *lateral paracore*. Whereas rostrally the median paracore comes to an end where the raphe area thins out into the posterior wall of the hypothalamus, the lateral paracore continues into the diencephalon. In fact, it is ambiguous whether the lateral hypothalamic area forms a rostral extension of the lateral paracore or a lateral extension of the core region.

The median and lateral paracores have several features in common:

1. As already mentioned, they are both directly continuous with the core region.
2. The constituent cells of the paracore regions contain different neuromediators, but monoamines predominate in both. In the median paracore numerous serotoninergic cells are found, but in the lateral paracore catecholaminergic neurons prevail. These include adrenergic elements, which are found exclusively in the most caudal part of the zone (Fig. 8); noradrenergic neurons, which occur throughout the length of the rhombencephalon (Figs. 6, 7); and dopaminergic neurons, concentrated in the mesencephalic portion of the lateral paracore (Fig. 5).
3. Both paracores lie clearly beyond the trajectories of the large, well-myelinated ascending and descending pathways.
4. Both paracores contain numerous thin, longitudinally oriented fibres. In the median paracore these fibres constitute the dorsal and ventral ascending serotoninergic pathways, as well as the dorsal and ventral descending serotoninergic pathways (Felten and Sladek 1983; Fig. 57). It is important to note that many of these ascending and descending fibres may contain one or even two additional neuromediators (see statement 9 and Table 3). In the lateral paracore the thin, longitudinally arranged fibres assemble in the longitudinal catecholamine bundle of Jones and Friedman (1983; Fig. 34), which tends to divide over its course into more or less separate dorsal and ventral components (Felten and Sladek 1983; Fig. 57). The catecholamine (mostly noradrenergic) fibres in the bundle just discussed are joined by many different kinds of peptidergic fibres. In the rat, axons containing VIP, neurotensin, LHRH, somatostatin, CRF, ACTH, α-MSH, endorphin and enkephalin are present in the area of the longitudinal catecholamine bundle throughout the brain stem. The same may hold true for the longitudinal serotoninergic

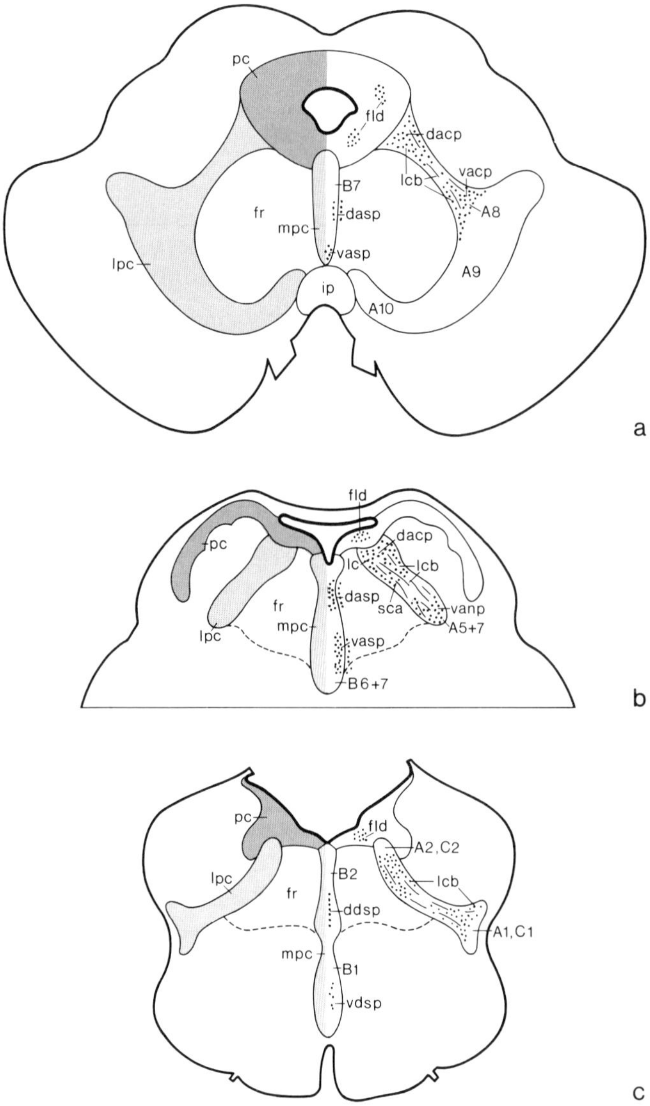

A1-A7	Noradrenergic cell groups	*lc*	Locus coeruleus (A6)
A8-A10	Dopaminergic cell groups	*lcb*	Longitudinal catecholamine bundle
B1-B7	Serotoninergic cell groups	*lpc*	Lateral paracore
C1, C2	Adrenergic cell groups	*mpc*	Medial paracore
dacp	Dorsal ascending catecholamine pathway	*pc*	[Paracrine?] core
ddsp	Dorsal descending serotonin pathway	*sca*	Subcoeruleus area (A6)
fld	Fasciculus longitudinalis dorsalis of Schütz	*vacp*	Ventral ascending catecholamine pathway
fr	Formatio reticularis	*vdsp*	Ventral descending serotonin pathway
ip	Interpeduncular nucleus		

Fig. 57 a–c. Diagrammatic frontal sections through the human brain stem at mesencephalic (**a**), rostral metencephalic (**b**) and myelencephalic levels (**c**) to show the positions of the [paracrine?] core, the median paracore and the (bilateral) lateral paracores. In the [paracrine?] core the position of the coarsest fibres in the fasciculus longitudinalis dorsalis is indicated; however, it is known that this region contains large numbers of thin, unmyelinated fibres. Further explanation is given in the text

fibre assemblies, although direct evidence for this assumption is lacking.
5. Both paracores contain sets of cells giving rise to networks of fibres that pervade virtually all grisea in the neuraxis, i.e. the serotoninergic cells in the median paracore and the noradrenergic cells in the lateral paracore.
6. Felten and colleagues (Felten and Sladek 1983; Felten et al. 1981; Cummings and Felten 1979) observed that tanycytes situated in the ventricular lining of the brain stem penetrate with their long, peripherally extending processes several cell masses in both the median paracore and the lateral paracore, where they directly contact the monoaminergic neurons present in these centres. They supposed that these specialized ependymal elements act as a transport link, providing a route for cerebrospinal fluidborne substances to reach receptors on the surface of the monoaminergic neurons.
7. Felten and colleagues (Felten and Crutcher 1979; Felten and Sladek 1983) also observed in primates only that in most monoaminergic centres included here in the median and lateral paracores, neurovascular appositions occur where the basement membrane of capillaries directly abuts on a monoaminergic soma or dendrite. They postulated that the neurons in question may be influenced directly by hormones or other substances in the blood.
8. Both paracores exchange numerous axons with the core region. It has already been discussed (see statement 13. and Fig. 37) that the periaqueductal grey and the adjacent nucleus raphes dorsalis project heavily to the nucleus raphes magnus, and that this projection consists of fibres containing several different neuromediators. Conversely, the periaqueductal grey receives afferents from several raphe nuclei, including the nucleus raphes dorsalis, the nucleus centralis superior, the nucleus raphes magnus and the nucleus raphes obscurus (Beitz 1982c; Marchand and Hagiro 1983). As regards the lateral paracore, the substantia nigra is known to project heavily to the periaqueductal grey, and the locus coeruleus and the medial parabrachial nucleus also distribute afferents to that mesencephalic core centre (Beitz 1982c).
9. Finally, it is remarkable that the core and the paracores together embrace the reticular formation (Fig. 57).

What is the function of the paracore regions? The connections with the core region suggest that they act in concert with the latter, and the fact that both of them are endowed with profusely branching axonal systems strongly suggests that they exert wide-ranging influences over neurons in virtually every major subdivision of the central nervous system. However, concerning the specific functions of these two all-pervading neuronal groups, I feel unable to add much to the functional comments on the serotoninergic and noradrenergic projections presented earlier (see pp. 39 and 25). Nevertheless, the following preliminary notes may give some idea concerning the activities in which the paracore regions are involved.
The complex formed by the Kölliker-Fuse nucleus and the most ventral part of the medial parabrachial nucleus is considered to represent the pontine respiratory or 'pneumotaxic' centre. This portion of the lateral paracore projects to some more caudal centres within the same zone, i.e. the ventral medullary respiratory centre, which is situated in the vicinity of the A1 and C1 cell groups, and the dorsal medullary respiratory centre, which lies in the ventrolateral portion of the solitary nucleus, in the same region where the A2 and C2 cell groups are also located (see Cohen 1979; Long and Duffin 1984; Fulwiler and Saper 1984). In addition, the Kölliker-Fuse nucleus projects directly to the intermediolateral cell column in the thoracic spinal cord. The names of the various centres reflect the strong electrophysiological evidence that these cell groups, and hence considerable parts of the caudal lateral paracore, are implicated in the control of respiration.
The caudal parts of the lateral paracore not only contain respiratory centres, but also

play an important role in cardiovascular control. Its superficial, ventrolateral portions contain chemosensitive neurons which directly monitor blood composition, particularly the concentration of CO_2 (Dampney and Moon 1980; Trouth et al. 1982). The dorsomedial and ventrolateral moieties of the caudal lateral paracore are intensively and reciprocally interconnected. As indicated in Fig. 57c, both of these moieties contain noradrenergic (A1, A2) as well as adrenergic cell groups (C1, C2). It is also worth mentioning that the ventrolateral moiety, like the dorsomedial one with its adjoining dorsal vagal complex, is particularly rich in neuropeptides. Within the ventrolateral moiety, populations of cells containing substance P, CCK, somatostatin, TRH, and enkephalin have been observed (see Mantyh and Hunt 1984a).

It has already been mentioned that the entity which I have tentatively designated as the core of the brain is involved most directly in processes aimed at the survival of the individual and of the species, and that these vital processes require the initiation of action patterns such as exploratory behaviour, foraging behaviour associated with food and water procurement, and reproductive activities. It will be obvious that locomotor responses are associated with all of these action patterns. Relying heavily on the very thoughtful papers of Kuypers (1982), Mogenson et al. (1983) and Swanson et al. (1984) I now suggest that a chain of core and paracore centres is involved in the initiation and execution of the behavioural patterns mentioned. This chain of centres includes the nucleus accumbens, the subpallidal region, the lateral preoptico-hypothalamic continuum, the TPc and adjacent cell groups, and the raphe nuclei, particularly the nucleus raphes magnus.

It is known that the nucleus accumbens is involved in the modulation of locomotor behaviour. This nucleus projects heavily to a subpallidal region that includes parts of the substantia innominata, the lateral preoptic area and the lateral hypothalamic area. These areas, in turn, send numerous fibres via the medial forebrain bundle to a region in the midbrain which corresponds roughly with the TPc. There is experimental evidence indicating that the locomotor responses evoked by the nucleus accumbens depend to a considerable degree on the integrity of the subpallidal region (Swerdlow et al. 1984). Moreover, it has been reported that locomotor stepping movements can be elicited by stimulating the medial forebrain bundle (Sinnamon et al. 1984), i.e. the channel along which the subpallidal region projects to the TPc, as mentioned. It is relevant that the nucleus accumbens is under the control of such typical core regions as the hippocampus and the amygdaloid complex (Fig. 55), and that it also receives a heavy (mainly dopaminergic) projection from the area tegmentalis ventralis, which forms part of the mesencephalic lateral paracore (Fig. 57: A10).

The TPc, which is situated in the caudolateral midbrain tegmentum, corresponds roughly with an area known as the mesencephalic locomotor region. The latter has received this name because its electrical stimulation in post-mammillary decerebrate cats (Shik et al. 1967) and rats (Skinner and Garcia-Rill 1984) induces coordinated locomotion on a treadmill. The TPc and its directly adjacent areas are caudally continuous with a zone designated as the pontomedullary locomotor strip (Shik and Yagodnitsyn 1979). This zone contains numerous noradrenergic elements, and it is important to note that it, as well as the TPc, falls entirely within the confines of the lateral paracore. In addition to the heavy projection from the subpallidal region, the TPc receives afferents from numerous other sources, including the motor cortex, the globus pallidus, the bed nucleus of the stria terminalis, the medial preoptic area, the parvocellular part of the paraventricular nucleus and the pars reticulata of the substantia nigra. It is also worth mentioning that the subpallidal region distributes fibres to the periaqueductal grey and that this typical core region in turn projects to the TPc. The descending efferents of the mesencephalic locomotor region terminate mainly in the rhombencephalic medial reticular formation and in the nucleus raphes magnus (Steeves and Jordan 1984).

I consider it likely that the nucleus raphes magnus and other (mainly serotoninergic) median paracore centres, and the medullary part of the lateral paracore, form a caudal continuation of the somatomotor control system described above. Kuypers (1982) pointed out that extensive serotoninergic and noradrenergic fibre systems project from the brain stem to the spinal anterior horn, and that the cells of origin of these fibre systems may be under limbic (and therefore core) control. He continued that, if activation of the serotoninergic and noradrenergic brain stem pathways to the spinal anterior horn produces an increase in the responsiveness of the motoneurons, as has been suggested in the literature, it would seem likely that these pathways are especially active under circumstances which require a high level of motor activity, for instance in fight and flight. Kuypers hypothesized that the serotoninergic and noradrenergic spinal projections constitute a component of the motor system which may be instrumental in providing motivational drive in the execution of movements. He extended this theory one step further by assuming that the increase in the responsiveness of the motoneurons induced by the descending projections discussed might under certain circumstances occur simultaneously with an activation of the brain stem projections to the spinal dorsal horn. The latter are known to suppress pain transmission (see p. 133 and Fig. 37). Such a coupling of inhibitory impulses toward crucial nociceptive centres and increase of the responsiveness of motoneurons would enable the individual to ignore painful stimuli during the execution of motor actions of vital priority.

CHAPTER 6

'Classical' Neuroanatomy, 'Chemical' Neuroanatomy and Neurobiology; Some Concluding Remarks

In this book an attempt has been made to summarize our present-day knowledge of the neuromediator-specified neuronal populations in the mammalian central nervous system, and to explore the relations between what are now called 'classical neuroanatomy' and 'chemical neuroanatomy'. Many uncertainties have been expressed and few definitive conclusions have been drawn. One of the major difficulties encountered was that opportunities for an immunohistochemical approach to the central nervous system offered themselves to neurosciences at a time when the facade, not to mention the building, of classical neuroanatomy was by no means finished. Moreover, the set of new techniques triggered a tremendous explosion of research which has led – as explosions usually do – to the generation of numberless fragments ('data'), but not in itself to any great gain in our understanding and insights. Times in which explosions occur do not invite reflexion, and it is no exaggeration to say that, at present, theory formation in neurobiology lags far behind. Yet, if any real progress is to be made, serious attempts are compellingly necessary to create a new framework within which classical neuroanatomy and chemical neuroanatomy will harmoniously converge towards a new neurobiology. Needless to say, the creation of such a new framework is far beyond the capability of a single individual. It should therefore be appreciated from the outset that the concluding comments which follow have only the intention of emphasizing a few aspects which might be relevant to such an enterprise.

Following some remarks on a few classical techniques and their importance for chemical neuroanatomy, I will briefly touch on the present state of hodology, the various forms of chemical transmission, the presumed importance of the extracellular space, and some possible roles of neuroglia, to terminate with a comment on the 'new' entities introduced in this book.

Nissl staining has been, since its development 100 years ago (see Kreuzberg 1984), one of the most powerful tools in neuroanatomy. Its application and the study of the results obtained with it together represent an indispensable first step in the analysis of any part of the central nervous system. Its strength is that by its use, the baffling intricacy of nervous tissue is comfortingly reduced to showing only the perikarya, leaving the feltwork of dendrites, axons and terminals out of sight. This single technique has led to the birth of a separate neuroanatomical discipline: cytoarchitectonics. The aim of this discipline is to subdivide the grey matter of the central nervous system into *adjacent* provinces, called grisea, which include cell masses or nuclei and fields within laminated structures or cortices. The boundaries between these entities are drawn on the basis of differences in size, shape and density of their neuronal perikarya. The delineation of grisea is not practiced as an esoteric activity; rather, it is hoped that by this approach units of biological significance will be found (see e.g. Olszewski and Baxter 1954). However, it is well known that in many parts of the brain the drawing of unambiguous cytoarchitectonic boundaries is extremely difficult, if not impossible. Nevertheless, in the many cytoarchitectonic atlases which are used as a frame of reference for various types of neurobiological research the boundaries are shown, but the reasons for the authors placing their boundaries at particular sites are rarely made explicit (see Swanson 1984).

Apart from the inherent subjectivity of the cytoarchitectonic approach per se, doubt should also be cast upon the assumption implicit in this subdiscipline that the grey matter consists entirely of adjacent portions, and is entirely built as a kind of three-dimensional jigsaw puzzle. Examination of Nissl preparations frequently reveals that particular populations of perikarya which are easily distinguishable by their size, shape or stainability do not respect the boundaries as drawn in cytoarchitectonic atlases. I now feel strongly that the relation between cytoarchitectonics and 'chemoarchitectonics' should not be a one-way affair. The 'chemoarchitectonist' should not finish up with an enumeration of the places where her/his results fit or do not fit with those of the cytoarchitectonist. Rather, the question should be posed: What may be the meaning of the incongruities found? If cytoarchitectonists and chemoarchitectonists can in this way enter into a dialogue, new insights may well result. Possibly it will appear that in a number of regions, at least, a pattern of mutually overlapping fields is a more accurate type of functional subdivision than the usual parcellations presented in cytoarchitectonic atlases.

The Golgi technique was, is, and is likely to remain the cornerstone of neuroanatomy. It is indispensable in clarifying interneuronal relationships and patterns of microcircuitry in general. An adequate explanation of many electron-microscopical data can be given only in light of parallel studies with the Golgi technique. Hence, the recent development of the combined Golgi–electron-microscopical technique was of paramount importance. It should also be pointed out that it is only with the aid of the Golgi technique that a complete typology of the neurons present in a given griseum can be made. Fortunately, immunochemistry frequently yields 'Golgi-like' images of neurons, and careful matching of the images thus obtained may well lead to highly important results. In such cases chemical neuroanatomy indeed adds a dimension (in this case, neuromediator specification) to the results of classical neuroanatomy.

Notwithstanding the existence of many different so-called tract-tracing techniques, hodology – i.e. the subdiscipline focussing on the structure of fibre paths –, is at present full of uncertainties. One of the reasons for this is that many of these techniques are in fact not 'tract tracing' at all, and reveal little or nothing concerning the course and composition of fibre systems. Also, an adequate vocabulary to describe the various types of connections has not yet been developed. In Fig. 58 I have attempted to bring together a few of these variations, which may be tentatively designated as follows:

a) The through-conducting, usually myelinated fibre.

b) The 'open-line' polysynaptic system, which at several synaptic interruptions is accessible to side inputs, as described by Ricardoh and Koh (1978, p. 20). Their elucidation is worth quoting in full; after indicating that such systems are accessible at each synaptic interruption to side inputs of related ('re-entrant' circuits) or of unrelated origins, they continue: 'The "open-line" componentry of such systems appears to reflect the need of many homeostatic functions to be guided by several rather than a single modality of afferent signals; simultaneously, it might serve as a device allowing selective and finely graded modulation of the impulse flow by re-entering circuits'.

c1) The 'open-line' fibre issuing collaterals and synaptic endings throughout its course. Fibres of this type have recently been observed by Swanson et al. (1984) to connect the subpallidal region with the pedunculopontine nucleus.

c2) The 'open-line' fibre issuing collaterals and synaptic endings in selected centres. The fact that the fibres indicated under c1) and c2) participate in the formation of classical synapses implies that their mode of secretion is neurocrine.

d1) The 'open-line', unmyelinated fibre provided with paracrine output sites throughout.
d2) The 'open-line', unmyelinated fibre possessing paracrine output sites in selected centres.

It should be appreciated that most long ascending and descending pathways of classical neuroanatomy are composed of type-a fibres, whereas in the 'open multi-neuromediator channels' postulated earlier in this work (see p. 128), axons or axonal chains of types b and c, and particularly type d, are thought to prevail. It is also worth noting that a particular 'open-line' fibre, emanating from a neuron containing and releasing more than one neuromediator, may well exert different influences in different centres, depending on the types of receptors present on the neurons in these centres (Swanson 1983).
Neurocrine (i.e. classical synaptic) neurotransmission has been repeatedly touched upon in this book, but only in a very simplified manner (see e.g. Fig. 50b). It is appropriate to emphasize that this mode of chemical transmission is in fact very complex, because each of the sequential steps in this process – (a) the manufacturing of the synthesizing enzyme(s) of the neuromediator(s), (b) the transport of these synthesizing enzymes, (c) the production of the neuromediator(s), (d) the release of the neuromediator(s), (e) the eventual reuptake of the neu-

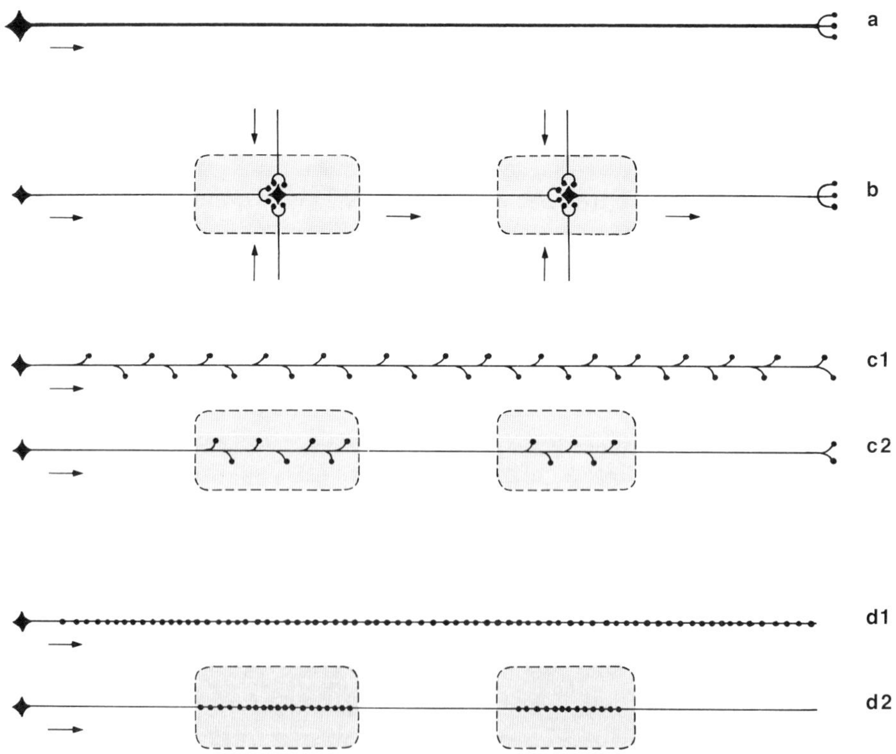

Fig. 58. Diagrammatic representation of different types of fibre connections: *a*, through-conducting fibre; *b*, open-line connection with afferent influences in several way stations; *c1*, open-line fibre with neurocrine outputs throughout; *c2*, open-line fibre with neurocrine outputs in selected centres; *d1*, open-line fibre with paracrine outputs throughout; *d2*, open-line fibre with paracrine outputs in selected centres

romediator(s), (f) the binding of the neuromediator(s) with their specific receptors, and (g) the degradation of the neuromediator(s) – is under the control of 'modulating' influences.

Paracrine chemical neurotransmission, which is by definition non-synaptic, has been postulated here to be present widely in the central nervous system, particularly in the entities tentatively designated as the core and paracore regions (Figs. 55–57). It should be emphasized once more that direct evidence for the occurrence of this process in the mammalian neuraxis is entirely lacking at present.

Neurocrine secretion and paracrine secretion are different processes; however, they have one important feature in common, i.e. exocytotic release of neuromediators (Fig. 50). There is also evidence suggesting that several modes of non-synaptic, non-exocytotic chemical neurotransmission play a modulatory role in the central nervous system (for a recent review, see Cooper and Meyer 1984).

In relation to the possible occurrence of paracrine secretion, it has been emphasized in this book that not only the synaptic clefts, but the entire extracellular space throughout the central nervous system is available for neuronal interactions, and thus potentially represents a most important communication channel (Fig. 50). With regard to this space it is appropriate to quote Watson (1974, p. 251), who stated: 'The histology of the brain has had two phases: in the first, light microscopists described spaces that were not there; in the second, electron microscopists denied the existence of spaces that were.' At present, several lines of evidence indicate that the extracellular space between neurons and glia cells adds up to about 20% of brain volume (Varon and Somjen 1979; Cragg 1979). Nicholson (1979) noted that this value can be equated with an average intercellular gap of 20 nm. However, it is quite possible that the width of this space varies from place to place. It is also important to note that, from a biophysical point of view, there are no fundamental objections against the proposition that molecules, including neuropeptide molecules, may well travel over a considerable distance, i.e. several hundred micra, via the neural extracellular space. Finally, I should like to mention that I concur with the opinion of Chan-Palay (1982) that the periventricular part of the neural extracellular space presumably communicates via ependymal elements with the ventricular system, which in its turn may also represent an important communication channel, acting as a vehicle for the widespread distribution of neuroactive substances. Interestingly, Rennels et al. (1985) most recently presented experimental evidence suggesting that in the central nervous system a circulation of fluid occurs via 'paravascular pathways', which include the perivascular spaces surrounding large penetrating vessels and the basal laminae around the intraparenchymal capillary network.

Little attention has been paid in this book to the functions of neuroglial cells. For general information on the established and possible roles of these elements the reader is referred to the able reviews of Watson (1974) and Varon and Somjen (1979). Suffice it to mention here that glia (a) regulates the composition of the extracellular fluid by active uptake of K^+ from and the release of Na^+, Cl^- and Ca^{++} into that fluid, and (b) may remove neuromediators, particularly amino acids, from the extracellular fluid. It is supposed that the clearance function last mentioned is exerted by velate glial processes which surround synapses or synapse complexes. Alternatively, such processes are considered to insulate the junctional structures they ensheath so as to prevent or reduce the diffusion of neuromediators. Conceivably, larger compartments could also be surrounded by glial processes. It is interesting in this context that the intriguing striosomes, which have a diameter of several hundred micra, are at least partly rimmed by glial septa (Graybiel 1984). Still larger compartments, for instance those shown diagrammatically in Fig. 58, could also be surrounded by glial processes, thus forming what might be termed 'microhomeostatic units'.

In conclusion I should like to emphasize that the hypothetical 'new' entities which I have introduced in this book, i.e. the [paracrine?] core and the median and lateral paracores, are to be taken merely as heuristic concepts with a view to learning something by paying special attention to both the classical *and* chemical neuroanatomical peculiarities of the regions involved: these entities are not meant as moulds into which the facts are to be fitted. The idea that in the central nervous system (a) there is a central core in which multipeptidergic transmission predominates, (b) that this is assisted by largely monoaminergic paracores, and (c) that the paracores in turn have reciprocal relations with superimposed structures, at least one of which (the cerebellum) operates almost entirely by amino acid neurotransmitters may be attractive in its simplicity. However, the well-established fact that the cerebral cortex contains numerous different neuropeptides should keep us thinking – as perhaps they do, indeed, keep us thinking!

Table 7. Abbreviations for amino acids

Ala	Alanine
Arg	Arginine
Asn	Asparagine
Asp	Aspartic acid
Cys	Cysteine
Gln	Glutamine
Glu	Glutamic acid
Gly	Glycine
(p-Gly	pyro-Glycine)
His	Histidine
Ile	Isoleucine
Leu	Leucine
Lys	Lysine
Met	Methionine
Phe	Phenylalanine
Pro	Proline
Ser	Serine
Thr	Threonine
Trp	Tryptophan
Tyr	Tyrosine
Val	Valine

References

Aamodt A, Aambø A, Walaas I, Søreide AJ, Fonnum F (1984) Autoradiographic demonstration of glutamate structures after stereotaxic injection of kainic acid in rat hippocampus. Brain Res 294:341–345

Abols IA, Basbaum AI (1981) Afferent connections of the rostral medulla of the cat: A neural substrate for midbrain-medullary interactions in the modulation of pain. J Comp Neurol 201:285–297

Adams JC (1983) Cytology of periolivary cells and the organization of their projections in the cat. J Comp Neurol 215:275–289

Adèr J-P, Room P, Postema F, Korf J (1980) Bilaterally diverging axon collaterals and contralateral projections from rat locus coeruleus neurons, demonstrated by fluorescent retrograde double labeling and norepinephrine metabolism. J Neural Transm 49:207–218

Aghajanian GK, Gallager DW (1975) Raphe origin of serotonergic nerves terminating in the cerebral ventricles. Brain Res 88:221–231

Agnati LF, Fuxe K, Zini I, Benfenati F, Hökfelt T, De Mey J (1982) Principles for the morphological characterization of transmitter-identified nerve cell groups. J Neurosci Meth 6:157–167

Agnati LF, Fuxe K, Calza L, Zini I, Hökfelt T, Steinbusch A, Verhofstad A (1984) A method for rostrocaudal integration of morphometric information from transmitter-identified cell groups. A morphometrical identification and description of 5-HT cell groups in the medulla oblongata of the rat. J Neurosci Meth 10: 83–101

Aizawa T, Greer MA (1981) Delineation of the hypothalamic area controlling thyrotropin secretion in the rat. Endocrinol 109:1731–1738

Akil H, Watson SJ, Berger PhA, Barchas JD (1978) Endorphins, β-LPH and ACTH: Biochemical, pharmacological and anatomical studies. Adv Biochem Psychoph 18:125–139

Akil H, Watson SJ, Young B, Lewis ME, Khachaturian H, Walker JM (1984) Endogenous opioids: Biology and function. Ann Rev Neurosci 7:223–255

Altschuler RA, Neises GR, Harmison GG, Wenthold RJ, Fex J (1981) Immunocytochemical localization of aspartate aminotransferase immunoreactivity in cochlear nucleus of the guinea pig. Proc Natl Acad Sci USA 78:6553–6557

Altschuler RA, Fex J, Parakkal MH, Eckenstein F (1984) Colocalization of enkephalin-like and choline acetyltransferase-like immunoreactivites in olivocochlear neurons of the guinea pig. J Histochem Cytochem 32:839–843

Amaral DG, Sinnamon HM (1977) The locus coeruleus: neurobiology of a central noradrenergic nucleus. Prog Neurobiol 9:147–196

Anand P, Gibson SJ, McGregor GP, Blank MA, Ghatei MA, Bacarese-Hamilton AJ, Polak JM, Bloom SR (1983) A VIP-containing system concentrated in the lumbosacral region of human spinal cord. Nature 305:143–145

Andén NE, Carlsson A, Dahlström A, Fuxe K, Hillarp NA, Larrson K (1964) Demonstration and mapping out of nigro-neostriatal dopamine neurons. Life Sciences 3:523–530

Anderson EG, Proudfit HK (1981) The functional role of the bulbospinal serotonergic nervous system. In: Jacobs BL, Gelperin A (eds) Serotonin Neurotransmission and Behavior. The MIT Press, London, pp 307–338

Anderson P, Eccles JC, Løyning Y (1964) Location of postsynaptic inhibitory synapses on hippocampal pyramids. J Neurophysiol 27:592–607

Andres KH, Von Düring M (1976) The synapse. In: Gispen WH (ed) Molecular and functional neurobiology. Elsevier, Amsterdam, pp 1–45

Anthony ELP, King JC, Stopa EG (1984) Immunocytochemical localization of LHRH in the median eminence, infundibular stalk, and neurohypophysis. Cell Tissue Res 236:5–14

Antonaccio MJ (1977) Neuropharmacology of central mechanisms governing the circulation. In: Antonaccio MJ (ed) Cardiovascular pharmacology. Raven Press, New York, pp 131–165

Antunes JL, Carmel PW, Zimmerman EA (1977) Projections from the paraventricular nucleus to the zona externa of the median eminence of the rhesus monkey: an immunohistochemical study. Brain Res 137:1–10

Araki M, McGeer PL, McGeer EG (1984) Retrograde HRP tracing combined with a pharmacohistochemical method for GABA transaminase for the identification of presumptive GABAergic projections to the habenula. Brain Res 304:271–277

Arendt T, Bigl V, Arendt A, Tennstedt A (1983) Loss of neurons in the nucleus basalis of Meynert in Alzheimer's disease, paralysis agitans and Korsakoff's disease. Acta Neuropathol 61:101–108

Arluison M, Conrath-Verrier M, Tauc M, Mailly Ph, Séraphin de la Manche I, Cesselin F, Bourgoin S, Hamon M (1983) Different localizations of met-enkephalin-like immunoreactivity in rat forebrain and spinal cord using hydrogen peroxide and triton X-100 light microscopic study. Brain Res Bull 11:555–571

Arluison M, Dietl M, Thibault J (1984) Ultrastructural morphology of dopaminergic nerve terminals and synapses in the striatum of the rat using tyrosine hydroxylase immunocytochemistry: A topographical study. Brain Res Bull 13:269–285

Armstrong DA, Saper CB, Levey AI, Wainer BH, Terry RD (1983) Distribution of cholinergic neurons in the rat brain: demonstrated by the immunohistochemical localization of choline acetyltransferase. J Comp Neurol 216:53–68

Armstrong DM, Ross CA, Pickel VM, Joh TH, Reis DJ (1982) Distribution of dopamine-, noradrenaline-, and adrenaline-containing cell bodies in the rat medulla oblongata: Demonstrated by the immunocytochemical localization of catecholamine biosynthetic enzymes. J Comp Neurol 212:173–187

Armstrong DM, Miller RJ, Beaudet A, Pickel VM (1984) Enkephalin-like immunoreactivity in rat area postrema: Ultrastructural localization and coexistence with serotonin. Brain Res 310:269–278

Arnold MB (1969) Emotion, motivation and the limbic system. Ann NY Acad Sci 159:1041–1058

Aronin N, DiFiglia M, Liotta AS, Martin JB (1981) Ultrastructural localization and biochemical features of immunoreactive leu-enkephalin in monkey dorsal horn. J Neurosci 1:561–577

Aronin N, DiFiglia M, Graveland GA, Schwartz WJ, Wu J-Y (1984) Localization of immunoreactive enkephalins in GABA synthesizing neurons of the rat neostriatum. Brain Res 300:376–380

Aston-Jones G, Shaver R, Dinan T (1984) Cortically projecting nucleus basalis neurons in rat are physiologically heterogeneous. Neurosci Lett 46:19–24

Axelsson S, Björklund A, Falck B, Lindvall O, Svensson LA (1973) Glyoxylic acid condensation: a new fluorescence method for the histochemical demonstration of biogenic monoamines. Acta Physiol Scand 87:57–62

Azmitia EC, Segal M (1978) An autoradiographic analysis of the differential ascending projections of the dorsal and median raphe nuclei in the rat. J Comp Neurol 179:641–668

Azmitia E, Gannon P (1983) The ultrastructural localization of serotonin immunoreactivity in myelinated and unmyelinated axons within the medial forebrain bundle of rat and monkey. J Neurosci 3:2083–2090

Baali-Cherif H, Arluison M, Tramu G (1984) Ultrastructural study of cholecystokinin-like immunoreactive nerve terminals in the rat nucleus accumbens. Neurosci Lett 49:331–335

Barber RP, Vaughn JE, Saito K, McLaughlin BJ, Roberts E (1978) GABAergic terminals are presynaptic to primary afferent terminals in the substantia gelatinosa of the rat spinal cord. Brain Res 141:35–55

Barber RP, Vaughn JE, Roberts E (1982) The cytoarchitecture of gabaergic neurons in rat spinal cord. Brain Res 238:305–328

Barbin G, Palacios JM, Garbarg M, Schwartz JC, Gaspar P, Javoy-Agid F, Agid Y (1980) L-histidine decarboxylase in the human brain: properties and localization. J Neurochem 35:400–406

Barchas JD, Akil H, Elliott GR, Holman RB, Watson SJ (1978) Behavioral neurochemistry: Neuroregulators and behavioral states. Science 200:964–973

Bargmann W, Scharrer E (1951) The site of the hormones of the posterior pituitary. Am Sci 39:255–259

Barry J (1977) Immunofluorescence study of LRF neurons in man. Cell Tissue Res 181:1–14

Barry J (1978) Septo-epithalamo-habenular LRF-reactive neurons in monkeys. Brain Res 151:183–187

Basbaum AI, Fields HL (1978) Endogenous pain control mechanisms: review and hypothesis. Ann Neurol 4:451–462

Basbaum AI, Fields HL (1979) The origin of descending pathways in the dorsolateral funiculus of the spinal cord of the cat and rat: pain modulation. J Comp Neurol 187:513–532

Basbaum AI, Fields HL (1984) Endogenous pain control systems: Brainstem spinal pathways and endorphin circuitry. Ann Rev Neurosci 7:309–338

Basbaum AI, Glazer EJ (1983) Immunoreactive vasoactive intestinal polypeptide is concentrated in the sacral spinal cord: A possible marker for pelvic visceral afferent fibers. Somatosens Res 1:69–82

Basbaum AI, Clanton CH, Fields HL (1978) Three bulbospinal pathways from the rostral medulla of the cat: an autoradiographic study of pain modulating systems. J Comp Neurol 178:209–224

Bates D, Weinshilboum RM, Campbell RJ, Sundt Jr, TM (1977) The effect of lesions in the locus coeruleus on the physiological responses of the cerebral blood vessels in cats. Brain Res 136:431–443

Beach TG, McGeer EG (1983) Neocortical substance P neurons in the baboon: An immunohistochemical finding. Neurosci Lett 41:265–270

Beach TG, McGeer EG (1984) The distribution of substance P in the primate basal ganglia: an immunohistochemical study of baboon and human brain. Neuroscience 13:29–52

Beal MF, Domesick VB, Martin JB (1983) Regional somatostatin distribution in the rat striatum. Brain Res 278:103–108

Beaudet A, Descarries L (1978) The monoamine innervation of rat cerebral cortex: synaptic and nonsynaptic axon terminals. Neurosci 3:851–860

Beaudet A, Sotelo C (1980) Synaptic remodeling of serotonin axon terminals in rat agranular cerebellum. Brain Res 206:305–329

Beckstead RM, Norgren R (1979) An autoradiographic examination of the central distribution of the trigeminal, facial, glossopharyngeal and vagal nerves in the monkey. J Comp Neurol 184:455–472

Beckstead RM, Domesick VB, Nauta WJH (1979) Efferent connections of the substantia nigra and ventral tegmental area in the rat. Brain Res 175:191–217

Beinfeld MC, Meyer DK, Brownstein MJ (1980) Cholecystokinin octapeptide in the rat hypothalamo-neurohypophysial system. Nature 288:376–378

Beinfeld MC, Meyer DK, Eskay RL, Jensen RT, Brownstein MJ (1981) The distribution of cholecystokinin immunoreactivity in the central nervous system of the rat as determined by radioimmunoassay. Brain Res 212:51–57

Beitz AJ (1982a) The sites of origin of brain stem neurotensin and serotonin projections to the rodent nucleus raphe magnus. J Neurosci 2:829–842

Beitz AJ (1982b) The nuclei of origin of brain stem enkephalin and substance P projections to the rodent nucleus raphe magnus. Neuroscience 7:2753–2768

Beitz AJ (1982c) The organization of afferent projections to the midbrain periaqueductal gray of the rat. Neuroscience 7:133–159

Beitz AJ, Shepard RD, Wells W (1983) The periaqueductal gray-raphe magnus projection contains somatostatin, neurotensin and serotonin but not cholecystokinin. Brain Res 261:132–137

Belcher G, Smock T, Fields HL (1982) Effects of intrathecal ACTH on opiate analgesia in the rat. Brain Res 247:373–377

Ben Ari Y, Le Gal La Salle G, Barbin G, Schwartz JC, Garbarg M (1977) Histamine synthesizing afferents within the amygdaloid complex and bed nucleus of the stria terminalis of the rat. Brain Res 138:285–294

Ben-Jonathan N, Oliver C, Weiner HJ, Mical RS, Porter JC (1977) Dopamine in hypophysial portal plasma of the rat during the estrous cycle and throughout pregnancy. Endocrinology 100:452–458

Bennett JP Jr, Snyder SH (1976) Angiotensin II binding to mammalian brain membranes. J Biol Chem 251:7423–7430

Bennett-Clarke C, Joseph SA (1982) Immunocytochemical distribution of LHRH neurons and processes in the rat: Hypothalamic and extrahypothalamic locations. Cell Tissue Res 221:493–504

Bennett-Clarke C, Romagno MA, Joseph SA (1980) Distribution of somatostatin in the rat brain: Telencephalon and diencephalon. Brain Res 188:473–486 (1980)

Berk ML, Finkelstein JA (1981) Afferent projections to the preoptic area and hypothalamic regions in the rat brain. Neuroscience 6:1601–1624

Bernardi G, Marciani MG, Morocutti C, Pavone F, Stanzione P (1978) The action of dopamine on rat caudate neurones intracellularly recorded. Neurosci Lett 8:235–240

Bigl V, Woolf NJ, Butcher LL (1982) Cholinergic projections from the basal forebrain to frontal, parietal, temporal, occipital and cingulate cortices: a combined fluorescent tracer and acetylcholinesterase analysis. Brain Res Bull 8:727–749

Bird ED, Spokes EGS, Iversen LL (1979) Increased dopamine concentration in limbic areas of brain from patients dying with schizophrenia. Brain 102:347–360

Björklund A, Nobin A (1973) Fluorescence histochemical and microspectrofluorometric mapping of dopamine and noradrenaline cell groups in the rat diencephalon. Brain Res 51:193–205

Björklund A, Skagerberg G (1979) Evidence for a major spinal cord projection from the diencephalic A11 dopamine cell group in the rat using transmitter-specific fluorescent retrograde tracing. Brain Res 177:170–175

Björklund A, Nobin A, Stenevi U (1973) The use of neurotoxic dihydroxytryptamines as tools for morphological studies and localized lesioning of central indolamine neurons. Z Zellforsch 145:479–501

Björklund A, Lindvall O, Nobin A (1975) Evidence of an incerto-hypothalamic dopamine neurone system in the rat. Brain Res 89:29–42

Blackstad TW, Fuxe K, Hökfelt T (1967) Noradrenaline nerve terminals in the hippocampal region of the rat and the guinea pig. Z Zellforsch 78:463–473

Blessing WW, Chalmers JP (1979) Direct projection of catecholamine (presumably dopamine) – containing neurons from hypothalamus to spinal cord. Neurosci Lett 11:35–40

Blessing WW, Furness JB, Costa M, West MJ, Chalmers JP (1981a) Projection of ventrolateral medullary (A1) catecholamine neurons toward nucleus tractus solitarii. Cell Tissue Res 220:27–40

Blessing WW, Goodchild AK, Dampney RAL, Chalmers JP (1981b) Cell groups in the lower brain stem of the rabbit projecting to the spinal cord, with special reference to catecholamine-containing neurons. Brain Res 221:35–55

Blessing WW, Jaeger CB, Ruggiero DA, Reis DJ (1982) Hypothalamic projections of medullary catecholamine neurons in the rabbit: A combined catecholamine fluorescence and HRP transport study. Brain Res Bull 9:279–286

Bloch B, Bugnon C, Fellmann D, Lenys D (1978) Immunocytochemical evidence that the same neurons in the human infundibular nucleus are stained with anti-endorphins and antisera of other related peptides. Neurosci Lett 10:147–152

Bloch B, Bugnon C, Fellmann D, Lenys D, Gouget A (1979) Neurons of the rat hypothalamus reactive with antisera against endorphins, ACTH, MSH and β-LPH. Cell Tissue Res 204:1–15

Bloom F, Battenberg E, Rossier J, Ling N, Leppaluoto J, Vargo ThM, Guillemin R (1977) Endorphins are located in the intermediate and anterior lobes of the pituitary gland, not in the neurohypophysis. Life Sci 20:43–48

Bloom F, Battenberg E, Rossier J, Ling N, Guillemin R (1978) Neurons containing β-endorphin in rat brain exist separately from those containing enkephalin: Immunocytochemical studies. Proc Natl Acad Sci USA 75:1591–1595

Bloom F, Battenberg ELF, Shibasaki T, Benoit R, Ling N, Guillemin R (1980) Localization of γ-melanocyte stimulating hormone (γ-MSH) immunoreactivity in rat brain and pituitary. Regulat Peptides 1:205–222

Bobillier P, Petitjean F, Salvert D, Ligier M, Sequin S (1975) Differential projections of the nucleus raphe dorsalis and nucleus raphe centralis as revealed by autoradiography. Brain Res 85:205–210

Bobillier P, Seguin S, Petitjean F, Salvert D, Touret M, Jouvet M (1976) The raphe nuclei of the cat brain stem: a topographical atlas of their efferent projections as revealed by autoradiography. Brain Res 113:449–486

Bobillier P, Seguin S, Degueurce A, Lewis BD, Pujol J-F (1979) The efferent connections of the nucleus raphe centralis superior in the rat as revealed by autoradiography. Brain Res 166:1–8

Bogerts B (1981) A brainstem atlas of catecholaminergic neurons in man, using melanin as a natural marker. J Comp Neurol 197:63–80

Bogerts B, Häntsch J, Herzer M (1983) A morphometric study of the dopamine-containing cell groups in the mesencephalon of normals, Parkinson patients, and schizophrenics. Biol Psychiat 18:951–969

Bohus B (1979) Neuropeptide influence on sexual and reproductive behavior. In: Zichella L, Pancheri P (eds) Psychoneuro-endocrinology in Reproduction. An interdisciplinary Approach. Developments in En-

docrinology, Vol 5, Elsevier/North Holland, Amsterdam, pp 111–120
Bohus B, De Wied D (1980) Pituitary-adrenal system hormones and adaptive behavior. In: Jones IC, Henderson JW (eds) General, Comparative and Clinical Endocrinology of the Adrenal Cortex. Academic Press, London, pp 265–347
Bolam JP (1984) Synapses of identified neurons in the neostriatum. In: Evered D, O'Connor M (eds) Functions of the basal ganglia; Ciba Foundation symposium 107. Pitman, London, pp 30–47
Bolam JP, Clarke DJ, Smith AD, Somogyi P (1983a) A type of aspiny neuron in the rat neostriatum accumulates (^{3}H)γ-aminobutyric acid: combination of Golgi-staining, autoradiography, and electron microscopy. J Comp Neurol 213:121–134
Bolam JP, Somogyi P, Takagi H, Fodor I, Smith AD (1983b) Localization of substance P-like immunoreactivity in neurons and nerve terminals in the neostriatum of the rat: a correlated light and electron microscopic study. J Neurocytol 12:325–344
Botticelli LJ, Cox BM, Goldstein A (1981) Immunoreactive dynorphin in mammalian spinal cord and dorsal root ganglia. Proc Natl Acad Sci USA 78:7783–7786
Bowker RM, Westlund KN, Coulter JD (1981a) Origins of serotonergic projections to the spinal cord in rat: An immunocytochemical-retrograde transport study. Brain Res 226:187–199
Bowker RM, Westlund KN, Coulter JD (1981b) Serotonergic projections to the spinal cord from the midbrain in the rat: An immunocytochemical and retrograde transport study. Neurosci Lett 24:221–226
Bowker RM, Westlund KN, Sullivan MC, Wilber JF, Coulter JD (1983) Descending serotonergic, peptidergic and cholinergic pathways from the raphe nuclei: A multiple transmitter complex. Brain Res 288:33–48
Braak H (1970) Über die Kerngebiete des menschlichen Hirnstammes. II. Die Raphekerne. Z Zellf 107:123–141
Bradbury AF, Smyth DG, Snell CR (1976a) Lipotropin: Precursor to two biologically active peptides. Biochem Biophys Res Co 69:950–956
Bradbury AF, Feldberg WF, Smyth DG, Snell CR (1976b) Lipotropin C-fragment: An endogenous peptide with potent analgesic activity. In: Kosterlitz HW (ed) Opiates and Endogenous Opioid Peptides. Elsevier/North Holland Biomedical Press, Amsterdam, pp 9–17
Bradbury AF, Smyth DG, Snell CR, Birdsall NJM, Hulme EC (1976c) C-fragment of lipotropin has a high affinity for brain opiate receptors. Nature 260:793–795
Bragin EO, Yeliseeva ZV, Vasilenko GF, Meizerov EE, Chuvin BT, Durinyan RA (1984) Cortical projections to the periaqueductal grey in the cat: A retrograde horseradish peroxidase study. Neurosci Lett 51:271–275
Brann MR, Emson PC (1980) Microiontophoretic injection of fluorescent tracer combined with simultaneous immunofluorescent histochemistry for the demonstration of efferents from the caudate-putamen projecting to the globus pallidus. Neurosci Lett 16:61–65
Brazeau P, Vale W, Burgus R, Ling N, Butcher M, Rivier J, Guillemin R (1973) Hypothalamic polypeptide that inhibits the secretion of immunoreactive pituitary growth hormone. Science 179:77–79
Broca P (1878) Anatomie comparée des circonvolutions cérébrales. Le grand lobe limbique et la scissure limbique dans la série des mammifères. Rev Anthrop 1:385–498
Brodal A (1981) Neurological Anatomy. Third Ed. Oxford University Press, New York, pp 1053
Brodal A, Taber E, Walberg F (1960) The raphe nuclei of the brain stem in the cat. II. Efferent connections. J Comp Neurol 114:239–259
Brooks MJ, Hubbard JI, Sirett NE (1983) Extracellular recording in rat area postrema in vitro and the effects of cholinergic drugs, serotonin and angiotensin II. Brain Res 261:85–90
Brown DA (1983) Slow cholinergic excitation – a mechanism for increasing neuronal excitability. Trends NeuroSci 6:303–307
Brown JC, Cook MA, Dryburgh JR (1973) Motilin, a gastric motor activity stimulating polypeptide: The complete amino acid sequence. Can J Biochem 51:533–537
Brown MR, Fisher LA, Rivier J, Spiess J, Rivier C, Vale W (1982) Corticotropin-releasing factor: effects on the sympathetic nervous system and oxygen consumption. Life Sci 30:207–210
Brownstein MJ, Palkovits M, Saavedra JM, Bassiri RM, Utiger RD (1974) Thyrotropin-releasing hormone in specific nuclei of rat brain. Science 185:267–269
Brownstein MJ, Utiger RD, Palkovits M, Kizer JS (1975) Effect of hypothalamic deafferentation on thyrotropin-releasing hormone levels in rat brain. Proc Natl Acad Sci USA 72:4177–4179
Brownstein MJ, Mroz EA, Tappaz ML, Leeman SE (1977) On the origin of substance P and glutamic acid decarboxylase (GAD) in the substantia nigra. Brain Res 135:315–323
Bryant MD, Bloom SR, Polak JM, Albuquerque RH, Modlin L, Pearse AGE (1976) Possible dual role for vasoactive intestinal peptide as gastrointestinal hormone and neurotransmitter substance. Lancet 1:991–993
Buck SH, Burks TF, Brown MR, Yamamura HI (1981) Reduction in basal ganglia and substantia nigra substance P levels in Huntington's disease. Brain Res 209:464–469
Bugnon C, Bloch B, Lenys D, Fellmann D (1979) Infundibular neurons of the human hypothalamus simultaneously reactive with antisera against endorphins, ACTH, MSH and β-LPH. Cell Tissue Res 199:177–196
Bugnon C, Fellmann D, Gouget A, Cardot J (1982) Corticoliberin in rat brain: immunocytochemical identification and localization of a novel neuroglandular system. Neurosci Lett 30:25–30
Buijs RM (1978) Intra- and extrahypothalamic vasopressin and oxytocin pathways in the rat. Pathways to the limbic system, medulla oblongata and spinal cord. Cell Tissue Res 192:423–435
Buijs RM, Swaab DF (1979) Immuno-electron microscopical demonstration of vasopressin and oxytocin synapses in the limbic system of the rat. Cell Tissue Res 204:355–365

Buijs RM, Swaab DF, Dogterom J, Leeuwen FW van (1978) Intra- and extrahypothalamic vasopressin and oxytocin pathways in the rat. Cell Tissue Res 186:423–433

Buijs RM, Vries GJ de, Leeuwen FW van, Swaab DF (1983) Vasopressin and oxytocin: distribution and putative functions in the brain. Prog Brain Res 60:115–122

Buijs RM, Geffard M, Pool CW, Hoorneman EMD (1984) The dopaminergic innervation of the supraoptic and paraventricular nucleus. A light and electron microscopical study. Brain Res 323:65–72

Buma P, Roubos EW Ultrastructural demonstration of non-synaptic release sites in the brains of the snail *Lymnaea stagnalis,* the insect *Periplaneta americana,* and the rat. Neuroscience, in press

Buma P, Roubos LW, Buijs RM (1984) Ultrastructural demonstration of exocytosis of neural, neuroendocrine secretions with an in vitro tannic acid (TARI-) method. Histochem 80:247–256

Burchanowski BJ, Sternberger LA (1980) Improved visualization of luteinizing hormone releasing hormone pathways by immunocytochemical staining of thick vibratome sections. J Histochem Cytochem 28:361–363

Burd GD, Davis BJ, Macrides F (1982a) Ultrastructural identification of substance P immunoreactive neurons in the main olfactory bulb of the hamster. Neuroscience 7:2697–2704

Burd GD, Marshall PE, Beal MF, Landis DMD, Martin JB (1982b) Effects of kainic and ibotonic acid on the neostriatal somatostatin system of the rat. Soc Neurosci Abstr 8:507

Burgus R, Dunn TF, Desideric D, Ward DN, Vale W, Guillemin R (1970) Characterization of ovine hypothalamic hypophysiotrophic TSH-releasing factor. Nature 226:321–325

Burnweit C, Forssmann WG (1979) Somatostatinergic nerves in the cervical spinal cord of the monkey. Cell Tissue Res 200:83–90

Burton H, Benjamin RM (1971) Central projections of the gustatory system. In: Beidler LM (ed) Handbook of Sensory Physiology, vol IV, pt 2, Springer, Berlin, pp 148–164

Butcher LL, Woolf NJ (1982) Cholinergic and Serotonergic Systems in the Brain and Spinal Cord: Anatomic Organization, Role in Intercellular Communication Processes and Interactive Mechanisms. Prog Brain Res 55:3–40

Cabot JB, Wild JM, Cohen DH (1979) Raphe inhibition of sympathetic preganglionic neurons. Science 203:184–186

Caffé AR, Leeuwen FW van (1983) Vasopressin-immunoreactive cells in the dorsomedial hypothalamic region, medial amygdaloid nucleus and locus coeruleus of the rat. Cell Tissue Res 233:23–33

Cajal, Ramon y (1911) Histologie du Systeme Nerveux de l'homme et des Vertebrés. Vol II. Maloine, Paris, pp 993

Calcutt CR (1976) The role of histamine in brain. Gen Pharmacol 7:15–25

Camacho A, Phillips MI (1981) Horseradish peroxidase study in rat of the neural connections of the organum vasculosum of the lamina terminalis. Neurosci Lett 25:201–204

Card JP, Moore RY (1984) The suprachiasmatic nucleus of the golden hamster: immunohistochemical analysis of cell and fiber distribution. Neuroscience 13:415–431

Carlton SM, Leichnetz GR, Young EG, Mayer DJ (1983) Supramedullary afferents of the nucleus raphe magnus in the rat: A study using the transcannula HRP gel and autoradiographic techniques. J Comp Neurol 214:43–58

Carraway R, Leeman SE (1973) The isolation of a new hypotensive peptide, neurotensin, from bovine hypothalami. J Biol Chem 248:6854–6861

Carraway R, Leeman SE (1975a) The amino acid sequence of a hypothalamic peptide, neurotensin. J Biol Chem 250:1907–1911

Carraway R, Leeman SE (1975b) The synthesis of neurotensin. J Biol Chem 250:1912–1918

Carraway R, Leeman SE (1976) Characterization of radioimmunoassayable neurotensin in the rat. J Biol Chem 251:7045–7052

Carstens E, Klump D, Zimmerman M (1980) Differential inhibitory effects of medial and lateral midbrain stimulation on spinal neuronal discharges to noxious skin heating in the cat. J Neurophysiol 43:332–342

Carstens E, Fraunhoffer M, Zimmerman M (1981) Serotoninergic mediation of descending inhibition from midbrain periaqueductal gray, but not reticular formation of spinal nociceptive transmission in the cat. Pain 10:149–167

Carter CJ (1982) Topographical distribution of possible glutamatergic pathways from the frontal cortex to the striatum and substantia nigra in rats. J Neuropharmacol 21:379–383

Carter DA, Fibiger HC (1977) Ascending projections of presumed dopamine-containing neurons in the ventral tegmentum of the rat as demonstrated by horseradish peroxidase. Neuroscience 2:569–576

Cash R, Ruberg M, Raisman R, Agid Y (1984) Adrenergic receptors in Parkinson's disease. Brain Res 322:269–275

Caverson MM, Ciriello J, Calaresu FR (1984) Paraventricular nucleus of the hypothalamus: an electrophysiological investigation of neurons projecting directly to intermediolateral nucleus in the cat. Brain Res 305:380–383

Cedarbaum JM, Aghajanian GK (1978) Afferent projections to the rat locus coeruleus as determined by a retrograde tracing technique. J Comp Neurol 178:1–16

Changaris DG, Keil LC, Severs WB (1978) Angiotensin II immunohistochemistry of the rat brain. Neuroendocrinology 25:257–274

Chan-Palay V (1977) Indoleamine neurons and their processes in the normal rat brain and in chronic diet-induced thiamine deficiency demonstrated by uptake of ^{3}H-serotonin. J Comp Neurol 176:467–494

Chan-Palay V (1978) The paratrigeminal nucleus. II. Identification and inter-relations of catecholamine axons, indoleamine axons, and substance P immunoreactive cells in the neuropil. J Neurocytol 7:419–442

Chan-Palay V (1982) Serotonin neurons and their axons in the raphe dorsalis of rat and rhesus monkey: Demonstration by high resolution autoradiography with ^{3}H-serotonin. In: Chan-Palay V, Palay SL (eds) Cy-

tochemical Methods in Neuroanatomy. AR Liss Inc, New York, pp 357–386
Chan-Palay V (1984) Purkinje cells of the cerebellum: Localization and function of multiple neuroactive substances. Exp Brain Res Suppl 9:129–144
Chan-Palay V, Jonsson G, Palay SL (1978) Serotonin and substance P coexist in neurons of the rat's central nervous system. Proc Natl Acad Sci USA 75:1582–1586
Chan-Palay V, Nilaver G, Palay SL, Beinfeld MC, Zimmerman EA, Wu J-Y, O'Donohue TL (1981) Chemical heterogeneity in cerebellar Purkinje cells: Existence and coexistence of glutamic acid decarboxylase-like and motilin-like immunoreactivities. Proc Natl Acad Sci USA 78:7787–7791
Chan-Palay V, Lin CT, Palay S, Yamamoto M, Wu J-Y (1982a) Taurine in the mammalian cerebellum: Demonstration by autoradiography with (^{3}H)taurine and immunocytochemistry with antibodies against the taurine-synthesizing enzyme, cysteine-sulfinic acid decarboxylase. Proc Natl Acad Sci USA 79:2695–2699
Chan-Palay V, Palay S, Wu J-Y (1982b) Sagittal cerebellar microband of taurine neurons: Immunocytochemical demonstration by using antibodies against the taurine-synthesizing enzyme cysteine sulfinic acid decarboxylase. Proc Natl Acad Sci USA 79:4221–4225
Chan-Palay V, Ito M, Tongroach P, Sakurai M, Palay SL (1982c) Inhibitory effects of motilin, somatostatin, leucine- and methionine enkephalin, and taurine on neurons of the lateral vestibular nucleus: their interaction with gamma-aminobutyric acid. Proc Natl Acad Sci USA 79:3355–3359
Charnay Y, Paulin C, Chauvialle J-A, Dubois PM (1983) Distribution of substance P-like immunoreactivity in the spinal cord and dorsal root ganglia of the human foetus and infant. Neuroscience 10:41–55
Charney DS, Redmond DE Jr (1983) Neurobiological mechanisms in human anxiety. Neuropharmacol 22:1531–1536
Chavkin C, James IF, Goldstein A (1982) Dynorphin is a specific endogenous ligand of the kappa opioid receptor. Science 215:413–415
Cheung AL, Goldstein A (1976) Failure of hypophysectomy to alter brain content of opioid peptides (endorphins). Life Sci 19:1005–1008
Chevalier G, Thierry AM, Shibazaki T, Féger J (1981) Evidence for a gabaergic inhibitory nigrotectal pathway in the rat. Neurosci Lett 21:67–70
Childs JA, Gale K (1983) Neurochemical evidence for a nigrotegmental GABAergic projection. Brain Res 258:109–114
Cho HJ, Ani YS, Shiosaka S, Inagaki S, Kubota Y, Kiyama H, Umegaki K, Tateishi K, Hashimura E, Hamaoka T, Tohyama M (1983) Ontogeny of cholecystokinin-8-containing neuron system of the rat: An immunohistochemical analysis. I. Forebrain and upper brainstem. J Comp Neurol 218:25–41
Cho HJ, Shiotani Y, Tateish K, Hashimura E, Hamaoka T, Tohyama M (1984) Cholecystokinin-8-like immunoreactivity in the rat anterior commissure revealed by immunocytochemistry. Brain Res 300:372–375
Clark AW, Parhad IM, Folstein SE, Whitehouse PJ, Hedreen JC, Price DL, Chase GA (1983) The nucleus basalis in Huntington's disease. Neurology 33:1262–1267
Clavier RM (1979) Afferent projections to the self-stimulation regions of the dorsal pons, including the locus coeruleus, in the rat as demonstrated by the horseradish peroxidase technique. Brain Res Bull 4:497–504
Clement-Jones V, Tomlin S, Rees L, McLoughlin L, Besser G, Wen H (1980) Increased β-endorphin but not met-enkephalin levels in human cerebrospinal fluid after acupuncture for recurrent pain. Lancet 2:946–948
Clineschmidt BV, McGuffin J (1977) Neurotensin administered intracisternally inhibits responsiveness of mice to noxious stimuli. Europ J Pharmcol 46:395–396
Cohen MI (1979) Neurogenesis of respiratory rhythm in the mammal. Phys Rev 59:1105–1173
Collins GGS (1979) Effect of chronic bulbectomy on the depth distribution of amino acid transmitter candidates in rat olfactory cortex. Brain Res 171:552–555
Collins GGS, Anson J, Probett GA (1981) Patterns of endogenous amino acid release from slices of rat and guinea-pig olfactory cortex. Brain Res 204:103–120
Commissiong JW (1981) Evidence that the noradrenergic coerulospinal projection decussates at the spinal level. Brain Res 212:145–151
Commissiong JW, Hellström SO, Neff NH (1978) A new project from locus coeruleus to the spinal ventral columns: histochemical and biochemical evidence. Brain Res 148:207–213
Cone RI, Weber E, Barchas JD, Goldstein A (1983) Regional distribution of dynorphin and neo-endorphin peptides in rat brain, spinal cord and pituitary. J Neuroscience 3:2146–2152
Conrad LCA, Leonard ChM, Pfaff DW (1974) Connections of the median and dorsal raphe nuclei in the rat: an autoradiographic and degeneration study. J Comp Neurol 156:179–206
Conrad LCA, Pfaff DW (1976) Efferents from medial basal forebrain and hypothalamus in the rat. II. An autoradiographic study of the anterior hypothalamus. J Comp Neurol 169:221–262
Conrath-Verrier M, Dietl M, Arluison M, Cesselin F, Bourgoin S, Hamon M (1983) Localization of met-enkephalin-like immunoreactivity within pain-related nuclei of cervical spinal cord, brainstem and midbrain in the cat. Brain Res Bull 11:587–604
Conrath-Verrier M, Dietl M, Tramu G (1984) Cholecystokinin-like immunoreactivity in the dorsal horn of the spinal cord of the rat: A light and electron microscopic study. Neuroscience 13:871–885
Contestabile A, Fonnum F (1983) Cholinergic and GABA-ergic forebrain projections to the habenula and nucleus interpeduncularis: surgical and kainic acid lesions. Brain Res 275:287–297
Cools AR (1980) Role of the neostriatal dopaminergic activity in sequencing and selecting behavioural strategies: Facilitation of processes involved in selecting the best strategy in a stressful situation. Behav Brain Res 1:361–378
Cools AR, Van den Bercken JHL, Horstink MWI, Van Spaendonck KPM, Berger HJC (1984a) Cognitive

and motor shifting aptitude disorder in Parkinson's disease. J Neurol Neurosur Ps 47:443–453
Cools AR, Japsers R, Schwartz M, Sontag K-H, Vrijmoed-De Vries M, Van den Bercken J (1984b) Basal ganglia and switching motor programs In: McKenzie JS, Kemm RE, Wilcock LN (eds) Basal Ganglia Structure and Function. Plenum Publ Corp, New York, pp 513–544
Coons AH (1958) Fluorescent antibody methods. In: Danielli JF (ed) General Cytochemical Methods. Academic Press, New York, pp 399–422
Cooper PE, Fernstrom MH, Rorstad OP, Leeman SE, Martin JB (1981) The regional distribution of somatostatin, substance P and neurotensin in human brain. Brain Res 218:219–232
Cooper JR, Meyer EM (1984) Possible mechanisms involved in the release and modulation of release of neuroactive agents. Neurochem Int 6:419–433
Cragg B (1979) Overcoming the failure of electronmicroscopy to preserve the brain's extracellular space. Trends Neurosci 2:159–161
Critchlow V, Rice RW, Abe K, Vale W (1978) Somatostatin content of the median eminence in female rats with lesion-induced disruption of the inhibitory control of growth hormone secretion. Endocrinology 103:817–825
Cropper EC, Eisenman JS, Azmitia EC (1984) An immunocytochemical study of the serotonergic innervation of the thalamus of the rat. J Comp Neurol 224:38–50
Crutcher KA, Bingham WG Jr (1978) Descending monoaminergic pathways in the primate spinal cord. Am J Anat 153:159–164
Cuello AC, Kanazawa I (1978) The Distribution of substance P immunoreactive fibers in the rat central nervous system. J Comp Neurol 178:129–156
Cuello AC, Paxinos G (1978) Evidence for a long leu-enkephalin striopallidal pathway in rat brain. Nature 271:178–180
Cuello AC, Sofroniew MV (1984) The anatomy of the CNS cholinergic neurons. Trends NeuroSci 7:74–78
Cuello AC, DelFiacco M, Paxinos G (1978) The central and peripheral ends of the substance P-containing sensory neurones in the rat trigeminal system. Brain Res 152:499–509
Cuello AC, DelFiacco M, Paxinos G, Somogyi P, Priestley JV (1981) Neuropeptides in striato-nigral pathways. J Neural Transm 51:83–96
Cullheim S, Kellerth J-O (1981) Two kinds of recurrent inhibition of cat spinal α-motoneurones as differentiated pharmacologically. J Physiol 312:209–224
Cummings JP, Felten DL (1979) A raphe dendrite bundle in the rabbit medulla. J Comp Neurol 183:1–24
Cummings S, Elde R, Ells J, Lindall A (1983) Corticotropin-releasing factor immunoreactivity is widely distributed within the central nervous system of the rat: An immunohistochemical study. J Neurosci 3:1355–1368
Cupédo RNJ, Weerd de H (1980) Serotonergic intraventricular axons in the habenular region. Phagocytosis after induced degeneration. Anat Embryol 158:213–226
Curtis DR, Johnson GAR (1974) Amino-acid transmitters in the mammalian central nervous system. Ergebn Physiol 69:97–188
Curtis DR, Watkins JC (1960) The excitation and depression of spinal neurons by structurally related amino acids. J Neurochem 6:117–141
Curtis DR, Hösli L, Johnston GAR, Johnston IH (1968) The hyperpolarization of spinal motoneurons by glycine and related amino acids. Exp Brain Res 5:235–258
Dahlgren N, Lindvall O, Sakabe T, Stenevi U, Siesjö B (1981) Cerebral blood flow and oxygen consumption in the rat brain after lesions of the noradrenergic locus coeruleus system. Brain Res 209:11–23 (1981)
Dahlström A, Fuxe K (1964) Evidence for the existence of monoamine-containing neurons in the central nervous system. I. Demonstration of monoamines in the cell bodies of brain stem neurons. Acta Physiol Scand 62, Suppl 232:1–55
Dahlström A, Fuxe K (1965) Evidence for the existence of monoamine neurons in the central nervous system. II. Experimentally induced changes in the intraneuronal amine levels of bulbospinal neuron systems. Acta Physiol Scand 64, Suppl. 247:1–36
Dahlström A, Fuxe K, Olson L, Ungerstedt U (1965) On the distribution and possible function of monoamine nerve terminals in the olfactory bulb of the rabbit. Life Sci 4:2071–2074
Dale HH (1935) Pharmacology and nerve-endings. Proc R Soc Lond B 28:319–322
Dalsgaard CJ, Vincent SR, Hökfelt T, Lundberg JM, Dahlström A, Schultzberg M, Dockray GJ, Cuello AC (1982a) Coexistence of cholecystokinin- and substance P-like peptides in neurons of the dorsal root ganglia of the rat. Neurosci Lett 33:159–163
Dalsgaard C-J, Hökfelt T, Elfvin L-G, Terenius L (1982b) Enkephalin-containing sympathetic preganglionic neurons projecting to the inferior mesenteric ganglion: evidence from combined retrograde tracing and immunohistochemistry. Neuroscience 7:2019–2050
Dampney RAL, Moon EA (1980) Role of ventrolateral medulla in vasomotor response to cerebral ischemia. Am J Physiol 239:H349–H358
Davies P (1979) Neurotransmitter-related enzymes in senile dementia of the Alzheimer type. Brain Res 171:319–327
Davies P, Josef J, Thompson A (1981) Anterior to posterior variations in the concentration of somatostatin-like immunoreactivity in human basal ganglia. Brain Res Bull 7:365–368
Davies SW, McBean GJ, Roberts PJ (1984) A glutamatergic innervation of the nucleus basalis/substantia innominata. Neurosci Lett 45:105–110
Day TA, Blessing W, Willoughby JO (1980) Noradrenergic and dopaminergic projections to the medial preoptic area of the rat. A combined horseradish peroxidase/catecholamine fluorescence study. Brain Res 193:543–548
DeFeudis FV (1984) GABA and endocrine regulation – relation to neurologic-psychiatric disorders. Neurochem Int 6:1–16
De Lanerolle NC, LaMotte CC (1983) Ultrastructure of chemically defined neurons systems in the dorsal horn of the monkey. I. Substance P immunoreactivity. Brain Res 274:31–49
DelFiacco M, Cuello AC (1980) Substance P- and enke-

phalin-containing neurones in the rat trigeminal system. Neuroscience 5:803–815
DelFiacco M, Paxinos G, Cuello AC (1982) Neostriatal enkephalin-immunoreactive neurones project to the globus pallidus. Brain Res 231:1–17
DelFiacco M, Dessì ML, Levanti MC (1984) Topographical localization of substance P in the human post-mortem brainstem. An immunohistochemical study in the newborn and adult tissue. Neurosci 12:591–611
Dennis M, Seidah NG, Chrétien M (1983) Regional heterogeneity in the processing of pro-opiomelanocortin in rat brain. Life Sci 33:49–52
De Robertis E (1966) Synaptic complexes and synaptic vesicles as structural and biochemical units of the central nervous system. In: Rodahl K, Issekutz B Jr (eds) Nerve as a tissue, Hoeber, New York, pp 88–115
De Robertis E (1967) Ultrastructure and cytochemistry of the synaptic region. Science 156:907–914
De Robertis E, Bennett H (1954) Submicroscopic vesicular component in the synapse. Fed Proc 13:35
De Robertis E, Bennett HS (1955) Some features of the submicroscopic morphology of synapses in frog and earthworm. J Biophys Biochem Cytol 1:47–58
Descarries L, Beaudet A, Watkins KC (1975) Serotonin nerve terminals in adult rat neocortex. Brain Res 100:563–588
Descarries L, Watkins KC, Lapierre Y (1977) Noradrenergic axon terminals in the cerebral cortex of rat. III. Topometric ultrastructural analysis. Brain Res 133:197–222
Descarries L, Watkins KC, Garcia S, Beaudet A (1982) The serotonin neurons in nucleus raphe dorsalis of adult rat: A light and electron microscope radioautographic study. J Comp Neurol 207:239–254
De Vries GJ, Buijs RM (1983) The origin of the vasopressinergic and oxytocinergic innervation of the rat brain with special reference to the lateral septum. Brain Res 273:307–317
De Wied D (1964) Influence of anterior pituitary on avoidance learning and escape behavior. Am J Physiol 207:255–259
De Wied D (1976) Behavioral effects of intraventricularly administered vasopressin and vasopressin fragments. Life Sci 19:685–690
De Wied D (1977) Peptides and behavior. Life Sci 20:195–204
De Wied D (1982) Pro-opio(melano)cortin and brain homeostasis. Prog Brain Res 55:473–482
De Wied D (1983) Central actions of neurohypophysial hormones. Prog Brain Res 60:155–167
De Wied D, Gispen WH (1977) Behavioral effects of peptides. In: Gainer H (ed) Peptides in neurobiology. Plenum Press, New York, pp 397–448
De Wied D, Van Wimersma Greidanus TjB, Bohus B, Uraban I, Gispen WH (1976) Vasopressin and memory consolidation. Prog Brain Res 45:181–194
De Wied D, Gaffori O, Van Ree JM, De Jong W (1984) Central target for the behavioural effects of vasopressin neuropeptides. Nature 308:276–278
Di Chiara G, Porceddu ML, Morelli M, Mulas ML, Gessa GL (1979) Evidence for a gabaergic projection from the substantia nigra to the ventromedial thalamus and to the superior colliculus of the rat. Brain Res 176:273–284
Dierickx K, Vandesande F (1979) Immunocytochemical demonstration of separate vasopressin-neurophysin and oxytocin-neurophysin neurons in the human hypothalamus. Cell Tissue Res 196:203–212
DiFiglia M, Aronin N (1984a) Quantitative electron microscopic study of immunoreactive somatostatin axons in the rat neostriatum. Neurosci Lett 50:325–331
DiFiglia M, Aronin N (1984b) Immunoreactive leu-enkephalin in the monkey hypothalamus including observations on its ultrastructural localization in the paraventricular nucleus. J Comp Neurol 225:313–326
DiFiglia M, Aronin N, Martin JB (1982a) Light and electron microscopic localization of immunoreactive leu-enkephalin in the monkey basal ganglia. J Neurosci 2:303–320
DiFiglia M, Aronin N, Leeman SE (1982b) Light microscopic and ultrastructural localization of immunoreactive substance P in the dorsal horn of monkey spinal cord. Neuroscience 7:1127–1139
Difiglia M, Aronin N, Leeman SE (1984) Ultrastructural localization of immunoreactive neurotensin in the monkey superficial dorsal horn. J Comp Neurol 225:1–12
Dismukes K (1977) New look at the aminergic nervous system. Nature 269:557–558
Divac I (1975) Magnocellular nuclei of the basal forebrain project to neocortex, brain stem, and olfactory bulb. Review of some functional correlates. Brain Res 93:385–398
Divac I, Fonnum F, Storm-Mathisen J (1977) High affinity uptake of glutamate in terminals of corticostriatal axons. Nature (Lond) 266:377–378
Dockray GJ (1976) Immunocytochemical evidence of cholecystokinin-like peptides in brain. Nature 264:568–570
Dockray GJ, Gregory RA, Hutchison JB (1978) Isolation, structure and biological activity of two cholecystokinin octapeptides from sheep brain. Nature 274:711–713
Dodd J, Kelly JS (1979) Excitation of CA1 pyramidal neurones of the hippocampus by the tetra- and octapeptide C-terminal fragments of cholecystokinin. J Physiol – London 295:61P–62P
Dodd J, Kelly JS, Said SI (1979) Excitation of CA1 neurons of the rat hippocampus by the octacosapeptide, vasoactive intestinal polypeptide (VIP). Brit J Pharmacol 66:125P
Doris PA (1984) Vasopressin and central integrative processes. Neuroendocrinol 38:75–85
Dray A, Gonye TJ, Oakley NR, Tanner T (1976) Evidence for the existence of a raphe projection to the substantia nigra in rat. Brain Res 113:45–57
Dubé D, Lissitzky JC, Leclerc R, Pelletier G (1978) Localization of α-melanocyte-stimulating hormone in rat brain and pituitary. Endocrinology 102:1283–1291
Dubois B, Ruberg M, Javoy-Agid F, Ploska A, Agid Y (1983) A subcorticocortical cholinergic system is affected in Parkinson's disease. Brain Res 288:213–218
Dyer RG, Dyball REJ (1974) Evidence for a direct effect of LRF and TRF on single unit activity in the rostral hypothalamus. Nature 252:486–488

Eccles JC (1957) The Physiology of Nerve Cells. Johns Hopkins Press, Baltimore, pp 270

Eccles JC (1969) The Inhibitory Pathways of the Central Nervous System. Univ Press, Liverpool, pp 135

Eccles JC (1973) The understanding of the brain. McGraw-Hill Book Comp, New York, pp 238

Eccles JC, Ito M, Szentágothai J (1967) The Cerebellum as a neuronal machine. Springer Verlag, Berlin, pp 335

Eckenstein F, Baughman RW (1984) Two types of cholinergic innervation in cortex, one co-localized with vasoactive intestinal polypeptide. Nature 309:153–155

Eckenstein F, Sofroniew MV (1983) Identification of central cholinergic neurons containing both choline acetyltransferase and acetylcholinesterase and of central neurons containing only acetylcholinesterase. J Neurosci 3:2286–2291

Edley SM, Graybiel AM (1983) The afferent and efferent connections of the feline nucleus tegmenti pedunculopontinus, pars compacta. J Comp Neurol 217:187–215

Edvinsson L, Degueerce A, Duverger D, MacKenzie ET, Scatton B (1983) Central serotoninergic nerves project to the pial vessels of the brain. Nature 306:55–57

Edwards SB (1975) Autoradiographic studies of the projections of the midbrain reticular formation: descending projections of nucleus cuneiformis. J Comp Neur 161:341–358

Eipper BA, Mains RE (1980) Structure and biosynthesis of pro-adrenocorticotropin/endorphin and related peptides. Endocr Rev 1:1–27

Elde R (1979) Localization of hypophysiotropic peptides and other biologically active peptides within the brain. Ann Rev Physiol 41:587–602

Emson PC (1979a) Peptides as neurotransmitter candidates in the mammalian CNS. Prog Neurobiol 13:61–116

Emson PC (1979b) Distribution of putative neurotransmitters in the neocortex. Neuroscience 4:1–30

Emson PC (ed) (1983) Chemical neuroanatomy. Raven Press, New York, pp 560

Emson PC, Koob GF (1978) The origin and distribution of dopamine-containing afferents to the rat frontal cortex. Brain Res 142:249–267

Emson PC, Fahrenkrug J, Schaffalitzky de Muckadell OB, Jessell TM, Iversen LL (1978a) Vasoactive intestinal polypeptide (VIP) vesicular localization and potassium evoked release from rat hypothalamus. Brain Res 143:174–178

Emson PC, Jessell T, Paxinos G, Cuello AC (1978b) Substance P in the amygdaloid complex, bed nucleus and stria terminalis of the rat brain. Brain Res 149:97–105

Emson PC, Arregui A, Clement-Jones V, Sandberg BEB, Rossor M (1980a) Regional distribution of methionine-enkephalin and substance P-like immunoreactivity in normal human brain and in Huntington's disease. Brain Res 199:147–160

Emson PC, Lee CM, Rehfeld JF (1980b) Cholecystokinin octapeptide: Vesicular localization and calcium dependent release from rat brain in vitro. Life Sci 26:2157–2163

Emson PC, Corder R, Ratter SJ, Tomlin S, Lowry PhJ, Ress L, Arregui A, Rosser MN (1984) Regional distribution of pro-opiomelanocortin-derived peptides in the human brain. Neuroendocrinology 38:45–50

Engberg I, Ryall RW (1966) The inhibitory action of noradrenaline and other monoamines on spinal neurones. J Physiol Lond 185:298–322

Eskay RL, Long RT, Palkovits M (1983) Localization of immunoreactive thyrotropin releasing hormone in the lower brainstem of the rat. Brain Res 277:159–162

Fagg GE, Foster AC (1983) Amino acid neurotransmitters and their pathways in the mammalian central nervous system. Neuroscience 9:701–719

Falck B (1962) Observations on the possibilities of the cellular localization of monoamines by a fluorescence method. Acta Physiol Scand Suppl 197, 56:1–25

Falck B, Hillarp NA, Thieme G, Torp A (1962) Flurorescence of catecholamines and related compounds condensed with formaldehyde. J Histochem Cytochem 10:348–354

Fallon JH, Moore RY (1978a) Catecholamine innervation of the basal forebrain. III. Olfactory bulb, anterior olfactory nuclei, olfactory tubercle and piriform cortex. J Comp Neurol 180:533–544

Fallon JH, Moore RY (1978b) Catecholamine innervation of the basal forebrain. IV. Topography of the dopamine projection to the neostriatum. J Comp Neurol 180:545–580

Fallon JH, Koziell DA, Moore RY (1978) Catecholamine innervation of the basal forebrain. II. Amygdala, suprarhinal cortex and entorhinal cortex. J Comp Neurol 180:509–532

Fallon JH, Loughlin SE (1982) Monoamine innervation of the forebrain: Collateralization. Brain Res Bull 9:295–307

Fallon JH, Seroogy KB (1984) Visual and auditory pathways contain cholecystokinin: Evidence from immunofluorescence and retrograde tracing. Neurosci Lett 45:81–87

Fardin V, Oliveras JL, Besson JM (1984) Projections from the periaqueductal gray matter to the B_3 cellular area (Nucleus raphe magnus and nucleus reticularis paragigantocellularis) as revealed by the retrograde transport of horseradish peroxidase in the rat. J Comp Neurol 223:483–500

Farley IJ, Hornykiewicz O (1976) Noradrenaline in subcortical brain regions of patients with Parkinson's disease and control subjects. In: Birkmeyer W, Hornykiewicz O (eds) Advances in Parkinsonism, Roche, Basel, pp 178–185

Felix D, Phillips MI (1978) Effects of angiotensin II on central neurons. In: Ryall RW, Kelly JS (eds) Iontophoresis and Transmitter Mechanisms in the Mammalian Central Nervous System. Elsevier/North-Holland, Amsterdam, pp 104–106

Felten DL (1976) Catecholamine neurons in the squirrel monkey hypothalamus. J Neural Transm 39:269–280

Felten DL, Crutcher KA (1979) Neuronal-vascular relationships in the raphe nuclei, locus coeruleus, and substantia nigra in primates. Am J Anat 155:467–482

Felten DL, Cummings JP (1979) The raphe nuclei of the rabbit brain stem. J Comp Neurol 187:199–244

Felten DL, Harrigan P (1980) Serotonergic dendrite bundles in nuclei raphe dorsalis and centralis superior of the rabbit: A possible substrate for local control of serotonergic neurons. Neurosci Lett 16:275–280

Felten DL, Sladek Jr JR (1983) Monoamine distribution in primate brain. V. Monoaminergic nuclei: Anatomy, pathways and local organization. Brain Res Bull 10:171–284

Felten DL, Laties AM, Carpenter MB (1974) Monoamine-containing cell bodies in the squirrel monkey brain. Am J Anat 139:153–166

Felten DL, Harrigan P, Burnett BT, Cummings JP (1981) Fourth ventricular tanycytes: A possible relationship with monoaminergic nuclei. Brain Res Bull 6:427–436

Ferrara P, Houghton R, Li CH (1979) β-Endorphin: Characteristics of binding sites in the rat brain. Biochem Biophys Res Co 89:786–792

Fex J, Altschuler RA (1981) Enkephalin-like immunoreactivity of olivocochlear nerve fibers in cochlea of guinea pig and cat. Proc Natl Acad Sci USA 78:1255–1259

Feyrter F (1953) Über die peripheren endokrinen (parakrinen) Drüsen des Menschen. Wilhelm Maudrich Verlag, Wien, pp 231

Fibiger HC (1982) The organization and some projections of cholinergic neurons of the mammalian forebrain. Brain Res Rev 4:327–388

Fibiger HC, Lloyd KG (1984) Neurobiological substrates of tardive dyskinesia: the GABA hypothesis. TINS 7:462–464

Fibiger HC, Miller JJ (1977) An anatomical and electrophysiological investigation of the serotonergic projection from the dorsal raphe nucleus to the substantia nigra in the rat. Neuroscience 2:975–987

Fields HL, Basbaum AI, Clanton CH, Anderson SD (1977) Nucleus raphe magnus inhibition of spinal cord dorsal horn neurons. Brain Res 126:441–453

Fields HL, Basbaum AI (1978) Brainstem control of spinal pain transmission neurons. Annu Rev Physiol 40:193–221

Finley JCW, Lindström P, Petrusz P (1981a) Immunocytochemical localization of β-endorphin-containing neurons in the rat brain. Neuroendocrinology 33:28–42

Finley JCW, Maderdrut JL, Roger LJ, Petrusz P (1981b) The immunocytochemical localization of somatostatin-containing neurons in the rat central nervous system. Neuroscience 6:2173–2192

Finley JCW, Maderdrut JL, Petrusz P (1981c) The immunocytochemical localization of enkephalin in the central nervous system of the rat. J Comp Neurol 198:541–565

Fischer-Ferraro C, Nahmod VE, Goldstein DJ, Finkelman S (1981) Angiotensin and renin in rat and dog brain. J Exp Med 133:353–361

Fisher LA, Rivier J, Rivier C, Spiess J, Vale W, Brown M (1982) Corticotropin releasing factor (CRF): central effects on mean arterial pressure and heart rate in rats. Endocrinology 110:2222–2224

Florey E (1967) Neurotransmitters and modulators in the animal kingdom. Fed Proc 26:1164–1178

Fonnum F (1984) Glutamate: A neurotransmitter in mammalian brain. J Neurochem 42:1–11

Fonnum F, Walaas I (1978) The effect of intrahippocampal kainic acid injections and surgical lesions on neurotransmitters in hippocampus and septum. J Neurochem 31:1173–1181

Fonnum F, Storm-Mathisen J, Walberg F (1970) Glutamate decarboxylase in inhibitory neurons, a study of the enzyme in Purkinje cell axons and boutons in the cat. Brain Res 20:259–275

Fonnum F, Gottesfeld Z, Grofova I (1978) Distribution of glutamate decarboxylase, choline acetyltransferase and aromatic amino acid decarboxylase in the basal ganglia of normal and operated rats. Evidence for striatopallidal, striatoentopeduncular and striatonigral gabaergic fibres. Brain Res 143:125–138

Fonnum F, Lund Karlsen R, Malthe-Sørenssen D, Skrede KK, Walaas I (1979) Localization of neurotransmitters, particularly glutamate, in hippocampus, septum, nucleus accumbens and superior colliculus. Progr Brain Res 51:167–191

Fonnum F, Storm-Mathisen J, Divac I (1981) Biochemical evidence for glutamate as neurotransmitter in corticostriatal and corticothalamic fibres in rat brain. Neuroscience 6:863–873

Foote SL, Aston-Jones G, Bloom FE (1980) Impulse activity of locus coeruleus neurons in awake rats and monkeys is a function of sensory stimulation and arousal. Proc Natl Acad Sci USA 77:3033–3037

Foote SL, Bloom FE, Aston-Jones G (1983) Nucleus locus coeruleus: New evidence of anatomical and physiological specificity. Physiol Rev 63:844–914

Forssmann WG (1978) A new somatostatinergic system in the mammalian spinal cord. Neurosci Lett 10:293–297

Forssmann WG, Burnweit C, Shehab T, Triepel J (1979) Somatostatin-immunoreactive nerve cell bodies and fibers in the medulla oblongata et spinalis. J Histochem Cytochem 27:1391–1393

Frohman LA, Maeda K, Berelowitz M, Szabo M, Thominet J (1982) Effects of neurotensin on hypothalamic and pituitary hormone secretion. Ann NY Acad Sci 400:172–182

Fulwiler CE, Saper CB (1984) Subnuclear organization of the efferent connections of the parabrachial nucleus in the rat. Brain Res Rev 7:229–259

Fuxe K, Jonsson g (1974) Further mapping of central 5-hydroxytryptamine neurons: studies with the neurotoxic dihydroxytryptamines. In: Costa E, Gessa GL, Sandler M (eds) Advances in Biochemical Psychopharmacology, vol 10. Raven Press, New York, pp 1–12

Fuxe K, Ganten D, Hökfelt T, Bolme P (1976) Immunohistochemical evidence for the existence of angiotensin II-containing nerve terminals in the brain and spinal cord in the rat. Neurosci Lett 2:229–234

Fuxe K, Hökfelt T, Said SI, Mutt V (1977) Vasoactive intestinal polypeptide and the nervous system: immunohistochemical evidence for localization in central and peripheral neurons, particularly intracortical neurons of the cerebral cortex. Neurosci Lett 5:241–246

Fuxe K, Ganten D, Hökfelt T, Locatelli V, Poulsen K, Stock G, Rix E, Taugner R (1980) Renin-like immunocytochemical activity in the rat and mouse brain. Neurosci Lett 18:245–250

Fuxe K, Anderson K, Calza L, Agnati LF, Lang RE, Poulsen K, Hökfelt T, Bernardi P (1982) Immunocytochemical demonstration of angiotensin II and renin-like immunoreactive nerve cells in the hypothalamus. Angiotensin peptides as comodulators in vasopressin and oxytocin neurons and their regulation of various types of central catecholamine nerve terminal systems. Exp Brain Res Suppl 4, 208–232

Gale K, Hong JS, Guidotti A (1977) Presence of substance P and GABA in separate striatonigral neurons. Brain Res 136:371–375

Gall C (1984) The distribution of cholecystokinin-like immunoreactivity in the hippocampal formation of the guinea pig: Localization in the mossy fibers. Brain Res 306:73–83

Gall C, Moore RY (1984) Distribution of enkephalin, substance P, tyrosine hydroxylase, and 5-hydroxytryptamine immunoreactivity in the septal region of the rat. J Comp Neurol 225:212–227

Gall C, Brecha N, Karten HJ, Chang K-J (1981) Localization of enkepahlin-like immunoreactivity to identified axonal and neuronal populations of the rat hippocampus. J Comp Neurol 198:335–350

Gall C, Moore RY (1984) Distribution of enkephalin, substance P, tyrosine hydroxylase, and 5-hydroxytryptamine immunoreactivity in the septal region of the rat. J Comp Neurol 225:212–227

Ganten D, Marquez-Julio A, Granger P, Hayduk K, Karsunsky KP, Boucher R, Genest J (1971) Renin in dog brain. Am J Physiol 221:1733–1737

Garbarg M, Barbin G, Feger J, Schwartz JC (1974) Histaminergic pathway in rat brain evidence by lesions of the medial forebrain bundle. Science 186:833–835

Garver DL, Sladek JR Jr (1975) Monoamine distribution in primate brain. I. Catecholamine-containing perikarya in the brain stem of Macaca speciosa. J Comp Neur 159:289–304

Gatter KC, Powell TPS (1977) The projection of the locus coeruleus upon the neocortex in the macaque monkey. Neuroscience 2:441–445

Geeraedts LMG (1985) Personal communication

Geffard M, Buijs RM, Seguela Ph, Pool CW, Le Moal M (1984) First demonstration of highly specific and sensitive antibodies against dopamine. Brain Res 294:161–165

Gerfen CR (1984) The neostriatal mosaic: compartmentalization of corticostriatal input and striatonigral output systems. Nature 311:461–464

Giachetti A, Said SI, Reynolds RC, Koniges FC (1977) Vasoactive intestinal polypeptide in brain: localization in and release from isolated nerve terminals. Proc Natl Acadi Sci USA 74:3424–3428

Gibson SJ, Polak JM, Bloom SR, Wall PD (1981) The distribution of nine peptides in rat spinal cord with special emphasis on the substantia gelatinosa and on the area around the central canal (Lamina X). J Comp Neurol 201:65–79

Gilbert RFT, Emson PC, Hunt SP, Bennett GW, Marsden CA, Sandberg BEB, Steinbusch HWM, Verhofstad AAJ (1982) The effects of monoamine neurotoxins on peptides in the rat spinal cord. Neuroscience 7:69–87

Gillis RA, Helke CJ, Hamilton BL, Norman WP, Jacobowitz DM (1980) Evidence that substance P is a neurotransmitter of baro- and chemoreceptor afferents in nucleus tractus solitarius. Brain Res 181:476–481

Glazer EJ, Basbaum AI (1981) Immunohistochemical localization of leucine-enkephalin in the spinal cord of the cat: enkephalin-containing marginal neurons and pain modulation. J Comp Neurol 196:377–389

Glazer EJ, Basbaum AI (1984) Axons which take up (^{3}H) serotonin are presynaptic to enkephalin immunoreactive neurons in cat dorsal horn. Brain Res 298:386–391

Glazer EJ, Steinbusch H, Verhofstad A, Basbaum AI (1981) Serotonin neurons in nucleus raphe dorsalis and paragigantocellularis of the cat contain enkephalin. J Physiol Paris 77:241–245

Goedert M (1984) Neurotensin – a status report. Trends NeuroSci 7:3–5

Goedert M, Lightman SL, Nagy JI, Marley PD, Emson PC (1982) Neurotensin in the rat anterior pituitary gland. Nature 298:163–165

Goedert M, Mantyh PW, Hunt SP, Emson PC (1983) Mosaic distribution of neurotensin-like immunoreactivity in the cat striatum. Brain Res 274:176–179

Goldman-Rakic PS (1982) Cytoarchitectonic heterogeneity of the primate neostriatum: subdivision into island and matrix cellular compartments. J Comp Neurol 205:398–413

Goldstein A, Ghazarossian VE (1980) Immunoreactive dynorphin in pituitary and brain. Proc Natl Acad Sci USA 77:6207–6210

Goldstein A, Tachibana S, Lowney LI, Hunkapiller M, Hood L (1979) Dynorphin-(1-13), an extraordinarily potent opioid peptide. Proc Natl Acad Sci USA 76:6666–6670

Goldstein A, Fischli W, Lowney LI, Hunkapiller M, Hood L (1981) Porcine pituitary dynorphin: Complete amino acid sequence of the biologically active heptadecapeptide. Proc Natl Acad Sci USA 78:7219–7223

Goldstein M, Lew JY, Matsumoto Y, Hökfelt T, Fuxe K (1978) Localization and function of PNMT in the central nervous system. In: Lipton MA, DiMascio A, Killam KF (eds) Psychopharmacology: A Generation of Progress. Raven Press, New York, pp 261–269

Goldstein M, Pearson J, Sauter A, Ueta K, Asano T, Engel J, Passeltiner P, Hökfelt T, Fuxe K (1980) Distribution and characterization of PNMT and epinephrine in the CNS. In: Fuxe K, Goldstein M, Hökfelt B, Hökfelt T (eds) Central Adrenaline Neurons. Basic aspects and their role in cardiovascular functions. Pergamon Press, Oxford, pp 49–55

Goltermann NR, Rehfeld JF, Roigaard-Petersen H (1980) In vivo biosynthesis of cholecystokinin in rat cerebral cortex. J Biol Chem 255:6181–6185

Goodchild AK, Moon EA, Dampney RAL, Howe PRC (1984) Evidence that adrenaline neurons in the rostral ventrolateral medulla have a vasopressor function. Neurosci Lett 45:267–272

Gorski RA, Harlan RE, Jacobson CD, Shryne JE, Southam AM (1980) Evidence for the existence of a sexually dimorphic nucleus in the preoptic area of the rat. J Comp Neurol 193:529–539

Gottesfeld Z, Jacobowitz DM (1979) Cholinergic projections from the septal-diagnonal band area to the habenular nuclei. Brain Res 176:391–394

Graham LT, Shank RP, Werman R, Aprison MH (1967) Distribution of some synaptic transmitter suspects

in cat spinal cord: glutamic acid aspartic acid γ-aminobutyric acid, glycine and glutamine. J Neurochem 14:465–472
Gramsch C, Kleber G, Höllt V, Pasi A, Mehraein P, Herz A (1980) Pro-opiocortin fragments in human and rat brain: β-endorphin and α-MSH are the predominant peptides. Brain Res 192:109–119
Gray TS, Cassell MD, Nilaver G, Zimmerman EA, Williams TH (1984) The distribution and ultrastructure of VIP-immunoreactivity in the central nucleus of the rat amygdala. Neuroscience 11:399–408
Graybiel AM (1984) Neurochemically specified subsystems in the basal ganglia. In: Evered D, O'Connor M (eds) Functions of the basal ganglia; Ciba Foundation symposium 107, Pitman, London, pp 114–149
Graybiel AM, Ragsdale Jr CW (1978) Histochemically distinct compartments in the striatum of human monkey and cat demonstrated by acetylthiocholinesterase staining. Proc. Natl Acid Sci USA 75:5723–5726
Graybiel AM, Ragsdale Jr CW (1979) Fiber connections of the basal ganglia. In: Cuénod M, Kreutzberg GW, Bloom FE (eds) Development and Chemical Specificity of Neurons, Elsevier, Amsterdam, pp 239–283
Graybiel AM, Ragsdale CW Jr (1983) Biochemical anatomy of the striatum. In: Emson PC (ed) Chemical Neuroanatomy. Raven Press, New York, p 427–504
Graybiel AM, Ragsdale CW Jr, Yoneoka ES, Elde RP (1981) An immunohistochemical study of enkephalins and other neuropeptides in the striatum of the cat with evidence that the opiate peptides are arranged to form mosaic patterns in register with the striosomal compartments visible by acetylcholinesterase staining. Neuroscience 6:377–397
Greengard P (1978) Cyclic Nucleotides, Phosphorylated Proteins, and Neuronal Function. Raven Press, New York, pp 124
Greengard P (1979) Cyclic nucleotides, phosphorylated proteins, and the nervous system. Fed Proc 38:2208–2217
Greenwood RS, Godar SE, Reaves Jr TA, Hayward JN (1981) Cholecystokinin in hippocampal pathways. J Comp Neurol 203:335–350
Gronan RJ, York DH (1978) Effects of angiotensin II and acetylcholine on neurons in the preoptic area. Brain Res 154:172–177
Grossman A, Rees LH (1983) The neuroendocrinology of opioid peptides. Brit Med Bull 39:83–88
Groves PM (1980) Synaptic endings and their post-synaptic targets in neostriatum: synaptic specializations revealed from analysis of serial sections. Proc Natl Acad Sci USA 77:6926–6929
Grzanna R, Molliver ME, Coyle JT (1978) Visualization of central noradrenergic neurons in thick sections by the unlabeled antibody method: A transmitter-specified Golgi image. Proc Natl Acad Sci USA 75:2502–2506
Guillemin R (1978) Peptides in the brain: The new endocrinology of the neuron. Science 202:390–402
Guillemin R, Vargo Th, Rossier J, Minick S, Ling N, Rivier C, Vale W, Bloom F (1977) β-Endorphin and adrenocorticotropin are secreted concomitantly by the pituitary gland. Science 197:1367–1369
Gunne LM (1962) Relative adrenaline content in brain tissue. Acta Physiol Scand 56:324–333
Gurtu S, Pant KK, Sinha JN, Bhargava KP (1984) An investigation into the mechanism of cardiovascular responses elicited by electrical stimulation of locus coeruleus and subcoeruleus in the cat. Brain Res 301:59–64
Guth PS, Norris CH, Bobbin RP (1976) The pharmacology of transmission in the peripheral auditory system. Pharmacol Rev 28:95–125
Guy J, Leclerc R, Vaudry H, Pelletier G (1980) Identification of a second category of α-melanocyte-stimulating hormone (α-MSH) neurons in the rat hypothalamus. Brain Res 199:135–146
Guyenet PG (1980) The coeruleospinal noradrenergic neurons: Anatomical and electrophysiological studies in the rat. Brain Res 189:121–133
Guyenet PG, Crane JK (1981) Non-dopaminergic nigrostriatal pathway. Brain Res 243:291–305
Haas HL, Hösli L (1973) The depression of brain stem neurones by taurine and its interaction with strychnine and bicuculline. Brain Res 52:399–402
Haber S, Elde R (1981) Correlation between met-enkephalin and substance P immuno-reactivity in the primate globus pallidus. Neuroscience 6:1291–1297
Haber S, Elde R (1982) The distribution of enkephalin immunoreactive fibers and terminals in the monkey central nervous system: an immunohistochemical study. Neuroscience 7:1049–1095
Halász N, Shepherd GM (1983) Neurochemistry of the vertebrate olfactory bulb. Neuroscience 10:579–619
Halász N, Ljungdahl A, Hökfelt T, Johansson O, Goldstein M, Park D, Biberfeld P (1977) Transmitter histochemistry of the rat olfactory bulb. I. Immunohistochemical localization of monoamine synthesizing enzymes. Support for intrabulbar, periglomerular dopamine neurons. Brain Res 126:455–474
Halasz N, Ljungdahl A, Hökfelt T (1979) Transmitter histochemistry of the rat olfactory bulb. III. Autoradiographic localization of (^{3}H)GABA. Brain Res 167:221–240
Hamill GS, Lenn NJ (1984) The subnuclear organization of the rat interpeduncular nucleus: A light and electron microscopic study. J Comp Neurol 222:396–408
Hamill GS, Olschowka JA, Lenn NJ, Jacobowitz DM (1984) The subnuclear distribution of substance P, cholecystokinin, vasoactive intestinal peptide, somatostatin, leu-enkephalin, dopamine-β-hydroxylase, and serotonin in the rat interpeduncular nucleus. J Comp Neurol 226:580–596
Hancock MB, Fougerousse CL (1976) Spinal projections from the nucleus locus coeruleus and nucleus subcoeruleus in the cat and monkey as demonstrated by the retrograde transport of horseradish peroxidase. Brain Res Bull 1:229–234
Handelmann G, Meyer DK, Beinfeld MC, Oertel WH (1981) CCK-containing terminals in the hippocampus are derived from intrinsic neurons: an immunohistochemical and radioimmunological study. Brain Res 224:180–184
Hara Y, Shiosaka S, Senba E, Sakanaka M, Inagaki S, Takagi H, Kawai Y, Takatsuki K, Matsuzaki T, Tohyama M (1982) Ontogeny of the neurotensin-containing neuron system of the rat: Immunohistochemical analysis. I. Forebrain and diencephalon. J Comp Neurol 208:177–195
Hardy SGP, Leichnetz GR (1981) Cortical projections

to the periaqueductal gray in the monkey: a retrograde and orthograde horseradish peroxidase study. Neurosci Lett 22:97–101

Hartman BK (1973) The innervation of cerebral blood vessels by central noradrenergic neurons. In: Usdin E, Snyder SH (eds) Frontiers in Catecholamine Research, Pergamon Press, Oxford, pp 91–96

Harvey LT, Scholfield CN, Graham LT, Aprison MH (1975) Putative transmitters in denervated olfactory cortex. J Neurochem 24:445–449

Hattori T, McGeer PL, Fibiger HC, McGeer EG (1973) On the source of gaba-containing terminals in the substantia nigra. Electron microscopic autoradiographic and biochemical studies. Brain Res 54:103–114

Hayashi H, Takagi H, Takeda N, Kubota Y, Tohyama M, Watanabe T, Wada H (1984) Fine structure of histaminergic neurons in the caudal magnocellular nucleus of the rat as demonstrated by immunocytochemistry using histidine decarboxylase as a marker. J Comp Neurol 229:233–241

Haywood JR, Fink GD, Buggy J, Phillips MI, Brody MJ (1980) The area postrema plays no role in the pressor action of angiotensin in the rat. Am J Physiol 239:H108–H113

Hedreen JC, Chalmers JP (1972) Neuronal degeneration in rat brain induced by 6-hydroxydopamine: A histological and biochemical study. Brain Res 47:1–36

Heimer L, Wilson RD (1975) The subcortical projections of the allocortex: Similarities in the neural associations of the hippocampus, the piriform cortex, and the neocortex. In: Santini M (ed) Golgi Centennial Symposium. Proceedings, Raven Press, New York, pp 177–193

Helke CJ (1984) Studies on the origin and release of somatostatin-immunoreactivity in the nucleus of the solitary tract. Brain Res Bull 12:51–55

Hemmendinger LM, Moore RY (1984) Interpeduncular nucleus organization in the rat: Cytoarchitecture and histochemical analysis. Brain Res Bull 13:163–179

Hendry SHC, Jones EG (1981) Sizes and distributions of intrinsic neurons incorporating tritiated gaba in monkey sensory-motor cortex. J Neurosci 1:390–408

Hendry SHC, Valentino KL, Jones EG, Beinfeld MC (1982) An investigation of neurons displaying cholecystokinin-like immunoreactivity in the cerebral cortex of the monkey and rat. Soc Neurosci Abstr 8:585

Heritage AS, Grant LD, Stumpf WE (1976) Autoradiographic localization of estradiol in catecholamine neurons of the rat brain stem. Proc Soc Neurosci 2:672

Heritage AS, Stumpf WE, Sar M, Grant LD (1980) Brainstem catecholamine neurons are target sites for sex steroid hormones. Science 207:1377–1379

Herkenham M, Moon edley S, Stuart J (1984) Cell clusters in the nucleus accumbens of the rat, and the mosaic relationship of opiate receptors, acetylcholinesterase and subcortical afferent terminations. Neuroscience 11:561–593

Heuser JE, Reese TS, Dennis MJ, Jan Y, Jan L, Evans L (1979) Synaptic vesicle exocytosis captured by quick freezing and correlated with quantal transmitter release. J Cell Biol 81:275–300

Higgins GA, Hoffman GE, Wray S, Schwaber JS (1984) Distribution of neurotensin-immunoreactivity within baroreceptive portions of the nucleus of the tractus solitarius and the dorsal vagal nucleus of the rat. J Comp Neurol 226:155–164

Hikosaka O, Wurtz RH (1983a) Visual and oculomotor functions of monkey substantia nigra pars reticulata. I: Relation of visual and auditory responses to saccades. J Neurophysiol 49:1230–1253

Hikosaka O, Wurtz RH (1983b) Visual and oculomotor functions of monkey substantia nigra pars reticulata. II: Visual responses related to fixation of gaze. J Neurophysiol 49:1254–1267

Hikosaka O, Wurtz RH (1983c) Visual and oculomotor functions of monkey substantia nigra pars reticulata. III: Memory-contingent visual and saccade responses. J Neurophysiol 49:1268–1284

Hikosaka O, Wurtz RH (1983d) Visual and oculomotor functions of monkey substantia nigra pars reticulata. IV: Relation of substantia nigra to superior colliculus. J Neurophysiol 49:1285–1301

Hirose S, Yokosawa H, Inagami T (1978) Immunocytochemical identification of renin in rat brain and distinction from acid proteases. Nature 274:392–393

Ho RH, Berelowitz M (1984) Somatostatin $28_{1\text{-}14}$ immunoreactivity in primary afferent neurons of the rat spinal cord. Neurosci Lett 46:161–166

Hoffman DL, Krupp L, Schrag D, Nilaver G, Valiquette G, Kilcoyne MM, Zimmerman EA (1982) Angiotensin immunoreactivity in vasopressin cells in rat hypothalamus and its relative deficiency in homozygous brattleboro rats. Ann NY Acad Sci 394:135–141

Hoffman GE, Hayes TA (1979) Somatostatin neurons and their projections in dog diencephalon. J Comp Neurol 186:371–392

Hoffman WE, Phillips MI (1976) Regional study of cerebral ventricle sensitive sites to angiotensin II. Brain Res 110:313–330

Hökfelt T (1968) In vitro studies on central and peripheral monoamine neurons at the ultrastructural level. Z Zellforsch Mikrosk Anat 91:1–74

Hökfelt T, Fuxe K (1969) Cerebellar monoamine nerve terminals a new type of afferent fibers to the cortex cerebelli. Exp Brain Res 9:63–72

Hökfelt T, Fuxe K, Johansson O, Ljungdahl A (1974a) Pharmacohistochemical evidence of the existence of dopamine nerve terminals in the limbic cortex. Eur J Pharmacol 25:108–112

Hökfelt T, Fuxe K, Goldstein M, Johansson O (1974b) Immunohistochemical evidence for the existence of adrenaline neurons in the rat brain. Brain Res 66:235–251

Hökfelt T, Fuxe K, Johansson O, Jeffcoate S, White N (1975a) Distribution of thyrotropin-releasing hormone (TRH) in the central nervous system as revealed with immunohistochemistry. Europ J Pharmacol 34:289–392

Hökfelt T, Fuxe K, Johansson O, Jeffcoate S, White N (1975b) Thyrotropin releasing hormone (TRH)-containing nerve terminals in certain brain stem nuclei and in the spinal cord. Neurosci Lett 1:133–139

Hökfelt T, Kellerth J-O, Nilsson G, Pernow B (1975c) Experimental immunohistochemical studies on the localization and distribution of substance P in cat primary sensory neurons. Brain Res 100:235–252

Hökfelt T, Johansson O, Fuxe K, Goldstein M, Park

D (1976a) Immunohistochemical studies on the localization and distribution of monoamine neuron systems in the rat brain. I. Tyrosine hydroxylase in the mes- and diencephalon. Med Biol 54:427–453

Hökfelt T, Elde R, Johansson O, Luft R, Nilsson G, Arimura A (1976b) Immuno histochemical evidence for separate populations of somatostatin-containing and substance P-containing primary afferent neurons in the rat. Neuroscience 1:131–136

Hökfelt T, Ljungdahl A, Terenius L, Elde R, Nilsson G (1977a) Immunohistochemical analysis of peptide pathways possible related to pain and analgesia: Enkephalin and substance P. Proc Natl Acad Sci USA 74:3081–3085

Hökfelt T, Elde R, Johansson O, Terenius L, Stein L (1977b) The distribution of enkephalin-immunoreactive cell bodies in the rat central nervous system. Neurosci Lett 5:25–31

Hökfelt T, Elde R, Johansson O, Ljungdahl A, Schultzberg M, Fuxe K, Goldstein M, Nilsson G, Pernow B, Terenius L, Ganten D, Jeffcoate L, Rehfeld J, Said S: Distribution of peptide-containing neurons. In: Lipton MA, DiMaschio A, Killam KF (eds) Psychopharmacology: A Generation of Progress. Raven Press, New York, pp 39–66

Hökfelt T, Elde R, Fuxe K, Johansson O, Ljungdahl A, Goldstein M, Luft R, Efendic S, Nilsson G, Terenius L, Ganten D, Jeffcoate SL, Rehfeld J, Said S, Perez de la Mora M, Possani L, Tapia R, Teran L, Palacios R (1978b) Aminergic and peptidergic pathways in the nervous system with special reference to the hypothalamus. In: Reichlin S, Baldessarini RJ, Martin JB (eds) The Hypothalamus. Raven Press, New York, pp 69–135

Hökfelt T, Ljungdahl A, Steinbusch H, Verhofstad A, Nilsson G, Brodin E, Pernow B, Goldstein M (1978c) Immunohistochemical evidence of substance-P like immunoreactivity in some 5-hydroxytryptamine containing neurons in the rat central nervous system. Neuroscience 3:517–538

Hökfelt T, Phillipson O, Goldstein M (1979a) Evidence for a dopaminergic pathway in the rat descending from the A11 cell group to the spinal cord. Acta Physiol Scand 107:393–395

Hökfelt T, Terenius L, Kuypers HGJM, Dann O (1979b) Evidence for enkephalin immunoreactive neurons in the medulla oblongata projecting to the spinal cord. Neurosci Lett 14:55–60

Hökfelt T, Johansson O, Ljungdahl A, Lundberg JM, Schultzberg M (1980a) Peptidergic neurones. Nature 284:515–521

Hökfelt T, Rehfeld JF, Skirboll L, Ivemark B, Goldstein M, Markey K (1980b) Evidence for coexistence of dopamine and CCK in meso-limbic neurones. Nature 285:476–478

Hökfelt T, Skirboll L, Rehfeld JF, Goldstein M, Markey K, Dann O (1980c) A subpopulation of mesencephalic dopamine neurons projecting to limbic areas contains a cholecystokinin-like peptide: evidence from immunohistochemistry combined with retrograde tracing. Neuroscience 5:2093–2124

Hökfelt T, Goldstein M, Fuxe K, Johansson O, Verhofstad A, Steinbusch H, Penke B, Vargas J (1980d) Histochemical identification of adrenaline containing cells with special reference to neurons. In: Fuxe K, Goldstein M, Hökfelt B, Hökfelt T (eds) Central Adrenaline Neurons. Basic aspects and their role in cardiovascular functions. Pergamon Press, Oxford, pp 19–47

Hökfelt T, Everitt BJ, Theodorsson-Norheim E, goldstein M (1984a) Occurrence of neurotensinlike immunoreactivity in subpopulations of hypothalamic, mesencephalic, and medullary catecholamine neurons. J Comp Neurol 222:543–559

Hökfelt T, Johansson O, Goldstein M (1984b) Chemical anatomy of the brain. Science 225:1326–1334

Holaday JW, Loh HH, Li CH (1978) Unique behavioral effects of β-endorphin and their relationship to thermoregulation and hypothalamic function. Life Sci 22:1525–1536

Hong JS, Costa E, Yang H-YT (1976) Effects of habenular lesions on the substance P content of various brain regions. Brain Res 118:523–525

Hong JS, Yang H-YT, Racagni G, Costa E (1977) Projections of substance P containing neurons from neostriatum to substantia nigra. Brain Res 122:541–544

Hoorneman EMD, Buijs RM (1982) Vasopressin fiber pathways in the rat brain following suprachiasmatic nucleus lesioning. Brain Res 243:235–241

Hoover DB, Baisden RH (1980) Localization of putative cholinergic neurons innervating the anteroventral thalamus. Brain Res Bull 5:519–524

Hoover DB, Jacobowitz DM (1979) Neurochemical and histochemical studies of the effect of a lesion of the nucleus cuneiformis on the cholinergic innervation of discrete areas of the rat brain. Brain Res 170:113–122

Hornykiewicz O (1978) Historical aspects and frontiers of Parkinson's disease research. Adv Exp Med Biol 90:1–20

Hornykiewicz O (1984) Neurochemical pathology of dyskinesias. In: Bruyn GW, Roos RAC, Buruma OJS (eds) Actua Sandoz, vol 7, Dyskinesias, Sandoz b.v., Uden, pp 39–47

Hosoya Y, Matsushita M (1979) Identification and distribution of the spinal and hypophyseal projection neurons in the paraventricular nucleus of the rat. A light and electron microscopic study with the horseradish peroxidase method. Exp Brain Res 35:315–331

Houser CR, Vaughn JE, Barber RP, Roberts E (1980) GABA neurons are the major cell type of the nucleus reticularis thalami. Brain Res 200:341–354

Howe PRC, Costa M, Furness JB, Chalmers JP (1980) Simultaneous demonstration of phenylethanolamine N-methyltransferase immunofluorescent and catecholamine fluorescent nerve cell bodies in the rat medulla oblongata. Neuroscience 5:2229–2238

Howe PRC, Moon E, Dampney RAL (1983) Distribution of serotonin nerve cells in the rabbit brainstem. Neurosci Lett 38:125–130

Hubbard JE, DiCarlo V (1973) Fluorescence histochemistry of monoamine-containing cell bodies in the brain stem of the squirrel monkey (Saimiri sciureus). I. The locus caeruleus. J Comp Neur 147:553–566

Hubbard JE, DiCarlo V (1974a) Fluorescence histochemistry of monoamine-containing cell bodies in the brain stem of the squirrel monkey (Saimiri sciureus). II. Catecholamine-containing groups. J Comp Neur 153:369–384

Hubbard JE, DiCarlo V (1974b) Fluorescence histochemistry of monoamine-containing cell bodies in the brain stem of the squirrel monkey (Saimiri sciureus). III. Serotonin-containing groups. J Comp Neur 153:385–398

Hubbard JI, Llinas R, Quastel DMJ (1969) Electrophysiological Analysis of Synaptic Transmission. Arnold E Ltd, London, pp 372

Hughes J (1975) Isolation of an endogenous compound from the brain with pharmacological properties similar to morphine. Brain Res 88:295–308

Hughes J, Smith TW, Kosterlitz HW, Fothergill LA, Morgan BA, Morris HR (1975) Identification of two related pentapeptides from the brain with potent opiate agonist activity. Nature 258:577–579

Ibata Y, Watanabe K, Kinoshita H, Kubo S, Sano Y (1979) The location of LH-RH neurons in the rat hypothalamus and their pathways to the median eminence. Cell Tissue Res 198:381–395

Ibata Y, Kawakami F, Fukui K, Obata-Tsuto HL, Tanaka M, Kubo T, Okamura H, Morimoto N, Yanaihara C, Yanaihara N (1984) Light and electron microscopic immunocytochemistry of neurotensin-like immunoreactive neurons in the rat hypothalamus. Brain Res 302:221–230

Inagaki S, Parent A (1984) Distribution of substance P and enkephalin-like immunoreactivity in the substantia nigra of rat, cat and monkey. Brain Res Bull 13:319–329

Innis RB, Snyder SH (1980) Distinct cholecystokinin receptors in brain and pancreas. Proc Natl Acad Sci USA 77:6917–6921

Innis RB, Corrêa FMA, Uhl GR, Schneider B, Snyder SH (1979) Cholecystokinin octapeptide-like immunoreactivity: histochemical localization in rat brain. Proc Natl Acad Sci USA 76:521–525

Isaacson RL (1975) The Limbic System. Plenum Press, New York, pp 292

Ishikawa M, Tanaka C (1977) Morphological organization of catecholamine terminals in the diencephalon of the rhesus monkey. Brain Res 119:43–55

Ito M (1984) The cerebellum and neural control. Raven Press, New York, pp 580

Iversen LL (1983) Neuropeptides – What next? Trends NeuroSci 6:293–294

Iversen LL, Iversen SD, Bloom F, Douglas C, Brown M, Vale W (1978) Calcium-dependent release of somatostatin and neurotensin from rat brain *in vitro*. Nature 273:161–163

Iversen LL, Lee CM, Gilbert RF, Hunt S, Emson PC (1980) Regulation of neuropeptide release. Proc R Soc Lond (Biol) 210:91–111

Jackson IMD, Reichlin S (1974) Thyrotropin-releasing hormone (TRH): Distribution in hypothalamic and extrahypothalamic brain tissues of mammalian and submammalian chordates. Endocrinol 95:854–862

Jacobowitz DM (1975) Fluorescence microscopic mapping of CNS norepinephrine systems in the rat forebrain. In: Stumpf WE, Grant LD (eds) Anatomical Neuroendocrinology, S Karger, Basel, pp 368–380

Jacobowitz DM, MacLean PD (1978) A brainstem atlas of catecholaminergic neurons and serotonergic perikarya in a pygmy primate (Cebuella pygmaea). J Comp Neurol 177:397–416

Jacobowitz DM, O'Donohue TL (1978) α-Melanocyte stimulating hormone: Immunohistochemical identification and mapping in neurons of rat brain. Proc Natl Acad Sci USA 75:6300–6304

Jaeger CB, Rugghiero DA, Albert VR, Park DH, Joh TH, Reis DJ (1984) Aromatic L-amino acid decarboxylase in the rat brain: immunocytochemical localization in neurons of the brain stem. Neuroscience 11:691–713

Jan YN, Jan LY (1983) A LHRH-like peptidergic neurotransmitter capable of 'action at a distance' in autonomic ganglia. Trends NeuroSci 6:320–325

Jancsó G, Hökfelt T, Lundberg JM, Kiraly E, Halász N, Nilsson G, Terenius L, Rehfeld J, Steinbusch H, Verhofstad A, Elde R, Said S, Brown MJ (1981) Immunohistochemical studies on the effect of capsaicin on spinal and medullary peptide and monoamine neurons using antisera to substance P, gastrin/CCK, somatostatin, VIP, enkephalin, neurotensin and 5-hydroxytryptamine. J Neurocytol 10:963–980

Jaspers R, Schwarz M, Sontag KH, Cools AR (1984) Caudate nucleus and programming behaviour in cats: role of dopamine in switching motor patterns. Behav Brain Res 14:17–28

Jegou S, Tonon MC, Guy J, Vaudry H, Pelletier G (1983) Biological and immunological characterization of α-melanocyte-stimulating hormone (α-MSH) in two neuronal systems of the rat brain. Brain Res 260:91–98

Jenkins JS, Ang VTY, Hawthorn J, Rossor MN, Iversen LL (1984) Vasopressin, oxytocin and neurophysins in the human brain and spinal cord. Brain Res 291:111–117

Jennes L, Stumpf WE, Kalivas PW (1982) Neurotensin: Topographical distribution in rat brain by immunohistochemistry. J Comp Neurol 210:211–224

Jessell TM, Iversen LL (1977) Opiate analgesics inhibit substance P release from rat trigeminal nucleus. Nature 268:549–551

Jessell TM, Emson PC, Paxinos G, Cuello AC (1978) Topographic projections of substance P and GABA pathways in the striatopallido-nigral system: A biochemical and immunohistochemical study. Brain Res 152:487–498

Johansson O, Hökfelt T (1980) Thyrotropin releasing hormone, somatostatin, and enkephalin: distribution studies using immunohistochemical techniques. J Histochem Cytochem 28:364–366

Johansson O, Hökfelt T, Pernow B, Jeffcoate SL, White N, Steinbusch HWM, Verhofstad AAJ, Emson PC, Spindel E (1981) Immunohistochemical support for three putative transmitters in one neuron: coexistence of 5-hydroxytryptamine, substance P- and thyrotropin releasing hormone-like immunoreactivity in medullary neurons projecting to the spinal cord. Neuroscience 6:1857–1881

Johansson O, Hökfelt T, Elde RP (1984) Immunohistochemical distribution of somatostatin-like immunoreactivity in the central nervous system of the adult rat. Neuroscience 13:265–339

Johnson JL, Aprison MH (1971) The distribution of glutamate and total free amino acids in thirteen specific regions of the cat CNS. Brain Res 26:141–148

Jones BE, Friedman L (1983) Atlas of catecholamine perikarya, varicosities and pathways in the brainstem of the cat. J Comp Neurol 215:382–396

Jones BE, Moore RY (1977) Ascending projections of the locus coeruleus in the rat. II. Autoradiographic study. Brain Res 127:23–53
Jones EG, Burton H, Saper CB, Swanson LW (1976) Midbrain, Diencephalic and Cortical Relationships of the Basal Nucleus of Meynert and Associated Structures in Primates. J Comp Neurol 167:385–418
Joseph SA (1980) Immunoreactive adrenocorticotropin in rat brain: A neuroanatomical study using antiserum generated against synthetic $ACTH^{1-39}$. Am J Anat 158:533–548
Joseph SA, Pilcher WH, Bennett-Clarke C (1983) Immunocytochemical localization of ACTH perikarya in nucleus tractus solitarius: Evidence for a second opiocortin neuronal system. Neurosci Lett 38:221–225
Jouvet M (1967) Mechanisms of the states of sleep: a neuropharmacological approach. Res Pub Ass Res Ment Dis 45:86–126
Jouvet M (1969) Biogenic amines and the states of sleep. Science 163:32–41
Joy MD, Lowe RD (1970) Evidence that the area postrema mediates the central cardiovascular response to angiotensin II. Nature 228:1303–1304
Jurna I, Zetler G (1981) Antinociceptive effect of centrally administered caerulein and cholecystokinin octapeptide (CCK-8). Eur J Pharmacol 73:323–331
Kahn D, Abrams GM, Zimmerman EA, Carraway R, Leeman SE (1980) Neurotensin neurons in the rat hypothalamus: An immunocytochemical study. Endocrinology 107:47–54
Kahn D, Hou-Yu A, Zimmerman EA (1982) Localization of neurotensin in the hypothalamus. Ann NY Acad Sci 400:117–131
Kakidani H, Furutani Y, Takahashi H, Noda M, Morimoto Y, Hirose T, Asai M, Inayama S, Nakanishi S, Numa S (1982) Cloning and sequence analysis of cDNA for porcine beta-neo-endorphin/dynorphin precursor. Nature 298:245–249
Kalia M, Mesulam MM (1980a) Brain stem projections of sensory and motor components of the vagus complex in the cat. I. The cervical vagus and nodose ganglion. J Comp Neurol 193:435–465
Kalia M, Mesulam MM (1980b) Brain stem projections of sensory and motor components of the vagus complex in the cat. II. Laryngeal, tracheobronchial, pulmonary, cardiac, and gastrointestinal branches. J Comp Neurol 193:467–508
Kalia M, Fuxe K, Agnati LF, Hökfelt T, Härfstrand A (1984a) Somatostatin produces apnea and is localized in medullary respiratory nuclei: a possible role in apneic syndromes. Brain Res 296:339–344
Kalia M, Fuxe K, Hökfelt T, Johansson O, Lang R, Ganten D, Cuello C, Terenius L (1984b) Distribution of neuropeptide immunoreactive nerve terminals within the subnuclei of the nucleus of the tractus solitarius of the rat. J Comp Neurol 222:409–444
Kalia M, Fuxe K, Hökfelt T, Harfstrand A, Lang RE, Ganten D (1984c) Distribution of neurophysin II immunoreactive nerve fibers within the subnuclei of the nucleus of the tractus solitarius of the rat. Brain Res 321:71–82
Kalivas PW, Miller JS (1984) Neurotensin neurons in the ventral tegmental area project to the medial nucleus accumbens. Brain Res 300:157–160
Kalivas PW, Jennes L, Nemeroff ChB, Prange AJ Jr (1982) Neurotensin: Topographical distribution of brain sites involved in hypothermia and antinociception. J Comp Neurol 210:225–238
Kanazawa I, Emson PC, Cuello AC (1977) Evidence for the existence of substance P-containing fibres in striato-nigral and pallido-nigral pathways in rat brain. Brain Res 119:447–453
Kanazawa I, Mogaki S, Muramoto O, Kuzuhara S (1980) On the origin of substance P-containing fibres in the entopeduncular nucleus and the substantia nigra of the rat. Brain Res 184:481–485
Kangawa K, Minamino N, Chino N, Sakakibara S, Matsuo H (1981) The complete amino acid sequence of alpha-neo-endorphin. Biochem Biophys Res Commun 99:871–878
Kapadia SE, De Lanerolle NC (1984) Populations of substance P, met-enkephalin and serotonin immunoreactive neurons in the interpeduncular nucleus of cat: Cytoarchitectonics. Brain Res 302:33–43
Karoum F, Commissiong JW, Neff NH, Wyatt RJ (1980) Biochemical evidence for uncrossed and crossed locus coeruleus projections to the spinal cord. Brain Res 196:237–241
Kasting NW, Veale WL, Cooper KE, Lederis K (1981) Vasopressin may mediate fibrile convulsions. Brain Res 213:327–333
Kasting NW, Veale WL, Cooper KE (1982) Vasopressin: a homeostatic effector in the febrile process. Neurosci Biobehav Rev 6:215–222
Kataoka K (1979) Regional distribution of immunoreactive neurotensin in monkey brain. Brain Res Bull 4:57–60
Katz B (1969) The Release of Neural Transmitter Substances. Liverpool University Press, Liverpool and Charles C Thomas, Springfield Illinois USA, pp 60
Kawaguchi S, Ono T (1973) Bicuculline and picrotoxin sensitive inhibition in interpositus neurones of cat. Brain Res 58:260–265
Kawai Y, Inagaki S, Shiosaka S, Shibasaki T, Ling N, Tohyama M, Shiotani Y (1984) The distribution and projection of γ-melanocyte stimulating hormone in the rat brain: An immunohistochemical analysis. Brain Res 297:21–32
Kawano H, Daikoku S (1981) Immunohistochemical demonstration of LHRH neurons and their pathways in the rat hypothalamus. Neuroendocrinol 32:179–186
Kawatani M, Lowe I, Morgan C, Nadelhaft I, Erdman S, deGroat WC (1982) Vasoactive intestinal polypeptide (VIP) in primary afferent projections to the sacral spinal cord of the cat. Soc Neurosci Abstr 8:585
Kawatani M, Lowe IP, Nadelhaft I, Morgan C, De Groat WC (1983) Vasoactive intestinal polypeptide in visceral afferent pathways to the sacral spinal cord of the cat. Neurosci Lett 42:311–316
Keefer DA, Stumpf WE (1975) Estrogen localization in the primate brain. In: Stumpf WE, Grant LD (eds) Anatomical Neuroendocrinology. S Karger, Basel, pp 153–165
Kelley AE, Domesick VB, Nauta WJH (1982) The amygdalostriatal projections in the rat. An anatomical study by anterograde and retrograde tracing methods. Neurosci 7:615–630
Khachaturian H, Watson SJ, Lewis ME, Coy D, Gold-

stein A, Akil H (1982) Dynorphin immunocytochemistry in the rat central nervous system. Peptides 3:941–954
Khachaturian H, Lewis ME, Hollt V, Watson SJ (1983a) Telencephalic enkephalinergic systems in the rat brain. J Neurosci 3:844–855
Khachaturian H, Lewis ME, Watson SJ (1983b) Enkephalin systems in diencephalon and brainstem of the rat. J Comp Neurol 220:310–320
Kievit J, Kuypers HGJM (1975) Basal forebrain and hypothalamic connections to frontal and parietal cortex in the rhesus monkey. Science 187:660–662
Kimura H, McGeer PL, Peng JH, McGeer EG (1981) The Central Cholinergic System Studied by Choline Acetyltransferase Immunohistochemistry in the Cat. J Comp Neurol 200:151–201
King GW (1980) Topology of ascending brainstem projections to nucleus parabrachialis in the cat. J Comp Neurol 191:615–638
King JC, Tobet SA, Snavely FL, Arimura AA (1982) LHRH immunopositive cells and their projections to the median eminence and organum vasculosum of the lamina terminalis. J Comp Neurol 209:287–300
King JC, Anthony ELP, Gustafson AW, Damassa DA (1984) Luteinizing hormone-releasing hormone (LH-RH) cells and their projections in the forebrain of the bat Myotis lucifugus lucifugus. Brain Res 298:289–301
Kiss JZ, Williams TH, Palkovits M (1984) Distribution and projections of cholecystokinin-immunoreactive neurons in the hypothalamic paraventricular nucleus of the rat. J Comp Neurol 227:173–181
Kita H, Oomura Y (1982) Evidence for a glycinergic cortico-lateral hypothalamic inhibitory pathway in the rat. Brain Res 235:131–136
Kitahama K, Sakai K, Jouvet M, Dubois P (1984) Mise en évidence de la projection directe des neurones immunoréactifs à l'ACTH sur le complexe du locus coeruleus chez le Chat. CR Acad Sc Paris 298:163–167
Kiyama H, Shiosaka S, Kubota Y, Cho HJ, Takagi H, Tateishi K, Hashimura E, Hamaoka T, Tohyama M (1983) Ontogeny of cholecystokinin-8 containing neuron system of the rat: an immunohistochemical analysis-II. Lower brain stem. Neuroscience 10:1341–1359
Klüver H, Bucy PC (1937) Psychic blindness and other symptoms following bilateral temporal lobectomy in rhesus monkeys. Amer J Physiol 119:352–353
Klüver H, Bucy PC (1939) Preliminary analysis of functions of the temporal lobes in monkeys. Arch Neurol Psychiat 42:979–1000
Knigge KM, Joseph SA (1981) Relationship of the central ACTH-immunoreactive opiocortin system to the median eminence and the pituitary gland of the rat. Cell Tissue Res 215:333–340
Knigge KM, Joseph SA (1984) Anatomy of the opioid-systems of the brain. Can J Neurol Sci 11:14–23
Knigge KM, Joseph SA, Nocton J (1981) Topography of the ACTH-immunoreactive neurons in the basal hypothalamus of the rat brain. Brain Res 216:333–341
Knowles WD, Phillips MI (1980) Angiotensin II responsive cells in the organum vasculosum lamina terminalis (OVTL) recorded in hypothalamic brain slices. Brain Res 197:256–259
Kobayashi RM, Brown M, Vale W (1977) Regional distribution of neurotensin and somatostatin in rat brain. Brain Res 126:584–588
Köhler Ch, Chan-Palay V (1983) Distribution of gamma aminobutyric acid containing neurons and terminals in the septal area. Anat Embryol 167:53–65
Köhler C, Eriksson LG (1984) An immunohistochemical study of somatostatin and neurotensin positive neurons in the septal nuclei of the rat brain. Anat Embryol 170:1–10
Köhler C, Swanson LW (1984) Acetylcholinesterase-containing cells in the lateral hypothalamic area are immunoreactive for α-melanocyte stimulating hormone (α-MSH) and have cortical projections in the rat. Neurosci Lett 49:39–43
Köhler Ch, Chan-Palay V, Wu J-Y (1984a) Septal neurons containing glutamic acid decarboxylase immunoreactivity project to the hippocampal region in the rat brain. Anat Embryol 169:41–44
Köhler C, Haglund L, Swanson LW (1984b) A diffuse αMSH-immunoreactive projection to the hippocampus and spinal cord from individual neurons in the lateral hypothalamic area and zona incerta. J Comp Neurol 223:501–514
Kölliker A (1896) Handbuch der Gewebelehre des Menschen. Vol 2-Nervensystem. Engelmann, Leipzig, pp 874
Kohno J, Shiosaka S, Shinoda K, Inagaki S, Tohyama M (1984) Two distinct strionigral substance P pathways in the rat: An experimental immunohistochemical study. Brain Res 308:309–317
Kojima M, Sano Y (1983) The organization of serotonin fibers in the anterior column of the mammalian spinal cord. Anat Embryol 167:1–11
Kojima M, Takeuchi Y, Goto M, Sano Y (1982) Immunohistochemical study on the distribution of serotonin fibers in the spinal cord of the dog. Cell Tissue Res 226:477–491
Kojima M, Takeuchi Y, Goto M, Sano Y (1983) Immunohistochemical study on the localization of serotonin fibers and terminals in the spinal cord of the monkey (Macaca fuscata). Cell Tissue Res 229:23–36
Kooy D vd, Hattori T (1980) Dorsal raphe cells with collateral projections to the caudate-putamen and substantia nigra: A fluorescent retrograde double labeling study in the rat. Brain Res 186:1–7
Kooy D van der, Hunt SP, Steinbusch HWM, Verhofstad AAJ (1981) Separate populations of cholecystokinin and 5-hydroxytryptamine-containing neuronal cells in the rat dorsal raphe, and their contribution to the ascending raphe projections. Neurosci Lett 26:25–30
Krayniak PF, Meibach RC, Siegel A (1981) Origin of brain stem and temporal cortical afferent fibers to the septal region in the squirrel monkey. Exp Neurol 72:113–121 (1981)
Kreuzberg GW (1984) 100 years of Nissl staining. Trends NeuroSci 7:236–237
Krieger DT (1983) Brain peptides: What, where, and why? Science 222:975–985
Krieger DT, Martin JB (1981a) Brain Peptides. New Engl J Med 304:876–885

Krieger DT, Martin JB (1981b) Brain peptides. New Engl J Med 304:944–951
Krieger DT, Liotta A, Brownstein MJ (1977a) Presence of corticotropin in brain of normal and hypophysectomized rats. Proc Natl Acad Sci USA 74:648–652
Krieger DT, Liotta A, Suda T, Palkovits M, Brownstein MJ (1977b) Presence of immunoassayable β-lipotropin in bovine brain and spinal cord: Lack of concordance with ACTH concentrations. Biochem Biophys Res Co 76:930–936
Krisch B (1978) Hypothalamic and extrahypothalamic distribution of somatostatin-immunoreactive elements in the rat brain. Cell Tissue Res 195:499–513
Krisch B (1979) Immunohistochemical results on the distribution of somatostatin in the hypothalamus and in limbic structures of the rat. J Histochem Cytochem 27:1389–1390
Krisch B (1981) Somatostatin-immunoreactive fiber projections into the brain stem and the spinal cord of the rat. Cell Tissue Res 217:531–552
Kromer LF, Moore RY (1976) Cochlear nucleus innervation by central norepinephrine neurons in the rat. Brain Res 118:531–537
Kromer LF, Moore RY (1980) Norepinephrine innervation of the cochlear nuclei by locus coeruleus neurons in the rat. Anat Embryol 158:227–244
Kubek MJ, Rea MA, Hodes ZI, Aprison MH (1983) Quantitation and characterization of thyrotropin-releasing hormone in vagal nuclei and other regions of the medulla oblongata of the rat. J Neurochem 40:1307–1313
Kubuta Y, Inagaki S, Shiosaka S, Cho HJ, Tateishi K, Hashimura E, Hamaoka T, Tohyama M (1983) The distribution of cholecystokinin octapeptide-like structures in the lower brain stem of the rat: An immunohistochemical analysis. Neuroscience 9:587–604
Kuhar MJ, Pert CB, Snyder SH (1973) Regional distribution of opiate receptor binding in monkey and human brain. Nature 245:447–450
Kupfermann I (1979) Modulatory actions of neurotransmitters. Ann Rev Neurosci 2:447–465
Kuriyama K, Haber B, Sisken B, Roberts E (1966) The γ-aminobutyric acid system in rabbit cerebellum. Proc Natl Acad Sci USA 55:846–852
Kuriyama K, Ida S, Nishimura C, Ohkuma S (1983) Distribution and function of taurine in nervous tissues: An introductory review. Progr Biol Clin Res 125:127–140
Kuypers HGJM (1982) A new look at the organization of the motor system. Progr Brain Res 57:381–403
Lackner KJ (1980) Mapping of monoamine neurones and fibres in the cat lower brainstem and spinal cord. Anat Embryol 161:169–195
Lamberts R, Vijayan E, Graf M, Mansky T, Wuttke W (1983) Involvement of preoptic-anterior hypothalamic GABA neurons in the regulation of pituitary LH and prolactin release. Exp Brain Res 52:356–362
LaMotte CC, De Lanerolle NC (1981) Human spinal neurons: innervation by both substance P and enkephalin. Neuroscience 6:713–723
LaMotte CC, De Lanerolle NC (1983a) Ultrastructure of chemically defined neuron systems in the dorsal horn of the monkey. II. Methionine-enkephalin immunoreactivity. Brain Res 274:51–63
LaMotte CC, De Lanerolle NC (1983b) Ultrastructure of chemically-defined neuron systems in the dorsal horn of the monkey. III. Serotonin immunoreactivity. Brain Res 274:65–77
Landas S, Phillips MI, Stamler JF, Raizada MK (1980) Visualization of specific angiotensin II binding sites in the brain by fluorescent microscopy. Science 210:791–793
Langevin H, Emson PC (1982) Distribution of substance P, somatostatin and neurotensin in the human hypothalamus. Brain Res 246:65–69
Larsson L-I, Rehfeld JF (1979) Localization and molecular heterogeneity of cholecystokinin in the central and peripheral nervous system. Brain Res 165:201–218
Larsson L-I, Rehfeld JF (1981) Pituitary gastrins occur in corticotrophs and melanotrophs. Science 213:768–770
Larsson L-I, Fahrenkrug J, Schaffalitzky de Muckadell O, Sundler F, Hakanson R, Rehfeld JF (1976) Localization of vasoactive intestinal polypeptide (VIP) to central and peripheral neurons. Proc Natl Acad Sci USA 73:3197–3200
Larsson L-I, Childers S, Snyder SH (1979) Met- and leu-enkephalin immunoreactivity in separate neurones. Nature 282:407–410
Lazarus LH, Brown MR, Perrin MH (1977) Distribution, localization and characteristics of neurotensin binding sites in the rat brain. Neuropharmacol 16:625–629
Lechan RM, Snapper SB, Jackson IMD (1983) Evidence that spinal cord thyrotropin-releasing hormone is independent of the paraventricular nucleus. Neurosci Lett 43:61–65
Leenen L, Meek J, Nieuwenhuys R (1982) Unmyelinated fibres in the pyramidal tract of the rat. A new view. Brain Res 256:297–301
Leeuwen F van, Caffé R (1983) Vasopressin-immunoreactive cell bodies in the bed nucleus of the stria terminalis of the rat. Cell Tissue Res 228:52–534
Leeuwen FW van, Swaab DF, Raay C de (1978) Immunoelectronmicroscopic localization of vasopressin in the rat suprachiasmatic nucleus. Cell Tissue Res 193:1–10
Léger L, Descarries L (1978) Serotonin nerve terminals in the locus coeruleus of adult rat: a radioautographic study. Brain Res 145:1–13
Léger L, Wiklund L, Descarries L, Persson M (1979) Description of an indolaminergic cell component in the cat locus coeruleus: A fluorescent histochemical and radioautographic study. Brain Res 168:43–56
Léger L, McRae-Dequerce A, Pujol J-F (1980) Origine de l'innervation sérotoninergique du locus coeruleus chez le rat. CR Acad Sci D NAT 290:807–810
Lehmann J, Fibiger HC (1979) Acetylcholinesterase and the cholinergic neuron. Life Sci 25:1939–1947
Lehtosalo JI, Uusitalo H, Stjernschantz J, Palkama A (1984) Substance P-like immunoreactivity in the trigeminal ganglion. A fluorescence, light and electron microscope study. Histochem 80:421–427
Leichnetz GR, Astruc J (1977) The cours of some prefrontal corticofugals to the pallidum, substantia innominata and amygdaloid complex in monkeys. Exp Neurol 54:104–109

Leichnetz GR, Watkins L, Griffin G, Murfin R, Mayer DJ (1978) The projection from nucleus raphe magnus and other brainstem nuclei to the spinal cord in the rat: A study using the HRP blue-reaction. Neurosci Lett 8:119–124

Lenn NJ, Hamill GS (1984) Subdivisions of the interpeduncular nucleus: A proposed nomenclature. Brain Res Bull 13:203–204

Léranth Cs, Frotscher M, Tömböl T, Palkovits M (1984) Ultrastructure and synaptic connections of vasoactive intestinal polypeptide-like immunoreactive nonpyramidal neurons and axon terminals in the rat hippocampus. Neurosci 12:531–542

Léranth G, Williams TH, Chretien M, Palkovits M (1980) Ultrastructural investigation of ACTH immunoreactivity in arcuate and supraoptic nuclei of the rat. Cell Tissue Res 210:11–19

Levitt P, Moore RY (1978) Noradrenaline neuron innervation of the neocortex in the rat. Brain Res 139:219–231

Levitt P, Moore RY (1979) Origin and organization of brainstem catecholamine innervation in the rat. J Comp Neurol 186:505–528

Levitt P, Rakic P, Goldman-Rakic P (1984) Region-specific distribution of catecholamine afferents in primate cerebral cortex: A fluorescence histochemical analysis. J Comp Neurol 227:23–36

Lew JY, Matsumoto Y, Pearson J, Goldstein M, Hökfelt T, Fuxe K (1977) Localization and characterization of phenylethanolamine-N-methyltransferase in the brain of various mammalian species. Brain Res 119:199–210

Lewicki JA, Fallon JH, Printz MP (1978) Regional distribution of angiotensinogen in rat brain. Brain Res 158:359–371

Lewis PR, Shute CCD (1967) The cholinergic limbic system: Projections to hippocampal formation, medial cortex, nuclei of the ascending cholinergic reticular system, and the subfornical organ and supra-optic crest. Brain 90:521–540

Lewis MS, Molliver ME, Morrison JH, Lidov HGW (1979) Complementarity of dopaminergic and noradrenergic innervation in anterior cingulate cortex of the rat. Brain Res 164:328–333

Li CH, Chung D (1976) Isolation and structure of an untriakontapeptide with opiate activity from camel pituitary glands. Proc Natl Acad Sci USA 73:1145–1148

Libet B (1979) Cyclic nucleotides as synaptic mediators. Life Sci 24:1043–1058

Lidov HGW, Grzanna R, Molliver ME (1980) The serotonin innervation of the cerebral cortex in the rat – An immunohistochemical analysis. Neuroscience 5:207–227

Lieberman A, Dziatolowsky M, Kupersmith M, Serby M, Goodgold A, Korein J, Goldstein M (1979) Dementia in Parkinson disease. Ann Neurol 6:355–359

Lightman SL, Todd K, Everitt BJ (1984) Ascending noradrenergic projections from the brainstem: Evidence for a major role in the regulation of blood pressure and vasopressin secretion. Exp Brain Res 55:145–151

Lind RW, Swanson LW (1984) Evidence for corticotropin releasing factor and leu-enkephalin in the neural projection from the lateral parabrachial nucleus to the median preoptic nucleus: a retrograde transport, immuno-histochemical double labeling study in the rat. Brain res 321:217–224

Lindvall O (1975) Mesencephalic dopamine afferents to the lateral septal nucleus of the rat. Brain Res 87:89–95

Lindvall O, Björklund A (1974a) The organization of the ascending catecholamine neuron systems in the rat brain as revealed by the glyoxylic acid fluorescence method. Acta Physiol Scand (Suppl) 412:1–48

Lindvall O, Björklund A (1974b) The glyoxylic acid fluorescence histochemical method: A detailed account of the methodology for the visualization of central catecholamine neurons. Histochemistry 39:97–127

Lindvall O, Björklund A (1978) Organization of catecholamine neurons in the rat central nervous system. In: Iversen LL, Iversen SD, Snyder SH (eds) Handbook of Psychopharmacology, Vol 9, Plenum Press, New York, pp 139–231

Lindvall O, Björklund A (1979) Dopaminergic innervation of the globus pallidus by collaterals from the nigrostriatal pathway. Brain Res 172:169–173

Lindvall O, Björklund A (1983) Dopamine- and norepinephrine-containing neuron systems: Their anatomy in the rat brain. In: Emson PC (ed) Clinical Neuroanatomy, Raven Press, New York, pp 229–255

Lindvall O, Stenevi U (1978) Dopamine and noradrenaline neurons projecting to the septal area in the rat. Cell Tissue Res 190:383–407

Lindvall O, Björklund A, Moore RY, Stenevi U (1974a) Mesencephalic dopamine neurons projecting to neocortex. Brain Res 81:325–331

Lindvall O, Björklund A, Nobin A, Stenevi U (1974b) The adrenergic innervation of the rat thalamus as revealed by the glyoxylic acid fluorescence method. J Comp Neur 154:317–348

Lindvall O, Björklund A, Divac I (1978) Organization of catecholamine neurons projecting to the frontal cortex in the rat. Brain Res 142:1–24

Lindvall O, Björklund A, Skagerberg G (1984) Selective histochemical demonstration of dopamine terminal systems in rat di- and telencephalon: New evidence for dopaminergic innervation of hypothalamic neurosecretory nuclei. Brain Res 306:19–30

Liotta AS, Gildersleeve D, Brownstein MJ, Krieger DT (1979) Biosynthesis in vitro of immunoreactive 31,000-dalton corticotropin/β-endorphin-like material by bovine hypothalamus. Proc Natl Acad Sci USA 76:1448–1452

Liposits Zs, Sétáló G (1980) Descending luteinizing hormone-releasing hormone (LH–RH) nerve fibers to the midbrain of the rat. Neurosci Lett 20:1–4

Ljungdahl A, Hökfelt T (1973) Autoradiographic uptake paterns of (^{3}H)GABA and (^{3}H)glycine in central nervous tissues with special reference to the cat spinal cord. Brain Res 62:587–595

Ljungdahl A, Hökfelt T, Halász N, Johansson O, Goldstein M (1977) Olfactory bulb dopamine neurons – the A15 catecholamine cell group. Acta Physiol Scand Suppl 452:31–35

Ljungdahl A, Hökfelt T, Nilsson G (1978) Distribution of substance P-like immunoreactivity in the central nervous system of the rat. I. Cell bodies and nerve terminals. Neuroscience 3:861–943

Llorens-Cortes C, Javoy-Agid F, Agid Y, Taquet H,

Schwartz JC (1984) Enkephalinergic markers in substantia nigra and caudate nucleus from parkinsonian subjects. J Neurochem 43:874–877
Loewy AD, McKellar S (1981) Serotonergic projections from the ventral medulla to the intermediolateral cell column in the rat. Brain Res 211:146–152
Loewy AD, Neil JJ (1981) The role of descending monoaminergic system in central of blood pressure. Fed Proc 40:2778–2785
Loewy AD, Gregorie EM, McKellar S, Baker RP (1979a) Electrophysiological evidence that the A5 catecholamine cell group is a vasomotor center. Brain Res 178:196–200
Loewy AD, McKellar S, Saper CB (1979b) Direct projections from the A5 catecholamine cell group to the intermediolateral cell column. Brain Res 174:309–314
Loewy AD, Wallach JH, McKellar S (1981) Efferent connections of the ventral medulla oblongata in the rat. Brain Res Rev 3:63–80
Loh HH, Tseng LF, Wei E, Li CH (1976) β-Endorphin is a potent analgesic agent. Proc Natl Acad Sci USA 73:2895–2898
Lombardini JB (1976) Regional and subcellular studies on taurine in the rat central nervous system. In: Huxtable RJ, Barbeau A (eds) Taurine. Raven Press, New York, pp 311–326
Long SE, Duffin J (1984) The medullary respiratory neurons: A review. Can J Physiol Pharmacol 62:161–182
Lorén I, Alumets J, Hakanson R, Sundler F (1979a) Distribution of gastrin and CCK-like peptides in rat brain. Histochemistry 59:249–257
Lorén I, Emson PC, Fahrenkrug J, Björklund A, Alumets J, Hakanson R, Sundler F (1979b) Distribution of vasoactive intestinal polypetide in the rat and mouse brain. Neuroscience 4:1953–1976
Loughlin SE, Foote SL, Fallon JH (1982) Locus coeruleus projections to cortex: topography, morphology, and collateralization. Brain Res Bull 9:287–294
Loy R, Koziell DA, Lindsey JD, Moore RY (1980) Noradrenergic innervation of the adult rat hippocampal formation. J Comp Neurol 189:699–710
Lundberg JM, Fahrenkrug J, Brimijoin S (1981) Characteristics of the axonal transport of vasoactive intestinal polypeptide (VIP) in nerves of the cat. Acta Physiol Scand 112:427–436
Lund Karlesen R, Fonnum F (1978) Evidence for glutamate as a neurotransmitter in the corticofugal fibres to the dorsal lateral geniculate body and the superior colliculus in rats. Brain Res 151:457–467
Lundberg JM, Hökfelt T (1983) Coexistence of peptides and classical neurotransmitters. Trends NeuroSci 6:325–333
Lundberg JM, Hökfelt T, Nilsson G, Terenius L, Rehfeld J, Elde R, Said S (1978) Peptide neurons in the vagus, splanchnic and sciatic nerves. Acta Physiol Scand 104:499–501
Lundberg JM, Hökfelt T, Schultzberg M, Uvnäs-Wallenstein K, Köhler C, Said SI (1979) Occurrence of vasoactive intestinal polypeptide (VIP)-like immunoreactivity in certain cholinergic neurons of the cat: evidence from combined immunohistochemistry and acetylcholinesterase staining. Neuroscience 4:1539–1559
Maciewicz R, Phipps BS, Grenier J, Poletti CE (1984) Edinger-Westphal nucleus: cholecystokinin immunocytochemistry and projections to spinal cord and trigeminal nucleus in the cat. Brain Res 299:139–145
MacKay AVP (1984) High dopamine in the left amygdala. Trends NeuroSci 7:107–108
MacLean PD (1952) Some psychiatric implications of physiological studies of frontotemporal portion of limbic system (visceral brain). Electroencephalogr Clin Neurophysiol 4:407–418
MacLean P (1958) The limbic system with respect to self-preservation and the preservation of the species. J Nerv Ment Dis 127:1–11
MacLean PD (1962) New findings relevant to the evolution of psychosexual functions of the brain. J Nerv Ment Dis 135:289–301
MacLean PD (1970) The Triune Brain, Emotion, and Scientific Bias. In: Schmitt FO (ed) The Neurosciences, 2nd study program, pp 336–349
Macleod RM, Lehmeyer JE (1974) Studies on the mechanism of the dopamine-mediated inhibition of prolactin secretion. Endocrinology 94:1077–1085
Macrides F, Davis BJ (1983) The olfactory bulb. In: Emson PC (ed) Chemical Neuroanatomy, Raven Press, New York, pp 391–426
Maderdrut JL, Yaksh TL, Petrusz P, Go VLW (1982) Origin and distribution of cholecystokinin-containing nerve terminals in the lumbar dorsal horn and nucleus caudalis of the cat. Brain Res 243:363–368
Maeda K, Frohman LA (1978) Dissociation of systemic and central effects of neurotensin on the secretion of growth hormone, prolactin, and thyrotrophin. Endocrinology 103:1903–1909
Magistretti PJ, Schorderet (1984) VIP and noradrenaline act synergistically to increase cyclic AMP in cerebral cortex. Nature 308:280–282
Magistretti PJ, Morrison JH, Shoemaker WJ, Sapin V, Bloom FE (1981) Vasoactive intestinal polypeptide induces glycogenolysis in mouse cortical slices: a possible regulatory mechanism for the local control of energy metabolism. Proc Natl Acad Sci USA 78:6535–6539
Mains RE, Eipper BA (1979) Synthesis and secretion of corticotropins, melanotropins and endorphins by rat intermediate pituitary cells. J Biol Chem 254:7885–7894
Mains RE, Eipper BA, Ling N (1977) Common precursor to corticotropins and endorphins. Proc Natl Acad Sci USA 74:3014–3018
Makara GB, Stark E, Karteszi M, Palkovitz M, Rappay G (1981) Effects of paraventricular lesions on stimulated ACTH release and CRF in stalk-median eminence of the rat. Am J Physiol E 240:441–446
Makino T, Yokokura T, Iizuka R (1978) Effect of neurotensin on pituitary gonadotropin release in vivo. Endocrinol Jpn 25:181–183
Manberg PJ, Youngblood WW, Nemeroff ChB, Rossor MN, Eversen LL, Prange AJ Jr, Kizer JS (1982) Regional distribution of neurotensin in human brain. J Neurochem 38:1777–1780
Mann DMA, Yates PO, Marcyniuk B (1984) A comparison of changes in the nucleus basalis and locus coeruleus in Alzheimer's disease. J Neurol Neurosur Ps 47:201–203

Mantyh PW (1982) Forebrain projections to the periaqueductal gray in the monkey, with observations in the cat and rat. J Comp Neurol 206:146–158
Mantyh PW (1983) Connections of midbrain periaqueductal gray in the monkey. I. Ascending efferent projections. J Neurophysiol 49:567–582
Mantyh PW, Hunt P (1984a) Evidence for cholecystokinin-like immunoreactive neurons in the rat medulla oblongata which project to the spinal cord. Brain Res 291:49–54
Mantyh PW, Hunt SP (1984b) Neuropeptides are present in projection neurones at all levels in visceral and taste pathways: from periphery to sensory cortex. Brain Res 299:297–311
Mantyh PW, Kemp A (1983) The distribution of putative neurotransmitters in the lateral geniculate nucleus of the rat. Brain Res 288:344–348
Mantyh PW, Peschanski M (1982) Spinal projections from the periaqueductal grey and dorsal raphe in the rat, cat and monkey. Neuroscience 7:2769–2776
Marchand ER, Riley JN, Moore RY (1980) Interpeduncular neucleus afferents in the rat. Brain Res 193:339–352
Marchand JE, Hagino N (1983) Afferents to the periaqueductal gray in the rat. A horseradish peroxidase study. Neuroscience 9:95–106
Marley PD, Emson PC, Hunt SP, Fahrenkrug J (1981) A long ascending projection in the rat brain containing vasoactive intestinal polypeptide. Neurosci Lett 27:261–266
Marley PD, Nagy JI, Emson PC, Rehfeld JF (1982) Cholecystokinin in the rat spinal cord: distribution and lack of effect of neonatal capsaicin treatment and rhizotomy. Brain Res 238:494–498
Marshall PE, Landis DMD, Zalneraitis EL (1983) Immunocytochemical studies of substance P and leucine-enkephalin in Huntington's disease. Brain Res 289:11–26
Martin GE, Bacino CB (1979) Action of intracerebrally injected β-endorphin on the rat's core temperature. Eur J Pharmacol 59:227–236
Martin GF, Cabana T, Ditirro FJ, Ho RH, Humbertson AO Jr (1982) Raphespinal projections in the north american opossum: Evidence for connectional heterogeneity. J Comp Neurol 208:67–84
Mason ST (1980) Noradrenaline and selective attention: A review of the model and the evidence. Life Sci 27:617–631
Mason ST (1981) Noradrenaline in the brain: Progress in theories of behavioral function. Progr Neurobiol 16:263–303
Mason ST, Fibiger HC (1979) Regional topography within noradrenergic locus coeruleus as revealed by retrograde transport of horseradish peroxidase. J Comp Neurol 187:703–724
Mason WD, Ho YW, Hatton GI (1984) Axon collaterals of supraoptic neurones: Anatomical and electrophysiological evidence for their existence in the lateral hypothalamus. Neurosci 11:169–182
Mauborgne A, Javoy-Agid F, Legrand JC, Agid Y, Cesselin F (1983) Decrease of substance P-like immunoreactivity in the substantia nigra and pallidum of parkinsonian brains. Brain Res 268:167–170
Mayeux R, Stern Y (1983) Intellectual dysfunction and dementia in Parkinson disease. Adv Neurol 38:211–227
Mayo W, Dubois B, Ploska A, Javoy-Agid F, Agid Y, Le Moal M, Simon H (1984) Cortical cholinergic projections from the basal forebrain of the rat, with special reference to the prefrontal cortex innervation. Neurosci Lett 47:149–154
McBride WJ, Aprison MH, Kusano K (1976a) Contents of several amino acids in the cerebellum, brain stem and cerebrum of the "staggerer", "weaver" and "nervous" neurologically mutant mice. J Neurochem 26:867–871
McBride WJ, Nadi NS, Altman J, Aprison MH (1976b) Effects of selective doses of X-irradiation on the levels of several amino acids in the cerebellum of the rat. Neurochem Res 1:141–152
McCann SM (1982) Physiology and pharmacology of LHRH and somatostatin. Annu Rev Pharmacol Toxicol 22:491–515
McCann SM, Vijayan E, Koenig J, Krulich L (1982) The effects of neurotensin on anterior pituitary hormone secretion. Ann NY Acad Sci 400:160–171
McCann SM, Lumpkin MD, Mizunuma H, Khorram O, Ottlecz A, Samson WK (1984) Peptidergic and dopaminergic control of prolactin release. Trends NeuroSci 7:127–131
McGeer EG, Singh EA (1980) Possible cortico-hypothalamic glycinergic tract. Neurosci Lett 17:85–87
McGeer PL (1984) The 12th JAF Stevenson Memorial Lecture. Aging, Alzheimer's disease, and the cholinergic system. Can J Physiol Pharmacol 62:741–754
McGeer PL, McGeer EG (1975) Evidence for glutamic acid decarboxylase-containing interneurons in the neostriatum. Brain Res 91:331–335
McGeer PL, McGeer EG (1964) Formation of adrenaline by brain tissue. Biochem Biophys Res Commun 17:502–507
McGeer PL, Hattori T, McGeer EG (1975) Chemical and autoradiographic analysis of γ-aminobutyric acid transport in Purkinje cells of the cerebellum. Exp Neurol 47:26–41
McGeer PL, McGeer EG, Scherer U, Singh K (1977) A glutamatergic corticostriatal path? Brain Res 128:369–373
McGeer PL, McGeer EG, Peng JH (1984a) Choline acetyltransferase: Purification and immunohistochemical localization. Life Sci 34:2319–2338
McGeer PL, McGeer EG, Suzuki J, Dolman CE, Nagai T (1984b) Aging, Alzheimer's disease and the cholinergic system of the basal forebrain. Neurology 34:741–745
McGeer EG, Staines WA, McGeer PL (1984c) Neurotransmitters in the basal ganglia. Can J Neurol Sci 11:89–99
McKellar S, Loewy AD (1982) Efferent projections of the A1 catecholamine cell group in the rat: An autoradiographic study. Brain Res 241:11–29
McLaughlin BJ, Barber R, Saito K, Roberts E, Wu J-Y (1975) Immunocytochemical localization of glutamate decarboxylase in rat spinal cord. J Comp Neurol 164:305–322
McLaughlin BJ, Wood JG, Saito K, Barber R, Vaughn JE, Roberts E, Wu J-Y (1980) The fine structural

localization of glutamate decarboxylase in synaptic terminals of rodent cerebellum. Brain Res 76:377–391

Mehler WR (1969) Some neurological species differences – A posteriori. Ann NY Acad Sci 167:424–468

Mehler WR (1981) The basal ganglia – circa 1982. Appl Neurophysiol 44:261–290

Mendelsohn FAO, Quirion R, Saavedra JM, Aguilera G, Catt KJ (1984) Autoradiographic localization of angiotensin II receptors in rat brain. Proc Natl Acad Sci USA 81:1575-1579

Merchenthaler I, Kovács G, Lovász G, Sétáló G (1980) The preoptico-infundibular LH–RH tract of the rat. Brain Res 198:63–74

Mesulam M-M, Hoesen GW van (1976) Acetylcholinesterase-rich projections from the basal forebrain of the rhesus monkey to neocortex. Brain Res 109:152–157

Mesulam M-M, Mufson EJ (1984) Neural inputs into the nucleus basalis of the substantia innominata (Ch4) in the rhesus monkey. Brain 107:253–274

Mesulam M-M, Mufson EJ, Levey AI, Wainer BH (1983a) Cholinergic Innervation of Cortex by the Basal Forebrain: Cytochemistry and Cortical Connections of the septal Area, Diagonal Band Nuclei, Nucleus Basalis (Substantia Innominata), and Hypothalamus in the Rhesus Monkey. J Comp Neurol 214:170–197

Mesulam M-M, Mufson EJ, Wainer BH, Levey AI (1983b) Central cholinergic pathways in the rat: an overview based on an alternative nomenclature (Ch1–Ch6). Neuroscience 10:1185–1201

Mesulam M-M, Mufson EJ, Levey AI, Wainer BH (1984a) Atlas of cholinergic neurons in the forebrain and upper brainstem of the macaque based on monoclonal choline acetyltransferase immunohistochemistry and acetylcholinesterase histochemistry. Neuroscience 12:669–686

Mesulam M-M, Rosen AD, Mufson EJ (1984b) Regional variations in cortical cholinergic innervation: chemoarchitectonics of acetylcholinesterase-containing fibers in the macaque brain. Brain Res 311:245–258

Meyer DK, Phillips MI, Eiden L (1982) Studies on the presence of angiotensin II in rat brain. J Neurochem 38:816–820

Meynert T (1872) Vom Gehirn der Säugetiere. In: Handbuch der Lehre von den Geweben des Menschen und der Tiere, Vol 2, Stricker S, Leipzig

Miachon S, Berod A, Leger L, Chat M, Hartman B, Pujol JF (1984) Identification of catecholamine cell bodies in the pons and pons-mesencephalon junction of the cat brain, using tyrosine hydroxylase and dopamine-β-hydroxylase immunohistochemistry. Brain Res 305:369–374

Micevych P, Elde R (1980) Relationship between enkephalinergic neurons and the vasopressin-oxytocin neuroendocrine system of the cat: an immunohistochemical study. J Comp Neurol 190:135–146

Michelakis AM, Yoshida H, Menzie J, Murakami K, Inagami T (1974) A radioimmunoassay for the direct measurement of renin in mice and its application to submaxillary gland and kidney studies. Endocrinology 94:1101–1105

Miletic V, Hoffert MJ, Ruda MA, Dubner R, Shigenaga Y (1984) Serotoninergic axonal contacts on identified cat spinal dorsal horn neurons and their correlation with nucleus raphe magnus stimulation. J Comp Neurol 228:129–141

Millan MH, Millan MJ, Herz A (1984) The hypothalamic paraventricular nucleus: Relationship to brain and pituitary pools of vasopressin and oxytocin as compared to dynorphin, β-endorphin and related opioid peptides in the rat. Neuroendocrinol 38:108–116

Millan MJ, Millan MH, Herz A (1983) Contribution of the supraoptic nucleus to brain and pituitary pools of immunoreactive vasopressin in the rat, and the interrelationships between these in the rat. Neuroendocrinol 36:310–319

Milner TA, Joh TH, Miller RJ, Pickel VM (1984) Substance P, neurotensin, enkephalin, and catecholamine-synthesizing enzymes: light microscopic localizations compared with autoradiographic label in solitary efferents to the rat parabrachial region. J Comp Neurol 226:434–447

Minagawa H, Shiosaka S, Inagaki S, Sakanaka M, Takatsuki K, Ishimoto I, Senba E, Kawai Y, Hara Y, Matsuzaki T, Tohyama M (1983) Ontogeny of neurotensin-containing neuron system of the rat: immunohistochemical analysis-II. Lower brain stem. Neuroscience 8:467–486

Miura M, Onai T, Takayama K (1983) Projections of upper structure to the spinal cardioacceleratory center in cats: an HRP study using a new microinjection method. J Auton Nerv Syst 7:119–139

Miselis RR, Shapiro RE, Hand PJ (1979) Subfornical organ efferents to neural systems for control of body water. Science 205:1022–1024

Mize RR, Spencer RF, Sterling P (1982) Two types of GABA-accumulating neurons in the superficial gray layer of the cat superior colliculus. J Comp Neurol 206:180–192

Mogenson GJ, Nielsen MA (1983) Evidence that an accumbens to subpallial GABAergic projection contributes to locomotor activity. Brain Res Bull 11:309–314

Mogenson GJ, Swanson LW, Wu M (1983) Neural projections from nucleus accumbens to globus pallidus, substantia innominata, and lateral preoptic-lateral hypothalamic area: An anatomical and electrophysiological investigation in the rat. J Neurosci 3:189–202

Molineaux CJ, Feuerstein G, Faden AL, Cox BM (1982) Distribution of immunoreactive dynorphin in discrete brain nuclei; Comparison with vasopressin. Neurosci Lett 33:179–184

Molliver ME, Grzanna R, Lidov HGW, Morrison JH, Olschowka JA (1982) Monoamine systems in the cerebral cortex. In: Chan-Palay V, Palay SL (eds) Cytochemical Methods in Neuroanatomy. Liss AR Inc, New York, pp 255–277

Moon HD, Hao Li C, Jennings BM (1973) Immunohistochemical and histochemical studies of pituitary β-lipotrophs. Anat Rec 175:529–538

Moore JK, Osen KK (1979) The cochlear nuclei in man. Am J Anat 154:393–418

Moore RY (1977) Organum vasculosum lamina terminalis: innervation by serotonin neurons of the midbrain raphe. Neurosci Lett 5:297–302

Moore RY (1978) Catecholamine innervation of the

basal forebrain. I. The septal area. J Comp Neur 177:665–684
Moore RY, Bloom FE (1978) Central catecholamine neuron systems: Anatomy and physiology of the dopamine systems. Ann Rev Neurosci 1:129–169
Moore RY, Bloom FE (1979) Central catecholamine neuron systems: Anatomy and physiology of the norepinephrine and epinephrine systems. Ann Rev Neurosci 2:113–168
Moore RY, Halaris AE (1975) Hippocampal innervation by serotonin neurons of the midbrain raphe in the rat. J Comp Neur 164:171–183
Moore RY, Halaris AE, Jones BE (1980) Serotonin neurons of the midbrain raphe: ascending projections. J Comp Neurol 180:417–438
Morgan Ch, Nadelhaft I, De Groat WC (1981) The distribution of visceral primary afferents from the pelvic nerve to Lissauer's tract and the spinal gray matter and its relationship to the sacral parasympathetic nucleus. J Comp Neurol 201:415–440
Morgane PJ, Jacobs MS (1979) Raphe projections to the locus coeruleus in the rat. Brain Res Bull 4:519–534
Morrell JI, Pfaff DW (1978) A neuroendocrine approach to brain function: Localization of sex steroid concentrating cells in vertebrate brains. Amer Zool 18:447–460
Morrison JH, Magistretti PJ (1983) Monoamines and peptides in cerebral cortex. Contrasting principles of cortical organization. Trends NeuroSci 6:146–151
Morrison JH, Grzanna R, Molliver ME, Coyle JT (1978) The distribution and orientation of noradrenergic fibers in neocortex of the rat: An immunofluorescence study. J Comp Neurol 181:17–40
Morrison JH, Molliver ME, Grzanna R, Coyle JT (1979) Noradrenergic innervation patterns in three regions of medial cortex: an immunofluorescence characterization. Brain Res Bull 4:849–857
Morrison JH, Molliver ME, Grzanna R, Coyle JT (1980) Noradrenergic innervation patterns in three regions of medial cortex an immunofluorescence characterization. Brain Res Bull 4:849–857
Morrison JH, Molliver ME, Grzanna R, Coyle JT (1981) The intra-cortical trajectory of the coeruleo-cortical projection in the rat: A tangentially organized cortical afferent. Neuroscience 6:139–158
Morrison JH, Foote SL, O'Connor D, Bloom FE (1982a) Laminar, tangential and regional organization of the noradrenergic innervation of monkey cortex: dopamine-β-hydroxylase immunohistochemistry. Brain Res Bull 9:309–319
Morrison JH, Foote SL, O'Connor D, Bloom FE (1982b) Laminar, tangential and regional organization of the noradrenergic innervation of monkey cortex: Dopamine-β-hydroxylase immunohistochemistry. Brain Res Bull 9:309–319
Morrison JH, Magistretti PJ, Benoit R, Bloom FE (1984) The distribution and morphological characteristics of the intracortical VIP-positive cell: An immunohistochemical analysis. Brain Res 292:269–282
Moss MS, Basbaum AI (1983) The fine structure of the caudal periaqueductal gray of the cat: Morphology and synaptic organization of normal and immunoreactive enkephalin-labeled profiles. Brain Res 289:27–43
Moss MS, Glazer EJ, Basbaum AI (1981) Enkephalin-immunoreactive perikarya in the cat raphe dorsalis. Neurosci Lett 21:33–37
Moss MS, Glazer EJ, Basbaum AI (1983) The peptidergic organization of the cat periaqueductal gray. I. The distribution of immunoreactive enkephalin-containing neurons and terminals. J Neurosci 3:603–616
Moss RL (1977) Role of hypophysiotrophic neurohormones in mediating neural and behavioral events. Fed Proc 36:1978–1983
Mroz EA, Brownstein MJ, Leeman SE (1976) Evidence for substance P in the habenulo-interpeduncular tract. Brain Res 113:597–599
Mroz EA, Brownstein MJ, Leeman SE (1977) Evidence for substance P in the striato-nigral tract. Brain Res 125:305–311
Mugnaini E (1965) Organization of cerebellar cortex. Exp Brain Res Suppl 1:8–19
Mugnaini E, Oertel WH, Wouterlood FF (1984a) Immunocytochemical localization of GABA neurons and dopamine neurons in the rat main and accessory olfactory bulbs. Neurosci Lett 47:221–226
Mugnaini E, Wouterlood FG, Dahl A-L, Oertel WH (1984b) Immunocytochemical identification of GABAergic neurons in the main olfactory bulb of the rat. Arch Ital Biol 122:83–112
Muller JE, Straus E, Yalow RS (1977) Cholecystokinin and its COOH-terminal octapeptide in the pig brain. Proc Natl Acad Sci USA 74:3035–3037
Mutt V, Said SI (1974) Structure of the porcine vasoactive intestinal octacosapeptide. Eur J Biochem 42:581–589
Nadi NS, McBride WJ, Aprison MH (1977) Distribution of several amino acids in regions of the cerebellum of the rat. J Neurochem 28:453–455
Nadler JV, Vaca KW, White WF, Lynch GS, Cotman CW (1976) Aspartate and glutamate as possible transmitters of excitatory hippocampal afferents. Nature, London 260:538–540
Nadler JV, White WF, Vaca KW, Perry BW, Cotman CW (1978) Biochemical correlates of transmission mediated by glutamate and aspartate. J Neurochem 31:147–155
Nagai T, Satoh K, Imamoto K, Maeda T (1981) Divergent projections of catecholamine neurons of the locus coeruleus as revealed by fluorescent retrograde double labeling technique. Neurosci Lett 23:117–123
Nagai T, McGeer PL, Peng JH, McGeer EG, Dolman CE (1983a) Choline acetyl-transferase immunohistochemistry in brains of Alzheimer's disease patients and controls. Neurosci Lett 36:195–199
Nagai T, McGeer PL, McGeer EG (1983b) Distribution of GABA-T-intensive neurons in the rat forebrain and midbrain. J Comp Neurol 218:220–238
Nair NPV, Bloom DM, Nestoros JN, Schwartz G (1983) Therapeutic efficacy of cholecystokinin in neuroleptic resistant schizophrenic subjects. Psychopharmacol Bull 19:134–136
Nair RMG, Barrett JF, Bowers CY, Schally AV (1970) Structure of porcine thyrotropin releasing hormone. Biochemistry 9:1103–1106
Nakanishi S, Inoue A, Kita T, Nakamura M, Chang ACY, Cohen SN, Numa S (1979) Nucleotide sequence of cloned cDNA for bovine corticotropin-β-lipotropin precursor. Nature 278:423–427

Nakano I, Hirano A (1983) Neuron loss in the nucleus basalis of Meynert in Parkinsonism-dementia complex of Guam. Ann Neurol 13:87–91
Nanopoulos D, Belin MF, Maitre G, Vincendon G, Pujol JF (1982) Immunocytochemical evidence for the existence of gaba-ergic neurons in the nucleus raphe dorsalis, possible existence of neurons containing serotonin and gaba. Brain Res 232:375–389
Nathanson JA (1977) Cyclic nucleotides and nervous system function. Physiol Rev 57:157–256
Nathanson JA, Greengard P (1977) Second messengers in the brain. Sci Am 37:108–119
Nauta HJW, Cuenod M (1982) Perikaryal cell labeling in the subthalamic nucleus following the injection of ^{3}H-aminobutyric acid into the pallidal complex: an autoradiographic study in cat. Neuroscience 7:2725–2734
Nauta WJH (1958) Hippocampal projections and related neural pathways to the midbrain in the cat. Brain 81:319–340
Nauta WJH (1973) Connections of the frontal lobe with the limbic system. In: Laitinen LV, Livingston KE (eds) Surgical Approaches in Psychiatry. Medical and Technical Publishing, Lancaster, pp 303–314
Nauta WJH, Domesick VB (1984) Afferent and efferent relationships of the basal ganglia. In: Evered D, O'Connor M (eds) Functions of the basal ganglia; Ciba Foundation symposium 107. Pitman, London, pp 3–29
Nauta WJH, Haymaker W (1969) Hypothalamic nuclei and fiber connections. In: Haymaker W, Anderson E, Nauta WJH (eds), The Hypothalamus. Thomas (ed), Springfield (Illinois), pp 136–209
Neckers LM, Schwartz JP, Wyatt RJ, Speciale SG (1979) Substance P afferents from the habenula innervate the dorsal raphe nucleus. Exp Brain Res 37:619–623
Nemeroff CB, Bissette G, Prange AJ Jr, Loosen PT, Barlow TS, Lipton MA (1977) Neurotensin: Central nervous system effects of a hypothalamic peptide. Brain Res 128:485–496
Nemeroff CB, Luttinger D, Prange AJ Jr (1983) Neurotensin and bombesin. In: Iverson LL, Iverson SD, Snyder SH (eds) Handbook of Psychopharmacology. Vol 16, Plenum Press, New York, pp 363–467
Nicholson C (1979) Brain-Cell Microenvironment as a Communication Channel. The Neurosciences Fourth Study Program, Schmitt FO, Worden G (eds). The MIT Press, Cambridge Mass., pp 457–476
Nicoll RA, Siggins GR, Ling N, Bloom FE, Guillemin R (1977) Neuronal actions of endorphins and enkephalins among brain regions. A comparative microiontophoretic study. Proc Natl Acad Sci USA 74:2584–2588
Nicoll RA, Schenker C, Leeman SE (1980) Substance P as a transmitter candidate. Ann Rev Neurosci 3:227–268
Nieoullon A, Kerkerian L, Dusticier N (1984) High affinity glutamate uptake in the red nucleus and ventrolateral thalamus after lesion of the cerebellum in the adult cat: Biochemical evidence for funtional changes in the deafferented structures. Exp Brain Res 55:409–419
Nieuwenhuys R (1984) Anatomy of the basal ganglia. In: Bruyn GW, Roos RAC, Buruma OJS (eds) Actua Sandoz, vol 7, Diskinesias. Sandoz, Uden, pp 15–25
Nieuwenhuys R, Voogd J, Van Huijzen Chr (1981) The Human Central Nervous System. Sec Ed. Springer, Berlin Heidelberg New York, pp 253
Nieuwenhuys R, Geeraedts LMG, Veening JG (1982) The medial forebrain bundle of the rat: I. General introduction. J Comp Neurol 206:49–81
Nieuwenhuys R, Cools AR (1984) Dorsal and ventral striatum; some anatomical and pharmacological comments. In: Bruyn GW, Roos RAC, Buruma OJS (eds) Actua Sandoz, vol 7, Diskinesias, Sandoz, Uden, pp 26–29
Nilaver G, Zimmerman EA, Defendini R, Liotta AS, Krieger DT, Brownstein MJ (1979) Adrenocorticotropin and β-lipotropin in the hypothalamus. J Cell Biol 81:50–58
Nobin A, Björklund A (1973) Topography of the monoamine neuron systems in the human brain as revealed in fetuses. Acta Physiol Scand Suppl 388:1–40
Nojyo Y, Sano Y (1978) Ultrastructure of the serotonergic nerve terminals in the suprachiasmatic and interpeduncular nuclei of rat brains. Brain Res 149:482–488
Nomura I, Senba E, Kubo T, Shiraishi T, Matsunaga T, Tohyama M, Shiotani Y, Wu J-Y (1984) Neuropeptides and γ-aminobutyric acid in the vestibular nuclei of the rat: an immunohistochemical analysis. I. Distribution. Brain Res 311:109–118
Norgren R (1974) Gustatory afferents to ventral forebrain. Brain Res 81:285–295
Norgren R (1978) Projections from the nucleus of the solitary tract in the rat. Neuroscience 3:207–218
North RA (1979) Opiates, opioid peptides and single neurones. Life Sci 24:1527–1546
Nygren L-G, Olson L (1977) A new major projection from locus coeruleus: the main source of noradrenergic nerve terminals in the ventral and dorsal columns of the spinal cord. Brain Res 132:85–93
Obata K, Ito M, Ochi R, Sato N (1967) Pharmacological properties of the postsynaptic inhibition by Purkinje cell axons and the action of γ-aminobutyric acid on Deiters neurones. Expl Brain Res 4:43–57
Obata K, Takeda K, Shinozaki H (1970) Further study on pharmacological properties of the cerebellar-induced inhibition of Deiters neurones. Expl Brain Res 11:327–342
Ochi J, Shimizu K (1978) Occurrence of dopamine-containing neurons in the midbrain raphe nuclei of the rat. Neurosci Lett 8:317–320
Oertel WH, Mugnaini E (1984) Immunocytochemical studies of GABAergic neurons in rat basal ganglia and their relations to other neuronal systems. Neurosci Lett 47:233–238
Ohta M, Oomura Y (1979) Inhibitory pathway from the frontal cortex to the hypothalamic ventromedial nucleus in the rat. Brain Res Bull 4:231–238
Okada Y, Shimada C (1975) Distribution of γ-aminobutyric acid (GABA) and glutamate decarboxylase (GAD) activity in the guinea pig hippocampus – microassay method for the determination of GAD activity. Brain Res 98:202–206
Okamoto K, Kimura H, Sakai Y (1983) Evidence for taurine as an inhibitory neurotransmitter in cerebellar stellate interneurons: selective antagonism by TAG (6-amino-methyl-4H,1,2,4-benzothiadiazine-1,1-dioxide). Brain Res 265:163–169

Okon E, Koch Y (1976) Localisation of gonadotropin-releasing and thyrotropin-releasing hormones in human brain by radioimmunoassay. Nature 263:345–347
Oliver C, Porter JC (1978) Distribution and characterization of α-melanocyte-stimulating hormone in the rat brain. Endocrinology 102:697–705
Oliver C, Eskay RL, Ben-Jonathan N, Porter JC (1974) Distribution and concentration of TRH in the rat brain. Endocrinol 95:540–546
Oliveras JL, Guilbaud G, Besson JM (1979) A map of serotoninergic structures involved in stimulation-produced analgesia in unrestrained freely moving cats. Brain Res 164:317–322
Olson L, Fuxe K (1971) On the projections from the locus coeruleus noradrenaline neurons: The cerebellar innervation. Brain Res 28:165–171
Olson L, Borens LO, Seiger A (1973) Histochemical demonstration and mapping of 5-hydroxytryptamine- and catecholamine-containing neuron systems in the human fetal brain. Z Anat Entwicklungsgesch 139:259–282
Olszewski J, Baxter D (1954) Cytoarchitecture of the human brain stem. S Karger, Basel, pp 199
Onteniente B, Geffard M, Calas A (1984) Ultrastructural immunocytochemical study of the dopaminergic innervation of the rat lateral septum with anti-dopamine antibodies. Neuroscience 13:385–393
Osamura RY, Komatsu N, Watanabe K, Nakai Y, Tanaka I, Imura H (1982) Immunohistochemical and immunocytochemical localization of γ-melanocyte stimulating hormone (γ-MSH)-like immunoreactivity in human and rat hypothalamus. Peptides 3:781–787
Osborne NN (1979) Is Dale's principle valid? Trends NeuroSci 2:73–75
Otsuka M, Konishi S (1983) Substance P – the first peptide neurotransmitter? Trends Neurosci 6:317–320
Ottersen OP (1981) Afferent connections to the amygdaloid complex of the rat with some observations in the cat. III. Afferents from the lower brain stem. J Comp Neurol 202:335–356
Ottersen OP, Storm-Mathisen J (1984) Glutamate- and GABA-containing neurons in the mouse and rat brain, as demonstrated with a new immunocytochemical technique. J Comp Neurol 229:374–392
Palacios JM, Wamsley JK, Kuhar MJ (1981) The distribution of histamine H_1-receptors in the rat brain: An autoradiographic study. Neuroscience 6:15–37
Palade GE, Palay SL (1954) Electron microscope observations of interneuronal and neuromuscular synapses. Anat Rec 118:335–336
Palay SL (1967) Principles of cellular organization in the nervous system. In: Quarton GC, Melnechuk T, Schmitt FO (eds) The Neurosciences. Rockefeller Univ Press, New York, pp 24–31
Palay SL (1982) Current status of neuroanatomical research in the cerebellum. Exp Brain Res Suppl 6:1–7
Palay SL, Chan-Palay V (1974) Cerebellar Cortex. Cytology and Organization. Springer, Berlin, pp 348
Palkovits M, Saavedra JM, Jacobowitz DM, Kizer JS, Zaborszky L, Brownstein MJ (1977) Serotonergic innervation of the forebrain: effect of lesions on serotonin and tryptophan hydroxylase levels. Brain Res 130:121–134
Palkovits M, Záborsky L, Feminger A, Mezey E, Fekete MIK, Herman JP, Kanyicska B, Szabó D (1980) Noradrenergic innervation of the rat hypothalamus: Experimental biochemical and electron microscopic studies. Brain Res 191:161–171
Palkovits M, Tapia-Arancibia L, Kordon C, Epelbaum J (1982a) Somatostatin connections between the hypothalamus and the limbic system of the rat brain. Brain Res 250:223–228
Palkovits M, Léránth Cs, Eiden LE, Rotsztejn W, Williams TH (1982b) Intrinsic vasoactive intestinal polypeptide (VIP)-containing neurons in the baroreceptor nucleus of the solitary tract in rat. Brain Res 244:351–355
Palkovits M, Eskay RL, Brownstein MJ (1982c) The course of thyrotropin-releasing hormone fibers to the median eminence in rats. Endocrinol 110:1526–1528
Palkovits M, Pattou E, Herman JP, Kordon C (1984a) Mapping of LH–RH containing projections to the mediobasal hypothalamus by differential deafferentation experiments. Brain Res 298:283–288
Palkovits M, Kiss JZ, Beinfeld MC, Brownstein MJ (1984b) Cholecystokinin in the hypothalamo-hypophyseal system. Brain Res 299:186–189
Panula P, Yang H-YT, Costa E (1984) Comparative distribution of bombesin/GRP- and substance-P-like immunoreactivities in rat hypothalamus. J Comp Neurol 224:606–617
Papez JW (1937) A proposed mechanism of emotion. Arch Neurol Psychiat 38:725–743
Parent A, Descarries L, Beaudet A (1981) Organization of ascending serotonin systems in the adult rat brain. A radioautographic study after intraventricular administration of (^{3}H)5-hydroxytryptamine. Neuroscience 6:115–138
Parent A, Csonka C, Etienne P (1984) The occurrence of large acetylcholinesterase-containing neurons in human neostriatum as disclosed in normal and Alzheimer-diseased brains. Brain Res 291:154–158
Parker CR Jr, Porter JC (1983) Regional localization and subcellular compartmentalization of thyrotropin-releasing hormone in adult human brain. J Neurochem 41:1614–1622
Pasquier DA, Gold MA, Jacobowitz DM (1980) Noradrenergic perikarya (A5–A7, subcoeruleus) projections to the rat cerebellum. Brain Res 196:270–275
Pasternak GW, Gintzler AR, Houghten RA, Ling GSF, Goodman RR, Spiegel K, Nishimura S, Johnson N, Recht LD (1983) Biochemical and pharmacological evidence for opioid receptor multiplicity in the central nervous system. Life Sci 33:167–173
Paton WDM (1958) Central and synaptic transmission in the nervous system (pharmacological aspects). Ann Rev Pharmacol 20:431–462
Paxinos G, Emson PC, Cuello AC (1978) The substance P projections to the frontal cortex and substantia nigra. Neurosci Lett 7:127–131
Paxinos G, Watson C (1982) The Rat Brain in Stereotaxic Coordinates. Academic Press, New York, pp 14+71 pl
Pearson J, Goldstein M, Markey K, Brandeis L (1983) Human brainstem catecholamine neuronal anatomy

as indicated by immunocytochemistry with antibodies to tyrosine hydroxylase. Neuroscience 8:3–32
Pelletier G, Dube D (1977) Electron microscopic immunohistochemical localization of α-MSH in the rat brain. Am J Anat 150:201–206
Pelletier G, Leclerc R (1979) Immunohistochemical localization of adrenocorticotropin in the rat brain. Endocrinology 104:1426–1433
Pelletier G, Leclerc R, Labrie F, Cote J, Chretien M, Lis M (1977) Immunohistochemical localization of β-lipotropic hormone in the pituitary gland. Endocrinology 100:770–776
Pelletier G, Leclerc R, Saavedra JM, Brownstein MJ, Vaudry H, Ferland L, Labrie F (1980) Distribution of β-lipotropin (β-LPH), adrenocorticotropin (ACTH) and α-melanocyte-stimulating hormone (α-MSH) in the rat brain. I. Origin of the extra-hypothalamic fibers. Brain Res 192:433–440
Pelletier G, Steinbusch HWM, Verhofstad AAJ (1981) Immunoreactive substance P and serotonin present in the same dense-core vesicles. Nature 293:71–72
Perry RH, Candy JM, Perry EK, Irving D, Blessed G, Fairbairn AF, Tomlinson BE (1982) Extensive loss of choline acetyltransferase activity is not reflected by neuronal loss in the nucleus of Meynert in Alzheimer's disease. Neurosci Lett 33:311–315
Perry RH, Candy JM, Perry EK, Thompson J, Oakley AE (1984) The substantia innominata and adjacent regions in the human brain: histochemical and biochemical observations. J Anat 138:713–732
Pert A, Dionne R, Ng L, Bragin E, Moody T, Pert C (1981) Alterations in rat central nervous system endorphins following transauricular electroacupuncture. Brain Res 224:83–93
Peterson RP (1966) Magnocellular neurosecretory centers in the rat hypothalamus. J Comp Neurol 128:181–190
Phillips MI (1978) Angiotensin in the Brain. Neuroendocrinology 25:354–377
Phillips MI (1983) New evidence for brain angiotensin and for its role in hypertension. Fed Proc 42:2667–2672
Phillips MI, Felix D (1976) Specific angiotensin II receptive neurons in the cat subfornical organ. Brain Res 109:531–540
Phillips MI, Weyhenmeyer J, Felix D, Ganten D, Hoffman WE (1979) Evidence for an endogenous brain renin-angiotensin system. Fed Proc 38:2260–2266
Phillips MI, Quinlan JT, Weyhenmeyer J (1980) An angiotensin-like peptide in the brain. Life Sci 27:2589–2594
Phillis JW, Kirkpatrick JR (1980) The actions of motilin, luteinizing hormone releasing hormone, cholecystokinin, somatostatin, vasoactive intestinal peptide, and other peptides on rat cerebral cortical neurons. Can J Physiol Pharm 58:612–623
Phillis JW, Kirkpatrick JR, Said SI (1978) Vasoactive intestinal polypeptide excitation of central neurons. Can J Physiol Pharmacol 56:337–340
Phipps BS, Maciewicz R, Sandrew BB, Poletti CE, Foote WE (1983) Edinger-Westphal neurons that project to spinal cord contain substance P. Neurosci Lett 36:125–131
Pickel VM, Joh TH, Reis DJ (1976) Monoamine-synthesizing enzymes in central dopaminergic, noradrenergic and serotonergic neurons. Immunocytochemical localization by light and electron microscopy. J Histochem Cytochem 24:792–806
Pickel VM, Joh TH, Reis DJ (1977) A serotonergic innervation of noradrenergic neurons in nucleus locus coeruleus: Demonstration by immunocytochemical localization of the transmitter specific enzymes tyrosine and tryptophan hydroxylase. Brain Res 131:197–214
Pickel VM, Sumai KK, Beckley SC, Miller RJ, Reis DJ (1980) Immunocytochemical localization of enkephalin in the neostriatum of rat brain: A light and electron microscopic study. J Comp Neurol 189:721–740
Pickel VM, Beckley SC, Joh TH, Reis DJ (1981) Ultrastructural immunocytochemical localization and tyrosine hydroxylase in the neostriatum. Brain Res 225:373–385
Pierce ET, Foote WE, Hobson JA (1976) The efferent connection of the nucleus raphe dorsalis. Brain Res 107:137–144
Pilcher WH, Joseph SA (1984) Co-localization of CRF-ir perikarya and ACTH-ir fibers in rat brain. Brain Res 299:91–102
Pinget M, Straus E, Yalow RS (1979) Release of cholecystokinin peptides from a synaptosome-enriched fraction fo rat cerebral cortex. Life Sci 25:339–342
Pioro EPJ, Trevor Hughes J, Cuello AC (1984) Loss of substance P and enkephalin immunoreactivity in the human substantia nigra after striato-pallidal infarction. Brain Res 292:339–347
Pique L, Bertagna X, Javoy-Agid F, Luton J-P, Agid Y (1981) Simultaneous measurement of β-endorphin and γ-lipotropin-like peptides in the human hypothalamus. Neuropeptides 2:99–108
Polak JM, Bloom SR (1982) The central and peripheral distribution of neurotensin. Ann NY Acad Sci 400:75–93
Polc P, Haefely W (1982) Benzodiazepines enhance the bicuculline-sensitive part of recurrent Renshaw inhibition in the cat spinal cord. Neurosci Lett 28:193–197
Pollard H, Llorens-Cortis C, Barbin G, Garbarg M, Schwartz JC (1978) Histamine and histidine decarboxylase in brain stem nuclei: distribution and decrease after lesions. Brain Res 157:178–181
Poulain P, Martin-Bouyer L, Beauvillain JC, Tramu G (1984) Study of the efferent connections of the enkephalinergic magnocellular dorsal nucleus in the guinea-pig hypothalamus using lesions, retrograde tracing and immunohistochemistry: evidence for a projection to the lateral septum. Neuroscience 11:331–343
Prange AJ Jr, Nemeroff ChB (1982) The manifold actions of neurotensin: A first synthesis. Ann NY Acad Sci 400:368–375
Price DL, Whitehouse PJ, Struble RG, Clark AW, Coyle JT, DeLong MD, Hedreen JC (1982) Basal forebrain cholinergic systems in Alzheimer's disease and related dementias. Neurosci Comment 1:84–92
Price JL, Amaral DG (1981) An autoradiographic study of the projections of the central nucleus of the monkey amygdala. J Neurosci 1:1242–1259
Price JL, Stern R (1983) Individual cells in the nucleus basalis-diagonal band complex have restricted ax-

onal projections to the cerebral cortex in the rat. Brain Res 269:252–356
Priestley JV, Kelly JS, Cuello AC (1979) Uptake of (^{3}H) dopamine in periglomerular cells of the rat olfactory bulb: an autoradiographic study. Brain Res 165:149–155
Printz MP, Ganten D, Unger T, Phillips MI (1982) The brain renin angiotensin system 1982. Exp Brain Res Suppl 4:3–52
Puizillout JJ, Gaudin-Chazal G, Sayadi A, Vigier D (1981) Serotoninergic mechanisms and sleep. J Physiol Paris 77:415–424
Pycock CJ, Kerwin RW (1981) The status of glycine as a supraspinal neurotransmitter. Life Sci 28:2679–2686
Pijnenburg AJJ, Van Rossum JM (1973) Stimulation of locomotor activity following injection of dopamine into the nucleus accumbens. J Pharm Pharmacol 25:1003–1005
Pijnenburg AJJ, Honig WMM, Van der Heyden JAM, Van Rossum JM (1976) Effects of chemical stimulation of the mesolimbic dopamine system upon locomotor activity. Eur J Pharmacol 35:45–58
Quik M, Iversen LL, Bloom SR (1978) Effect of vasoactive intestinal peptide (VIP) and other peptides on cAMP accumulation in rat brain. Biochem Pharmacol 27:2209–2213
Racagni G, Apud JA, Cocchi D, Locatelli V, Muller EE (1982) GABAergic control of anterior pituitary hormone secretion. Life Sci 31:823–838
Raichle ME, Hartman BK, Eichling JO, Sharpe LG (1975) Central noradrenergic regulation of cerebral blood flow and vascular permeability. Proc Nat Acad Sci USA 72:3726–3730
Randić M, Yu HH (1976) Effects of 5-hydroxytryptamine and bradykinin in cat dorsal horn neurones activated by noxious stimuli. Brain Res 111:197–203
Rasmussen GL (1946) The olivary peduncle and other fiber projections of the superior olivary complex. J Comp Neurol 84:141–220
Rasmussen GL (1953) Further observations of the efferent cochlear bundle. J Comp Neurol 99:61–74
Redmond DE Jr (1977) Alterations in the function of the nucleus locus coeruleus: A possible model for studies of anxiety. In: Hanin I, Usdin E (eds) Animal models in psychiatry and neurology. Pergamon Press, New York, pp 293–306
Redmond DE Jr, Huang YH (1979) New evidence for a locus coeruleus-norepinephrine connection with anxiety. Life Sci 25:2149–2162
Rees HD, Switz GM, Michael RP (1980) The estrogen-sensitive neural system in the brain of female cats. J Comp Neurol 193:789–804
Rehfeld JF (1978a) Immunochemical studies on cholecystokinin. I. Development of sequence-specific radioimmunoassays for porcine triaconta-triapeptide cholecystokinin. J Biol Chem 253:4016–4021
Rehfeld JF (1978b) Immunochemical studies on cholecystokinin. II. Distribution and molecular heterogeneity in the central nervous system small intestine of man and hog. J Biol Chem 253:4022–4030
Rehfeld JF (1980) Cholecystokinin. Trends NeuroSci 3:65–67
Rehfeld JF, Goltermann N, Larsson L-I, Emson PM, Lee CM (1979) Gastrin and cholecystokinin in central and peripheral neurons. Fed Proc 38:2325–2329
Reinhard JF Jr, Liebmann JE, Schlosberg AJ (1979) Serotonin neurons project to small blood vessels in the brain. Science 206:85–87
Renaud LP, Martin JB, Brazeau P (1975) Depressant action of TRH, LH–RH, and somatostatin on activity of central neurones. Nature 255:233–235
Rennels M, Gregory TF, Blaumanis OR, Fujimoto K, Grady PA (1985) Evidence for a 'Paravascular' fluid circulation in the mammalian central nervous system, provided by the rapid distribution of tracer protein throughout the brain from the subarachnoid space. Brain Res 326:47–63
Reynolds GP (1983) Increased concentrations and lateral asymmetry of amygdala dopamine in schizophrenia. Nature 305:527–529
Ribak ChE (1978) Aspinous and sparsely-spinous stellate neurons in the visual cortex of rats contain glutamic acid decarboxylase. J Neurocytol 7:461–478
Ribak CE (1981) The GABAergic neurons of the extrapyramidal system as revealed by immunocytochemistry. In: Di Chiara G, Gessa GL (eds) GABA and the basal ganglia. Raven Press, New York, pp 23–36
Ribak ChE, Vaughn JE, Saito K, Barber R, Roberts E (1976) Immunocytochemical localization of glutamate decarboxylase in rat substantia nigra. Brain Res 116:287–298
Ribak ChE, Vaughn JE, Saito K, Barber R, Roberts E (1977) Glutamate decarboxylase localization in neurons of the olfactory bulb. Brain Res 126:1–18
Ribak CE, Vaughn JE, Saito K (1978) Immunocytochemical localization of glutamic acid decarboxylase in neuronal somata following colchicine inhibition of axonal transport. Brain Res 140:315–332
Ribas JL (1977) Morphological evidence for a possible functional role of supraependymal nerves on ependyma. Brain Res 125:362–368
Ricardo JA, Koh ET (1978) Anatomical evidence of direct projections from the nucleus of the solitary tract to the hypothalamus, amygdala, and other forebrain structures in the rat. Brain Res 153:1–26
Richardson DB, Mogenson GJ (1981) Water intake elicited by injections of angiotensin II into preoptic area of rats. Am J Physiol 240:R70–R74
Rivier C, Brown M, Vale W (1977) Effect of neurotensin, substance P and morphine sulfate on the secretion of prolactin and growth hormone in the rat. Endocrinology 100:751–754
Roberts GW, Woodhams PL, Bryant MG, Crow TJ, Bloom SR, Polak JM (1980a) VIP in the rat brain: Evidence for a major pathway linking the amygdala and hypothalamus via the stria terminalis. Histochemistry 65:103–119
Roberts GW, Woodhams PL, Crow TJ, Polak JM (1980b) Loss of immunoreactive VIP in the bed nucleus following lesions of the stria terminalis. Brain Res 195:471–475
Roberts GW, Crow TJ, Polak JM (1981) Neurotensin: First report of a cortical pathway. Peptides 2 (Suppl I) p 37–43
Romagnano MA, Joseph SA (1983) Immunocytochemical localization of $ACTH_{1\text{-}39}$ in the brainstem of the rat. Brain Res 276:1–16

Rønnekleiv OK, Kelly MJ, Eskay RL (1984) Distribution of immunoreactive substance P neurons in the hypothalamus and pituitary of the rhesus monkey. J Comp Neurol 224:51–59
Room P, Postema F, Korf J (1981) Divergent axon collaterals of rat locus coeruleus neurons: Demonstration by a fluorescent double labeling technique. Brain Res 221:219–230
Ross CA, Armstrong DM, Ruggiero DA, Pickel VM, Joh TH, Reis DJ (1981) Adrenaline neurons in the rostral ventrolateral medulla innervate thoracic spinal cord: A combined immunocytochemical and retrograde transport demonstration. Neurosci Lett 25:257–262
Rossier J (1984) On the mapping of the cholinergic neurons by immunocytochemistry. Neurochem Int 6:183–184
Rossier J, Vargo TM, Minick S, Ling N, Bloom FE, Guillemin R (1977) Regional dissociation of β-endorphin and enkephalin contents in rat brain and pituitary. Proc Natl Acad Sci USA 74:5162–5165
Rossor MN, Svendsen C, Hunt SP, Mountjoy CQ, Roth M, Iversen LL (1982) The substantia innominata in Alzheimer's disease: an histochemical and biochemical study of cholinergic marker enzymes. Neurosci Lett 28:217–222
Rostène WH (1984) Neurobiological and neuroendocrine functions of the vasoactive intestinal peptide (VIP). Prog Neurobiol 22:103–129
Rotsztejn WH (1980) Neuromodulation in neuroendocrinology. Trends NeuroSci 3:67–70
Roubos EW, Buma P (1984) Non-synaptic release sites: Structural basis for the concept that neurons communicate by diffusely released messengers. Neurosci Lett Suppl 18:S 409
Rouzaire-Dubois B, Scarnati E, Hammond C, Crossman AR, Shibazaki T (1983) Microiontophoretic studies on the nature of the neurotransmitter in the subthalamo-entopeduncular pathway of the rat. Brain Res 271:11–20
Ruda MA (1982) Opiates and pain pathways: demonstration of enkephalin synapses on dorsal horn projection neurons. Science 215:1523–1525
Russchen FT, Price JL (1984) Amygdalostriatal projections in the rat. Topographical organization and fiber morphology shown using the lectin PHA-L as an anterograde tracer. Neurosci Lett 47:15–22
Rustioni A, Spreafico R, Cheema S, Cuénod M (1982) Selective retrograde labelling of somatosensory cortical neurons after ^{3}H-D-aspartate injections in the ventrobasal complex of rats and cats. Neuroscience 7:S 183
Rustioni A, Schmechel DE, Cheema S, Fitzpatrick D (1984) Glutamic acid decarboxylase-containing neurons in the dorsal column nuclei of the cat. Somatosens Res 1:329–358
Saavedra JM, Palkovits M, Brownstein MJ, Axelrod J (1974) Localisation of phenylethanolamine N-methyltransferase in the rat brain nuclei. Nature 248:695
Said SI, Mutt V (1970a) Polypeptide with broad biological activity: Isolation from small intestine. Science 169:1217–1218
Said SI, Mutt V (1970b) Potent peripheral and splanchnic vasodilator peptide from normal gut. Nature 225:863–864
Said SI, Rosenberg RN (1976) Vasoactive intestinal polypeptide: Abundant immunoreactivity in neural cell lines and normal nervous tissue. Science 192:907–908
Saito A, Sankaran H, Goldfine ID, Williams JA (1980) Cholecystokininreceptors in the brain: Characterization and distribution. Science 208:1155–1156
Saito K, Barber R, Wu J-Y, Matsuda T, Roberts E, Vaugn JE (1974) Immunohistochemical localization of glutamate decarboxylase in rat cerebellum. Proc Natl Acad Sci USA 71:269–273
Sakai K, Touret M, Salvert D, Leger L, Jouvet M (1977) Afferent projections to the cat locus coeruleus as visualized by the horseradish peroxidase technique. Brain Res 119:21–41
Sakamoto N, Takatusuji K, Shiosaka S, Tateish K, Hashimura E, Miura S, Hamaoka T, Tohyama M (1984) Cholecystokinin-8-like immunoreactivity in the pre- and post-central gyri of the human cerebral cortex. Brain Res 307:77–83
Sakanaka M, Shiosaka S, Takatsuki K, Inagaki S, Senba E, Kawai Y, Matsuzaki T, Tohyama M (1981a) Experimental immunohistochemical studies on the amygdalofugal peptidergic (substance P and somatostatin) fibers in the stria terminalis of the rat. Brain Res 221:231–242
Sakanaka M, Shiosaka S, Takatsuki K, Inagaki S, Takagi H, Senba E, Kawai Y, Hara Y, Iida H, Minagawa H, Matsuzaki T, Tohyama M (1981b) Evidence for the existence of a substance P-containing pathway from the nucleus laterodorsalis tegmenti (Castaldi) to the lateral septal area of the rat. Brain Res 230:351–355
Sakanaka M, Senba E, Shiosaka S, Takatsuki K, Inagaki S, Takagi H, Kawai Y, Hara Y, Tohyama M (1982a) Evidence for the existence of an enkephalin-containing pathway from the area just ventrolateral to the anterior hypothalamic nucleus to the lateral septal area of the rat. Brain Res 239:240–244
Sakanaka M, Shiosaka S, Takatsuki K, Inagaki S, Hara Y, Kawai Y, Senba E, Tohyama M (1982b) Origins of substance P-containing fibers in the lateral septal area of young rats: immunohistochemical analysis of experimental manipulations. J Comp Neurol 212:268–277
Sakanaka M, Shiosaka S, Takatsuki K, Tohyama M (1983) Evidence for the existence of a substance P-containing pathway from the nucleus laterodorsalis tegmenti (Castaldi) to the medial frontal cortex of the rat. Brain Res 259:123–126
Sakumoto T, Tohyama M, Satoh K, Kimoto Y, Kinugasa T, Tanizawa O, Kurachi K, Shimizu N (1978) Afferent fiber connections from lower brain stem to hypothalamus studied by the horseradish peroxidase method with special reference to noradrenaline innervation. Exp Brain Res 31:81–94
Salt TE, Hill RG (1983) Neurotransmitter candidates of somatosensory primary afferent fibres. Neuroscience 10:1083–1103
Sandoval ME, Cotman CW (1978) Evaluation of glutamate as a neurotransmitter of cerebellar parallel fibers. Neuroscience 3:199–206
Saper CB (1982) Convergence of autonomic and limbic connections in the insular cortex of the rat. J Comp Neurol 210:163–173

Saper CB, Loewy AD (1980) Efferent connections of the parabrachial nucleus in the rat. Brain Res 197:291–317
Saper CB, Petito CK (1982) Correspondence of melanin-pigmented neurons in human brain with A1–A14 catecholamine cell groups. Brain 105:87–101
Saper CB, Swanson LW, Cowan WM (1979) Some efferent connections of the rostral hypothalamus in the squirrel monkey (Saimiri sciureus) and cat. J Comp Neurol 184:205–242
Sar M, Stumpf WE (1975) Distribution of androgen-concentrating neurons in rat brain. In: Stumpf WE, Grant LD (eds) Anatomical Neuroendocrinology. S Karger, Basel, pp 120–133
Satoh K, Tohyama M, Yamamoto K, Sakumoto T, Shimizu N (1977) Noradrenaline innervation of the spinal cord studied by the horseradish peroxidase method combined with monoamine oxidase staining. Exp Brain Res 30:175–186
Sawchenko PE, Swanson LW (1981) Central noradrenergic pathways for the integration of hypothalamic neuroendocrine and autonomic responses. Science 214:685–687
Sawchenko PE, Swanson LW (1982a) The organization of noradrenergic pathways from the brainstem to the paraventricular and supraoptic nuclei in the rat. Brain Res Rev 4:275–325
Sawchenko PE, Swanson LW (1982b) Immunohistochemical identification of neurons in the paraventricular nucleus of the hypothalamus that project to the medulla or to the spinal cord in the rat. J Comp Neurol 205:260–272
Scatton B, Serrano A, Rivot JP, Nishikawa T (1984) Inhibitory GABAergic influence on striatal serotonergic transmission exerted in the dorsal raphe as revealed by in vivo voltammetry. Brain Res 305:343–352
Scharrer B (1970) General Principles of Neuroendocrine Communication. The Neurosciences, Sec Study Program, Schmitt FO (ed). The Rockefeller Univ Press, New York, pp 519–529
Schmitt FO (1984) Molecular regulators of brain function: A new view. Neuroscience 13:991–1001
Schofield SPM, Dixson AF (1982) Comparative anatomy of brain monoaminergic neurons in new world and old world monkeys. Am J Primatol 2:3–19
Schofield SPM, Everitt BJ (1981) The organization of indoleamine neurons in the brain of the rhesus monkey (Macaca mulatta). J Comp Neurol 197:369–383
Schofield SPM, Everitt BJ (1982) The organisation of catecholamine-containing neurons in the brain of the rhesus monkey (Macaca mulatta). J Anat 132:391–418
Schrøder HD (1984) Somatostatin in the caudal spinal cord: An immunohistochemical study of the spinal centers involved in the innervation of pelvic organs. J Comp Neurol 223:400–414
Schulman JA (1983) Chemical neuroanatomy of the cerebellar cortex. In: Emson PC (ed) Chemical Neuroanatomy. Raven Press, New York, pp 209–228
Schwanzel-Fukuda M, Morrell JI, Pfaff DW (1984) Localization of forebrain neurons which project directly to the medulla and spinal cord of the rat by retrograde tracing with wheat germ agglutinin. J Comp Neurol 226:1–20
Schwartz JC (1975) Histamine as a transmitter in brain. Life Sci 17:503–518
Schwartz JC (1977) Histaminergic mechanisms in brain. Ann Rev Pharmacol Toxicol 17:325–339
Schwartz JC (1979) Minireview: histamine receptors in brain. Life Sci 25:895–912
Schwartz JC, Pollard H, Quach TT (1980a) Histamine as a neurotransmitter in mammalian brain: neurochemical evidence. J Neurochem 35:26–33
Schwartz JC, Barbin G, Duchemin AM, Garbarg M, Palacios JM, Quach TT, Rose C (1980b) Histamine receptors in the brain: Characterization by binding studies and biochemical effects. In: Pepeu G, Kuhar MJ, Enna SJ (eds) Receptors for Neurotransmitters and Peptide Hormones. Raven Press, New York, pp 169–182
Schwartzberg DG, Nakane PK (1981) Pro ACTH-Endorphin antigenicities in medullary neurons of the rat. Soc Neurosci Abstr 7:224
Scott FH (1905) On the metabolism and action of nerve cells. Brain 28:506–526
Seguela P, Geffard M, Buijs RM, Le Moal M (1984) Antibodies against γ-aminobutyric acid: Specificity studies and immunocytochemical results. Proc Natl Acad Sci USA 81:3888–3892
Seizinger BR, Höllt V, Herz A (1981) Evidence for the occurrence of the opioid octapeptide dynorphin (1–8) in the neurointermediate pituitary of rats. Biochem Biophys Res Commun 102:197–205
Senba E, Tohyama M (1983) Reticulofacial enkephalinergic pathway in the rat: an experimental immunohistochemical study. Neuroscience 10:831–839
Senba E, Shiosaka S, Hara Y, Inagaki S, Kawai Y, Takatsuki K, Sekanaka M, Iida H, Takagi H, Minagawa H, Tohyama M (1982) Ontogeny of the leucine-enkephalin neuron system of the rat: Immunohistochemical analysis. I. Lower brainstem. J Comp Neurol 205:341–359
Sernia C, Reid IA (1980) Release of angiotensinogen by rat brain in vitro. Brain Res 192:217–225
Seybold VS, Elde RP (1982) Neurotensin immunoreactivity in the superficial laminae of the dorsal horn of the rat: I. Light microscopic studies of cell bodies and proximal dendrites. J Comp Neurol 205:89–100
Sheppard MC, Kronheim S, Pimstone BL (1979) Effect of substance P, neurotensin and the enkephalins on somatostatin release from the rat hypothalamus in vitro. J Neurochem 32:647–649
Shibasaki T, Ling N, Guillemin R (1980) A radioimmunoassay for γ-melanocyte stimulating hormone. Life Sci 26:1781–1785
Shibasaki T, Ling N, Guillemin R, Silver M, Bloom F (1981) The regional distribution of γ_3-melanotropin-like peptides in bovine brain is correlated with adrenocorticotropin immunoreactivity but not with β-endorphin. Regulat. Peptides 2:43–52
Shik ML, Severin FV, Orlovsky GN (1967) Structures of the brain responsible for evoked locomotion. Sechenov physiol. J USSR 53:1125–1132 Quoted from Steeves and Jordan (1984)
Shik ML, Yagodnitsyn AD (1979) The pontobulbar 'locomotor strip'. Neurophysiol 9:95–97 Quoted from Steeves and Jordan 1984
Shimatsu A, Kato Y, Matsushita N, Katakami H, Yanaihara N, Imura H (1982) Effects of glucagon, neu-

rotensin, and vasoactive intestinal polypeptide on somatostatin release from perifused rat hypothalamus. Endocrinology 110:2113–2117
Shimizu N, Ohnishi S, tohyama M, Maeda T (1974) Demonstration by degeneration silver method of the ascending projection from the locus coeruleus. Exp Brain Res 20:181–192
Shinoda K, Inagaki S, Shiosaka S, Kohno J, Tohyama M (1984) Experimental immunohistochemical studies on the substance P neuron system in the lateral habenular nucleus of the rat: Distribution and origins. J Comp Neurol 222:578–588
Shiosaka S, Tohyama M (1984) Evidence for an α-melanocyte stimulating hormonergic hippocampal commissural connection in the rat, revealed by a double-labeling technique. Neurosci Lett 49:213–216
Shiosaka S, Takatsuki K, Sakanaka M, Inagaki S, Takagi H, Senba E, Kawai Y, Iida H, Minagawa H, Hara Y, Matsuzaki T, Tohyama M (1982) Ontogeny of somatostatin-containing neuron system of the rat: immunohistochemical analysis. II. Forebrain and diencephalon. J Comp Neurol 204:211–224
Shipley MT, Adamek GD (1984) The connections of the mouse olfactory bulb: A study using orthograde and retrograde transport of wheat germ agglutinin conjugated to horsradish peroxidase. Brain Res Bull 12:669–688
Shivers BD, Harlan RE, Morrell JI, Pfaff DW (1983) Immunocytochemical localization of luteinizing hormone-releasing hormone in male and female rat brains. Neuroendocrinol 36:1–12
Shute CCD, Lewis PR (1967) The ascending cholinergic reticular system: neocortical, olfactory and subcortical projections. Brain 90:497–520
Silverman AJ (1984) Luteinizing hormone releasing hormone containing synapses in the diagonal band and preoptic area of the guinea pig. J Comp Neurol 227:452–458
Silverman AJ, Krey LC (1978) The luteinizing hormone-releasing hormone (LH–RH) neuronal networks of the guinea pig brain. I. Intra- and extra-hypothalamic projections. Brain Res 157:233–246
Silverman AJ, Zimmerman EA (1983) Magnocellular neurosecretory system. Ann Rev Neurosci 6:357–380
Silverman AJ, Antunes JL, Ferin M, Zimmerman EA (1977) The distribution of luteinizing hormone-releasing hormone (LHRH) in the hypothalamus of the Rhesus monkey. Light microscopic studies using immunoperoxidase technique. Endocrinology 101:134–142
Silverman AJ, Antunes JL, Abrams GM, Nilaver G, Thau R, Robinson JA, Ferin M, Krey LC (1982) The luteinizing hormone-releasing hormone pathways in rhesus (Macaca mulatta) and pigtailed (Macaca nemestrina) monkeys: New observations on thick, unembedded sections. J Comp Neurol 211:309–317
Simerly RB, Swanson LW, Gorski RA (1984) Demonstration of a sexual dimorphism in the distribution of serotonin-immunoreactive fibers in the medial preoptic nucleus of the rat. J Comp Neurol 225:151–166
Simon H, Le Moal M, Galey D, Cardo B (1976) Silver impregnation of dopaminergic systems after radiofrequency and 6-OHDA lesions of the rat ventral tegmentum. Brain Res 115:215–231
Simon H, Le Moal M, Stinus L, Calas A (1979) Anatomical relationships between the ventral mesencephalic tegmentum-A10 region and the locus coeruleus as demonstrated by anterograde and retrograde tracing techniques. J Neural Trans 44:77–86
Simpson JB (1981) The circumventricular organs and the central actions of angiotensin. Neuroendocrinology 32:248–256
Sims KB, Hoffman DL, Said S, Zimmerman EA (1980) Vasoactive intestinal polypeptide (VIP) in mouse and rat brain: An immunocytochemical study. Brain Res 186:165–183
Sinnamon HM, Lee SH, Adams DB, Stopford CK (1984) Locomotor stepping elicited by electrical stimulation of the lateral hypothalamus requires an ipsilateral descending pathway. Physiol & Behav 33:209–215
Sirett NE, McLean AS, Bray JJ, Hubbard JI (1977) Distribution of angiotensin II receptors in rat brain. Brain Res 122:299–312
Sirett NE, Bray JJ, Hubbard JI (1981) Localization of immunoreactive angiotensin II in the hippocampus and striatum of rat brain. Brain Res 217:405–411
Skagerberg G, Björklund A, Lindvall O, Schmidt RH (1982) Origin and termination of the diencephalo-spinal dopamine system in the rat. Brain Res Bull 9:237–244
Skagerberg G, Lindvall O, Björklund A (1984) Origin, course and termination of the mesohabenular dopamine pathway in the rat. Brain Res 307:99–108
Skinner RD, Garcia-Rill E (1984) The mesencephalic locomotor region (MLR) in the rat. Brain Res 323:385–389
Skirboll L, Hökfelt T, Rehfeld J, Cuello AC, Dockray G (1982) Coexistence of substance P- and cholecystokinin-like immunoreactivity in neurons of the mesencephalic periaqueductal central gray. Neurosci Lett 28:35–39
Sladek JR Jr, Walker P (1977) Serotonin-containing neuronal perikarya in the primate locus coeruleus and subcoeruleus. Brain Res 134:359–366
Sladek JR Jr, Garver DL, Cummings JP (1982) Monoamine distribution in primate brain. IV. Indoleamine-containing perikarya in the brain stem of Macaca arctoides. Neuroscience 7:477–493
Smock T, Fields HL (1980) $ACTH_{1-24}$ blocks opiate-induced analgesia in the rat. Brain Res 212:202–206
Snyder SH (1980) Brain peptides as neurotransmitters. Science 209:976–983
Sofroniew MV (1979) Immunoreactive β-endorphin and ACTH in the same neurons of the hypothalamic arcuate nucleus in the rat. Amer J Anat 154:283–289
Sofroniew MV (1983) Vasopressin and oxytocin in the mammalian brain and spinal cord. Trends Neurosci 6:467–472
Sofroniew MV, Weindl A (1978) Projections from the parvocellular vasopressin- and neurophysin-containing neurons of the suprachiasmatic nucleus. Am J Anat 153:391–430
Sofroniew MV, Weindl A (1980) Identification of parvocellular vasopressin and neurophysin neurons in the suprachiasmatic nucleus of a variety of mammals including primates. J Comp Neurol 193:659–675
Sofroniew MV, Weindl A, Schinko I, Wetzstein R (1979)

The distribution of vasopressin-, oxytocin-, and neurophysin-producing neurons in the Guinea pig brain. Cell Tiss Res 196:367–384
Sofroniew MV, Eckenstein F, Cuello AC (1985) Central cholinergic neurons visualized by immunohistochemical detection of choline acetyltransferase. In: Paxinos G, Watson C (eds) The Rat Nervous System. Academic Press, Sydney, in press
Somogyi P, Cowey A, Halász N, Freund TF (1981) Vertical organization of neurones accumulating ^{3}H-GABA in visual cortex of rhesus monkey. Nature 294:761–763
Somogyi P, Priestley JV, Cuello AC, Smith AD, Takagi H (1982) Synaptic connections of enkephalin-immunoreactive nerve terminals in the neostriatum: a correlated light and electron microscopic study. J Neurocytol 11:779–807
Speciale SG, Crowley WR, O'Donohue TL, Jacobowitz DM (1978) Forebrain catecholamine projections of the A5 cell group. Brain Res 154:128–133
Staines WmA, Nagy JI, Vincent SR, Fibiger HC (1980) Neurotransmitters contained in the efferents of the striatum. Brain Res 194:391–402
Steeves JD, Jordan LM (1984) Autoradiographic demonstration of the projections from the mesencephalic locomotor region. Brain Res 307:263–276
Steinbusch HWM (1981) Distribution of serotonin-immunoreactivity in the central nervous system of the rat. Cell bodies and terminals. Neuroscience 6:557–618
Steinbusch HWM, Mulder AH (1984) Immunohistochemical localization of histamine in neurons and mast cells in the rat brain. In: Björklund A, Hökfelt T, Kuhar MJ (eds) The Handbook of Chemical Neuroanatomy, vol 2: Classical transmitters and transmitter receptors in the CNS pt2. Elsevier, Amsterdam, pp 126–139
Steinbusch HWM, Nieuwenhuys R (1981) Localization of serotonin-like immunoreactivity in the central nervous system and pituitary of the rat, with special references to the innervation of the hypothalamus. Adv Exp Med Biol 133:7–36
Steinbusch HWM, Nieuwenhuys R (1983) The raphe nuclei of the rat brainstem: A cytoarchitectonic and immunohistochemical study. In: Emson PC (ed) Clinical Neuroanatomy. Raven Press, New York, pp 131–207
Steinbusch HWM, Verhofstad AAJ, Joosten HWJ (1978) Localization of serotonin in the central nervous system by immunohistochemistry: description of a specific and sensitive technique and some applications. Neuroscience 3:811–819
Steindler DA (1981) Locus coeruleus neurons have axons that branch to the forebrain and cerebellum. Brain Res 223:367–373
Sternberger LA (1979) Immunocytochemistry. J Wiley & Sons, New York, pp 354
Sternberger LA (1980) What is brain? An immunocytochemical neuroendocrine approach. Acta Histochem Cytochem 13:66–70
Stevens JR (1973) An anatomy of schizophrenia? Arch Gen Psychiat 29:177–189
Stevens JR (1979) Schizophrenia and dopamine regulation in the mesolimbic system. Trends Neuro Sci 2:102–105
Stevens RT, Hodge CJ Jr, Apkarian AV (1982) Kölliker-Fuse nucleus: the principal source of pontine catecholaminergic cells projecting to the lumbar spinal cord of cat. Brain Res 239:589–594
Stopa EG, King JC, Lydic R, Schoene WC (1984) Human brain contains vasopressin and vasoactive intestinal polypeptide neuronal subpopulations in the suprachiasmatic region. Brain Res 297:159–163
Storm-Mathisen J (1972) Glutamate decarboxylase in the rat hippocampal region after lesions of the afferent fibre systems. Evidence that the enzyme is localized in intrinsic neurones. Brain Res 40:215–235
Storm-Mathisen J, Fonnum F (1971) Quantitative histochemistry of glutamate decarboxylase in the rat hippocampal region. J Neurochem 18:1105–1111
Storm-Mathisen J, Woxen Opsahl M (1978) Aspartate and/or glutamate may be transmitters in hippocampal efferents to septum and hypothalamus. Neurosci Lett 9:65–70
Storm-Mathisen J, Leknes AK, Torbjørg Bore A, Line Vaaland J, Edminson P, Haug F-MS, Ottersen OP (1983) First visualization of glutamate and GABA in neurones by immunocytochemistry. Nature 301:517–520
Strahlendorf HK, Barnes CD (1983) Control of substantia nigra pars reticulata neurons by the nucleus accumbens. Brain Res Bull 11:259–263
Studler JM, Reibaud M, Tramu G, Blanc G, Glowinski J, Tassin JP (1984) Pharmacological study on the mixed CCK8/DA mesonucleus accumbens pathway: Evidence for the existence of storage sites containing the two transmitters. Brain Res 298:92–97
Stumpf WE (1970) Estrogen neurons in the periventricular brain. Am J Anat 129:207–218
Stumpf WE (1975) The Brain: An endocrine gland and hormone target. In: Stumpf WE, Grant LD (eds) Anatomical Neuroendocrinology. S Karger, Basel, pp 2–8
Stumpf WE, Sar M (1975) Hormone-architecture of the mouse brain with ^{3}H-estradiol. In: Stumpf WE, Grant LD (eds) Anatomical Neuroendocrinology. S Karger, Basel, pp 82–103
Stumpf WE, Sar M (1977) Steroid hormone target cells in the periventricular brain: relationship to peptide hormone producing cells. Fed Proc 36:1973–1977
Stumpf WE, Sar M (1978) Anatomical distribution of estrogen, androgen, progestin, corticosteroid and thyroid hormone target sites in the brain of mammals: Phylogeny and ontogeny. Amer Zool 18:435–445
Stumpf WE, Sar M, Keefer DA (1975) Atlas of estrogen target cells in rat brain. In: Stumpf WE, Grant LD (eds) Anatomical Neuroendocrinology. S Karger, Basel, pp 104–119
Sugimoto T, Hattori T (1984) Organization and efferent projections of nucleus tegmenti pedunculopontinus pars compacta with special reference to its cholinergic aspects. Neuroscience 11:931–946
Sugimoto T, Takada M, Kaneko T, Mizuno N (1984) Substance P-positive thalamo-caudate neurons in the center median-parafascicular complex in the cat. Brain Res 323:181–184
Swaab DF (1982) Neuropeptides. Their distribution and function in the brain. Prog Brain Res 55:97–122
Swanson LW (1976a) The locus coeruleus: a cytoarchi-

tectonic, Golgi and immunohistochemical study in the albino rat. Brain Res 110:39–56
Swanson LW (1976b) An autoradiographic study of the efferent connections of the preoptic region in the rat. J Comp Neurol 167:227–256
Swanson LW (1977) Immunohistochemical evidence for a neurophysin-containing autonomic pathway arising in the paraventricular nucleus of the hypothalamus. Brain Res 128:346–353
Swanson LW (1982) The projections of the ventral tegmental area and adjacent regions: A combined fluorescent retrograde tracer and immunofluorescence study in the rat. Brain Res Bull 9:321–353
Swanson LW (1983) Neuropeptides – new vistas on synaptic transmission. Trends NeuroSci 6:294–295
Swanson LW (1984) Review of "The Rat Brain in Stereotaxic Coordinates" by G. Paxinos and C. Watson. Trends NeuroSci 7:53
Swanson LW, Hartman BK (1975) The central adrenergic system, an immunofluorescence study of the location of cell bodies and their efferent connections in the rat utilizing dopamine-β-hydroxylase as a marker. J Comp Neur 163:467–506
Swanson LW, Hartman BK (1980) Biochemical specificity in central pathways related to peripheral and intracerebral homeostatic function. Neurosci Lett 16:55–60
Swanson LW, Kuypers HGJM (1980) The paraventricular nucleus of the hypothalamus: cytoarchitectonic subdivisions and the organization of projections to the pituitary, dorsal vagal complex and spinal cord as demonstrated by retrograde fluorescence double-labeling methods. J Comp Neurol 194:555–570
Swanson LW, McKellar S (1979) The distribution of oxytocin- and neurophysin-stained fibers in the spinal cord of the rat and monkey. J Comp Neurol 188:87–106
Swanson LW, Mogenson GJ (1981) Neural mechanisms for the functional coupling of autonomic, endocrine and somatomotor responses in adaptive behavior. Brain Res Rev 3:1–34
Swanson LW, Sawchenko PE (1983) Hypothalamic integration: organization of the paraventricular and supraoptic nuclei. Ann Rev Neurosci 6:269–324
Swanson LW, Connelly MA, Hartman BK (1977) Ultrastructural evidence for central monoaminergic innervation of blood vessels in the paraventricular nucleus of the hypothalamus. Brain Res 136:166–173
Swanson LW, Kucharczyk J, Mogenson J (1978) Autoradiographic evidence for pathways from the medial preoptic area to the midbrain involved in the drinking response to angiotensin II. J Comp Neurol 178:645–660
Swanson LW, Sawchenko PE, Bérod A, Hartman BK, Helle KB, Vanorden DE (1981) An immunohistochemical study of the organization of catecholaminergic cells and terminal fields in the paraventricular and supraoptic nuclei of the hypothalamus. J Comp Neurol 196:271–285
Swanson LW, Sawchenko PE, River J, Vale WW (1983) Organization of ovine corticotropin-releasing factor immunoreactive cells and fibers in the rat brain: an immunohistochemical study. Neuroendocrinology 36:165–186
Swanson LW, Mogenson GJ, Gerfen CR, Robinson P (1984) Evidence for a projection from the lateral preoptic area and substantia innominata to the mesencephalic locomotor region in the rat. Brain Res 295:161–178
Swerdlow NR, Swanson LW, Koob GF (1984) Substantia innominata: critical link in the behavioral expression of mesolimbic dopamine stimulation in the rat. Neurosci Lett 50:19–24
Szabadi E (1978) Functionally opposite receptors on neurons. Life Sci 23:1889–1898
Taber E, Brodal A, Walberg F (1960) The raphe nuclei of the brain stem in the cat. I. Normal topography and cytoarchitecture and general discussion. J Comp Neur 114:161–187
Taber Pierce E, Foote WE, Hobson JA (1976) The efferent connection of the nucleus raphe dorsalis. Brain Res 107:137–144
Taber Pierce E, Hoddevik GH, Walberg F (1977) The cerebellar projection from the raphe nuclei in the cat as studied with the method of retrograde transport of horseradish peroxidase. Anat Embryol 152:73–87
Tachibana S, Araki K, Ohya S, Yoshida S (1982) Isolation and structure of dynorphin, an opioid peptide, from porcine duodenum. Nature 295:339–340
Takagi H, Somogyi P, Somogyi J, Smith AD (1983) Fine structural studies on a type of somatostatin-immunoreactive neuron and its synaptic connections in the rat neostriatum. A correlated light and electron microscopic study. J Comp Neurol 214:1–16
Takagi H, Mizuta H, Inagaki S, Tateishi K, Hamaoka T (1984) The occurrence of cholecystokinin-like immunoreactive neurons in the rat neostriatum: light and electron microscopic analysis. Brain Res 309:346–349
Takahashi Y, Satoh K, Sakumoto T, Tohyama M, Shimuzu N (1979) A major source of catecholamine terminals in the nucleus tractus solitarii. Brain Res 172:372–377
Takatsuki K, Kawai Y, Sakanaka M, Shiosaka S, Senba E, Tohyama M (1983) Experimental and immunohistochemical studies concerning the major origins of the substance P-containing fibers in the lateral lemniscus and lateral parabrachial area of the rat, including the fiber pathways. Neurosci 10:57–71
Takeuchi Y, Hopkins DA (1984) Light and electron microscopic demonstration of hypothalamic projection to the parabrachial nuclei in the cat. Neurosci Lett 46:53–58
Takeuchi Y, Sano Y (1983) Immunohistochemical demonstration of serotonin nerve fibers in the neocortex of the monkey (Macaca fuscata). Anat Embryol 166:155–168
Takeuchi Y, Kimura H, Sano Y (1982a) Immunohistochemical demonstration of serotonin-containing nerve fibers in the cerebellum. Cell Tissue Res 226:1–12
Takeuchi Y, Kimura H, Sano Y (1982b) Immunohistochemical demonstration of the distribution of serotonin neurons in the brainstem of the rat and cat. Cell Tissue Res 224:247–267
Takeuchi Y, Kimura T, Matsuura T, Sano Y (1982c) Immunohistochemical demonstration of the organization of serotonin neurons in the brain of the monkey (Macaca fuscata). Acta Anat 114:106–124

Tanaka C, Ishikawa M, Shimada S (1982) Histochemical mapping of catecholaminergic neurons and their ascending fiber pathways in the rhesus monkey brain. Brain Res Bull 9:255–270

Tanaka I, Nakai Y, Jingami H, Fukata J, Nakao K, Oki S, Nakanishi S, Numa S, Imura H (1980) Existence of γ-melanotropin (γ-MSH)-like immunoreactivity in bovine and human pituitary glands. Biochem Biophys Res Co 94:211–217

Taquet H, Javoy-Agid F, Cesselin F, Hamon M, Legrand JC, Agid Y (1982) Microtopography of methionine-enkephalin, dopamine and noradrenaline in the ventral mesencephalon of human control and Parkinsonian brains. Brain Res 235:303–314

Taylor KM (1975) Brain histamine. In: Iverson LL, Iverson SD, Snyder SH (eds) Handbook of Psychopharmacology, Vol 3. Plenum Press, New York, pp 327–379

Taylor KM, Snyder SH (1972) Isotope microassay of histamine, histidine, histidine decarboxylase, and histamine methyltransferase in brain tissue. J Neurochem 19:1343–1358

Tennyson VM, Heikkila R, Mytilineou C, Coté L, Cohen G (1974) 5-Hydroxydopamine 'tagged' neuronal boutons in rabbit neostriatum: interrelationship between vesicles and axonal membrane. Brain Res 82:341–348

Tenovuo O, Rinne UK, Viljanen MK (1984) Substance P immunoreactivity in the post-mortem Parkinsonian brain. Brain Res 303:113–116

Ter Horst GJ, Luiten PGM, Kuipers F (1984) Descending pathways from hypothalamus to dorsal motor vagus and ambiguus nuclei in the rat. J Auton Nerv Syst 11:59–75

Thangnipon W, Storm-Mathisen J (1981) K^+-evoked Ca^{2+}-dependent release of D-(^{3}H)aspartate from terminals of the cortico-pontine pathway. Neurosci Lett 23:181–186

Thierry AM, Deniau JM, Herve D, Chevalier G (1980) Electrophysiological evidence for non-dopaminergic mesocortical and mesolimbic neurons in the rat. Brain Res 201:210–214

Tohyama M, Maeda T, Shimizu N (1974) Detailed noradrenaline pathways of locus coeruleus neuron to the cerebral cortex with use of 6-hydroxydopa. Brain Res 79:139–144

Tohyama M, Shiosaka S, Sakanaka M, Takagi H, Senba E, Saitoh Y, Takahashi Y, Sakumoto T, Shimizu N (1980) Detailed pathways of the raphe dorsalis neuron to the cerebral cortex with use of horseradish peroxidase-3,3′-5,5′tetramethyl benzidine reaction as a tool for the fiber tracing technique. Brain Res 181:433–439

Tramu G, Beauvillain JC, Croix D, Leonardelli J (1981) Comparative immunocytochemical localization of enkephalin and somatostatin in the median eminence, hypothalamus and adjacent areas of the guinea-pig brain. Brain Res 215:235–255

Trouth CO, Odek-Ogunde M, Holloway JA (1982) Morphological observations of superficial medullary CO_2-chemosensitive areas. Brain Res 246:35–46

Tseng L-F, Loh HH, Li CH (1976) β-Endorphin as a potent analgesic by intravenous injection. Nature 263:239–240

Uchizono K (1975) Excitation and inhibition. Synaptic morphology. Igaku S, Tokyo, pp 218

Udenfriend S, Kilpatrick DL (1983) Biochemistry of the enkephalins and enkephalin containing peptides. Arch Biochem Biophys 221:309–323

Ueda H, Katayama S, Kato R (1972) Area postrema angiotensin-sensitive site in brain. In: Ganong WF, Assaykeen T (eds) Control of Renin Secretion. Plenum, New York, pp 109–116

Uhl GR (1982) Distribution of neurotensin and its receptor in the central nervous system. Ann NY Acad Sci 400:132–149

Uhl G, Snyder S (1976) Regional and subcellular distributions of brain neurotensin. Life Sci 19:1827–1832

Uhl GR, Snyder SH (1977) Neurotensin receptor binding, regional and subcellular distributions favor transmitter role. Europ J Pharmacol 41:89–91

Uhl GR, Snyder SH (1979) Neurotensin: a neuronal pathway projecting from amygdala through stria terminalis. Brain Res 161:522–526

Uhl GR, Kuhar MJ, Snyder SH (1977) Neurotensin: Immunohistochemical localization in rat central nervous system. Proc Natl Acad Sci USA 74:4059–4063

Uhl GR, Kuhar MJ, Snyder SH (1978) Enkephalin-containing pathway: amygdaloid efferents in the stria terminalis. Brain Res 149:223–228

Uhl GR, Goodman RR, Kuhar MJ, Childers SR, Snyder SH (1979a) Immunohistochemical mapping of enkephalin containing cell bodies, fibers and nerve terminals in the brain stem of the rat. Brain Res 166:75–94

Uhl GR, Goodman RR, Snyder SH (1979b) Neurotensin-containing cell bodies, fibers and nerve terminals in the brain stem of the rat: Immunohistochemical mapping. Brain Res 167:77–91

Uhl GR, Whitehouse PJ, Price DL, Tourtelotte WW, Kuhar MJ (1984) Parkinson's disease: depletion of substantia nigra neurotensin receptors. Brain Res 308:186–190

Ungerstedt U (1971) Stereotaxic mapping of the monoamine pathways in the rat brain. Acta Physiol Scand Suppl 367:1–49

Vale W, Rivier C, Brown M (1977) Regulatory peptides of the hypothalamus. Ann Rev Physiol 39:473–527

Vale W, Spiess J, Rivier C, River J (1981) Characterization of a 41-residue ovine hypothalamic peptide that stimulates secretion of corticotropin and -endorphin. Science 213:1394–1397

Van der Gugten J, Palkovits M, Wijnen HJLM, Versteeg DHG (1976) Regional distribution of adrenaline in rat brain. Brain Res 107:171–174

Vanderhaeghen JJ, Signeau JC, Gepts W (1975) New peptide in the vertebrate CNS reacting with gastrin antibodies. Nature 257:604–605

Vanderhaeghen JJ, Lotstra F, De Mey J, Gilles C (1980) Immunohistochemical localization of cholecystokinin- and gastrin-like peptides in the brain and hypophysis of the rat. Proc Natl Acad Sci USA 7:1190–1194

Van de Kar LD, Lorens SA (1979) Differential serotonergic innervation of individual hypothalamic nuclei and other forebrain regions by the dorsal and median midbrain raphe nuclei. Brain Res 162:45–54

Van der Kooy D, Kuypers HGJM (1979) Fluorescent retrograde double labeling: Axonal branching in the

ascending raphe and nigral projections. Science 204:873–875
Van der Kooy D, Coscina DV, Hattori T (1981) Is there a non-dopaminergic nigrostriatal pathway? Neuroscience 6:345–357
Van der Kooy D, Koda LY, McGinty JF, Gerfen CR, Bloom FE (1984) The organization of projections from the cortex, amygdala, and hypothalamus to the nucleus of the solitary tract in rat. J Comp Neurol 224:1–24
Vandesande F, Dierickx K (1979) The activated hypothalamic magnocellular neurosecretory system and the one neuron – one neuronhypophysial hormone concept. Cell Tissue Res 200:29–33
Van Dongen PAM (1980) The noradrenergic locus coeruleus. Behavioral effects of intra-cerebral injections and a survey of its structure, function and pathology. Thesis. Meppel, Krips Repro, The Netherlands, pp 365
Van Dongen PAM (1981a) The central noradrenergic transmission and the locus coeruleus: A review of the data and their implications for neurotransmission and neuromodulation. Prog Neurobiol 16:117–143
Van Dongen PAM (1981b) The Human locus coeruleus in neurology and psychiatry. Prog Neurobiol 17:97–139
Van Houten M, Schiffrin EL, Mann JFE, Posner BI, Boucher R (1980) Radioautographic localization of specific binding sites for blood-borne angiotensin II in the rat brain. Brain Res 186:480–485
Van Praag HM (1977) New evidence of serotonin-deficient depressions. Neuropsychobiol 3:56–63
Van Praag HM, Korf J (1974) Serotonin metabolism in depression: (1974) Clinical application of the probenecid test. Int Pharmacopsychiat 9:35–51
Van Wimersma Greidanus TjB, De Wied D (1976) Dorsal hippocampus: a site of action of neuropeptides on avoidance behavior? Pharmacol Biochem Behav 5, Suppl 1:29–33
Van Wimersma Greidanus TjB, Croiset G, Bakker E, Bouman H (1979a) Amygdaloid lesions block the effect of neuropeptides (vasopressin, $ACTH_{4,10}$) on avoidance behavior. Physiol Behav 22:291–295
Van Wimersma Greidanus TjB, Croiset G, Schuiling GA (1979b) Fornix transection: discrimination between neuropeptide effects on attention and memory. Brain Res Bull 4:625–629
Varon SS, Somjen GG (1979) Neuron-glia interactions. Neurosci Res Prog Bull 17:1–183
Veening JG, Cornelissen FM, Lieven PAJM (1980) The topical organization of the afferents to the caudatoputamen of the rat. A horseradish peroxidase study. Neuroscience 5:1253–1268
Veening JG, Swanson LW, Sawchenko PE (1984) The organization of projections from the central nucleus of the amygdala to brainstem sites involved in central autonomic regulation: A combined retrograde transport-immunohistochemical study. Brain Res 303:337–357
Vijayan E, McCann SM (1979) *In vivo* and *in vitro* effects of substance P and neurotensin on gonadotropin and prolactin release. Endocrinology 105:64–68
Vijayan E, McCann SM (1980) Effects of substance P and neurotensin on growth hormone and thyrotropin release *in vivo* and *in vitro*. Life Sci 26:321–327
Villani L, Contestabile A, Fonnum F (1983) Autoradiographic labeling of the cholinergic habenulo-interpeduncular projection. Neurosci Lett 42:261–266
Vincent SR, Johansson O (1983) Striatal neurors containing both somatostatin- and avian pancreatic polypeptide (APP)-like immunoreactivities and NADPH-diaphorase activity. A light and electron microscopic study. J Comp Neurol 217:264–270
Vincent SR, Satoh K (1984) Corticotropin-releasing factor (CRF) immunoreactivity in the dorsolateral pontine tegmenturm further studies on the micturation reflex system. Brain Res 308:387–391
Vincent SR, Hattori T, McGeer EG (1978) The nigrotectal projection: a biochemical and ultrastructural characterization. Brain Res 151:159–164
Vincent SR, Staines WA, McGeer EG, Fibiger HC (1980) Transmitters contained in the efferents of the habenula. Brain Res 195:479–484
Vincent SR, Hökfelt T, Christensson I, Terenius L (1982) Dynorphin-immunoreactive neurons in the central nervous system of the rat. Neurosci Lett 33:185–190
Vincent SR, Hökfelt T, Skirboll LR, Wu J-Y (1983a) Hypothalamic γ-aminobutyric acid neurons project to the neocortex. Science 220:1309–1311
Vincent SR, Johansson O, Hökfelt T, Skirboll L, Elde RP, Terenius L, Kimmel J, Goldstein M (1983b) NADPH-diaphorase: A selective histochemical marker for striatal neurons containing both somatostatin- and avian pancreatic polypeptide (APP)-like immunoreactivities. J Comp Neurol 217:252–263
Vincent SR, Satoh K, Armstrong DM, Fibiger HC (1983c) Substance P in the ascending cholinergic reticular system. Nature 306:688–691
Vincent SR, Dalsgaard CJ, Schultzberg M, Hökfelt T, Christensson I, Terenius L (1984) Dynorphin-immunoreactive neurons in the autonomic nervous system. Neuroscience 11:973–987
Vogt M (1954) The concentration of sympathin in different parts of the central nervous system under normal conditions and after the administration of drugs. J Physiol 123:451–481
Voigt MM, Wang RY (1984) In vivo release of dopamine in the nucleus accumbens of the rat: modulation by cholecystokinin. Brain Res 296:189–193
Voshart K, Van der Kooy D (1981) The organization of the efferent projections of the parabrachial nucleus to the forebrain in the rat: a retrograde fluorescent double-labeling study. Brain Res 212:271–286
Wainer BH, Levey AI, Mufson EJ, Mesulam MM (1984) Cholinergic systems in mammalian brain identified with antibodies against choline acetyltransferase. Neurochem In 6:163–182
Walaas I (1981) Biochemical evidence for overlapping neocortical and allocortical glutamate projections tot the nucleus accumbens and rostral caudatoputamen in the rat brain. Neuroscience 6:399–405
Walaas I, Fonnum F (1980a) Biochemical evidence for γ-aminobutyrate containing fibres from the nucleus accumbens to the substantia nigra and ventral tegmental area in the rat. Neuroscience 5:63–72
Walaas I, Fonnum F (1980b) Biochemical evidence for glutamate as a transmitter in hippocampal efferents to the basal forebrain and hypothalamus in rat brain. Neuroscience 5:1691–1698

Wamsley JK, Zarbin MA, Young WS III, Kuhar MJ (1982) Distribution of opiate receptors in the monkey brain: an autoradiographic study. Neuroscience 7:595–613
Wang RY (1981) Dopaminergic neurons in the rat ventral tegmental area. I. Identification and characterization. Brain Res Rev 3:123–140
Wassef M, Berod A, Sotelo C (1981) Dopaminergic dendrites in the pars reticulata of the rat substantia nigra and their striatal input. Combined immunocytochemical localization of tyrosine hydroxylase and anterograde degeneration. Neuroscience 6:2125–2139
Watanabe T, Taguchi Y, Hayashi H, Tanaka J, Shiosaka S, Tahyama M, Kubota H, Terano Y, Wada H (1983) Evidence for the presence of a histaminergic neuron system in the rat brain: An immunohistochemical analysis. Neurosci Lett 39:249–254
Watanabe T, Taguchi Y, Shiosaka S, Tanaka J, Kubota H, Terano Y, Tohyama M, Wada H (1984) Distribution of the histaminergic neuron system in the central nervous system of rats: A fluorescent immunohistochemical analysis with histidine decarboxylase as a marker. Brain Res 295:13–25
Watkins L, Mayer D (1982) Organization of endogenous opiate and non-opiate pain control systems. Science 216:1185–1192
Watson SJ, Barchas JD, Hao Li C (1977) β-Lipotropin; Localization of cells and axons in rat brain by immunocytochemistry. Proc Natl Acad Sci USA 74:5155–5158
Watson SJ, Richard III, ChW, Barchas JD (1978) Adrenocorticotropin in rat brain: Immunocytochemical localization in cells and axons. Science 200:1180–1181
Watson SJ, Akil H (1980) α-MSH in rat brain: occurrence within and outside of β-endorphin neurons. Brain Res 182:217–223
Watson SJ, Akil H, Ghazarossian VE, Goldstein A (1981) Dynorphin immunocytochemical localization in brain and peripheral nervous system: Preliminary studies. Proc Natl Acad Sci USA 78:1260–1263
Watson SJ, Akil H, Fischli W, Goldstein A, Zimmerman E, Nilaver G, Van Wimersma Greidanus TB (1982a) Dynorphin and vasopressin: Common localization in magnocellular neurons. Science 216:85–87
Watson SJ, Khachaturian H, Akil H (1982b) Comparison of the distribution of dynorphin systems and enkephalin systems in brain. Science 218:1134–1136
Watson SJ, Khachaturian H, Taylor L, Fischli W, Goldstein A, Akil H (1983) Pro-dynorphin peptides are found in the same neurons throughout rat brain: Immunocytochemical study. Proc Natl Acad Sci USA 80:891–894
Watson WE (1974) Physiology of Neuroglia. Physiol Rev 54:245–271
Weber E, Evans CJ, Barchas JD (1982) Predominance of the amino-terminal hectapeptide fragment of dynorphin in rat brain regions. Nature 299:77–79
Weight FF (1979) Modulation of synaptic excitability. Fed Proc 38:2078–2079
Weight FF, Salmoiraghi GC (1966) Responses of spinal cord interneurons to acetylcholine, norepinephrine and serotonin administered by microelectrophoresis. J Pharmacol Exp Ther 153:420–427
Weiner RI, Ganong WF (1978) Role of brain monoamines and histamine in regulation of anterior pituitary secretion. Physiol Rev 58:905–976
Welch KMA, Gaudet R, Wang TPF, Chabi E (1977) Transient cerebral ischemia and brain serotonin: Relevance to migrane. Headache 17:145–147
Wenthold RJ (1978) Glutamic acid and aspartic acid in subdivisions of the cochlear nucleus after auditory nerve lesion. Brain Res 143:544–548
Wenthold RJ, Gulley RL (1977) Aspartic acid and glutamic acid levels in the cochlear nucleus after auditory nerve lesion. Brain Res 138:111–123
Wessendorf MW, Proudfit HK, Anderson EG (1981) The identification of serotonergic neurons in the nucleus raphe magnus by conduction velocity. Brain Res 214:168–173
Westlund KN, Coulter JD (1980) Descending projections of the locus coeruleus and subcoeruleus/medial parabrachial nuclei in monkey: Axonal transport studies and dopamine-β-hydroxylase immunocytochemistry. Brain Res Rev 2:235–264
Westlund KN, Bowker RM, Ziegler MG, Coulter JD (1983) Noradrenergic projections to the spinal cord of the rat. Brain Res 263:15–31
Westlund KN, Bowker RM, Ziegler MG, Coulter JD (1984) Origins and terminations of descending noradrenergic projections to the spinal cord of monkey. Brain Res 292:1–16
Weyhenmeyer JA, Phillips MI (1982) Angiotensin-like immunoreactivity in the brain of the spontaneously hypertensive rat. Hypertension 4:514–523
White JS, Warr WB (1983) The dual origins of the olivocochlear bundle in the albino rat. J Comp Neurol 219:203–214
Whitehouse PJ, Price DL, Clark AW, Coyle JT, DeLong R (1981) Alzheimer Disease: Evidence for Selective Loss of Cholinergic Neurons in the Nucleus Basalis. Ann Neurol 10:122–126
Whitehouse PJ, Price DL, Struble RG, Clark AW, Coyle JT, DeLong MR (1982) Alzheimer's Disease and Senile Dementia: Loss of Neurons in the Basal Forebrain. Science 215:1237–1239
Whitehouse PJ, Hedreen JC, White CL, Price DL (1983) Basal forebrain neurons in the dementia of Parkinson disease. Ann Neurol 13:143–148
Whittaker VP (1965) The application of subcellular fractionation techniques to the study of brain function. Prog Biophys 15:39–96
Wiklund L, Léger L, Persson M (1980) Monoamine cell distribution in the cat brain stem. A fluorescence histochemical study with quantification of indolaminergic and locus coeruleus cell groups. J Comp neurol 203:613–647
Wiklund L, Toggenburger G, Cuénod M (1982) Aspartate: Possible neurotransmitter in cerebellar climbing fibers. Science 216:78–79
Wilkin GP, Csillag A, Balázs R, Kingsbury AE, Wilson JE, Johnson AL (1981) Localization of high affinity (^{3}H)glycine transport sites in the cerebellar cortex. Brain Res 216:11–33
Willer J, Dehen H, Cambier J (1981) Stress-induced analgesia in humans: endogenous opioids and naloxone-reversible depression of pain reflexes. Science 212:689–691
Winokur A, Beckman AL (1978) Effects of thyrotropin-

releasing hormone. Norepinephrine, and acetylcholine on the activity of neurons in the hypothalamus, septum and cerebral cortex of the rat. Brain Res 150:205–209
Winokur A, Utiger RD (1974) Thyrotropin-releasing hormone: Regional distribution in rat brain. Science 185:265–267
Witkin JW, Paden ChM, Silverman AJ (1982) The luteinizing hormone-releasing hormone (LHRH) systems in the rat brain. Neuroendocrinol 35:429–438
Woodhams PL, Roberts GW, Polak JM, Crow TJ (1983) Distribution of neuropeptides in the limbic system of the rat: The bed nucleus of the stria terminalis, septum and preoptic area. Neuroscience 8:677–703
Woolf NJ, Butcher LL (1981) Cholinergic neurons in caudate-putamen complex proper are intrinsically organized: A combined Evans Blue and acetylcholinesterase analysis. Brain Res Bull 7:487–507
Yang HYT, Neff NH (1972) Distribution and properties of angiotensin converting enzyme of rat brain. J Neurochem 19:2443–2450
Yezierski RP, Bowker RM, Kevetter GA, Westlund KN, Coulter JD, Willis WD (1982) Serotoninergic projections to the caudal brain stem: a double label study using horseradish peroxidase and serotonin immunocytochemistry. Brain Res 239:258–264
Young AB, Oster-Granite ML, Herndon RM, Snyder SH (1974) Glutamic acid: selective depletion by viral induced granule cell loss in hamster cerebellum. Brain Res 73:1–13
Young AB, Bromberg MB, Penney JB Jr (1981) Decreased glutamate uptake in subcortical areas deafferented by sensorimotor cortical ablation in the cat. J Neurosc 1:241–249
Young EA, Walker JM, Houghten R, Akil H (1983) (^{3}H)Dynorphin a binds selectively to guinea pig brain. Eur J Pharmacol 91:327–328
Záborszky L, Beinfeld MC, Palkovits M, Heimer L (1984) Brainstem projection to the hypothalamic ventromedial nucleus in the rat: A CCK-containing long ascending pathway. Brain Res 303:225–231
Zaczek R, Hedreen JC, Coyle JT (1979) Evidence for a hippocampal-septal glutamatergic pathway in the rat. Exp Neurol 65:145–156
Zamir N, Palkovits M, Brownstein MJ (1983) Distribution of immunoreactive dynorphin in the central nervous system of the rat. Brain Res 280:81–93
Zamir N, Palkovits M, Brownstein MJ (1984a) Distribution of immunoreactive dynorphin $A_{1\text{-}18}$ in discrete nuclei of the rat brain: Comparison with dynorphin A. Brain Res 307:61–68
Zamir N, Palkovits M, Weber E, Mezey E, Brownstein MJ (1984b) A dynorphinergic pathway of leu-enkephalin production in rat substantia nigra. Nature 307:643–645
Zech M, Bogerts B, Tutsch J, Lesch A (1983) Reduction of met-enkephalin in the globus pallidus of patients suffering from Huntington's disease and schizophrenia. Neurosci Lett Suppl 14:S413
Zetler G (1980) Analgesia and ptosis caused by caerulein and cholecystokinin octapeptide (CCK-8). Neuropharmacology 19:415–422
Zimmerman EA, Liotta A, Krieger DT (1978) β-Lipotropin in brain: Localization in hypothalamic neurons by immunoperoxidase technique. Cell Tissue Res 186:393–398

Subject Index

Page numbers in **heavy type** indicate illustrations

Acetylcholine (ACh) 1, 6, 7–11, **7**
amygdaloid complex (amygdala) and 10
Alzheimer's disease and 10
anterior olfactory nucleus and 11
anterior thalamic nucleus and 9
arcuate nucleus and 11
ascending reticular activating system and 9
basal forebrain and 9
centrolateral nucleus of thalamus and 11
centromedian–parafascicular complex of thalamus and 11
cornu ammonis and 10
degradation of 7, **7**
descending corticocochlear projection and 9
dorsal striatum and 11
dorsal tegmental pathway of Shute and Lewis and 9
entorhinal cortex and 10
fascia dentata and 10
fasciculus olivocochlearis or bundle of Rasmussen and 7, 9, 10
fornix and 10
globus pallidus and 9
hippocampus and 10, 11
insula and 10
interpeduncular nucleus and 9
intralaminar thalamic nuclei and 9
lateral geniculate body and 9
lateral hypothalamus and 10
lateral parabrachial nucleus and 9
lateral thalamic nuclei and 9
limbic areas and 10
medial geniculate body and 9
medial habenular nucleus and 10
medial parabrachial nucleus and 9
medial septal nucleus and 9
α-motoneurons and 7
γ-motoneurons and 7
neocortex and 10
neostriatum and 11
nucleus accumbens and 11
nucleus basalis of Meynert and 9, 11
nucleus laterodorsalis tegmenti and 9
nucleus tegmenti pedunculopontinus, pars compacta and 11
olfactory bulb and 10
olfactory tubercle and 11
orbitofrontal cortex and 10
paralimbic areas and 10
pedunculopontine tegmental nucleus and 9
periolivary nuclei and 7
peripeduncular nucleus and 10
pontine taste area and 10
preganglionic autonomic neurons and 7
prepiriform cortex and 10
pretectal area and 9
rhombencephalic medial reticular formation and 9
senile dementia of the Alzheimer type and 10
septal region and 9, 10
spinal dorsal horn and 11
stria medullaris and 9
subthalamic nucleus and 9, 11
superior colliculus and 9
synthesis of 7, **7**
tractus habenulointerpeduncularis and 9
ventral amygdalofugal pathway and 10
ventral pallidum and 10
ventral striatum and 11
ventral tegmental area of Tsai and 9
Acetylcholinesterase (AChE) 7, 10
histochemistry 7
staining 5
Action potential 3
Adenosine 3′, 5′ monophosphate (cyclic AMP) 3, 59
Adenylate cyclase 3
Adrenaline 1, 6, 11, **12**, 30–32
amygdala and 32
area tegmentalis ventralis and 32
basal ganglia and 32
blood pressure and 32
cell groups C1–C3 and 30–32
dorsomedial nucleus and 32
habenula and 32
intermediolateral nucleus and 32
lateral hypothalamic area and 32
locus coeruleus and 32
medial longitudinal fascicle and 30
median eminence and 32
midline nuclei of thalamus and 32
nucleus accumbens and 32
nucleus dorsalis of the vagal nerve and 32
nucleus intermediolateralis and 32
nucleus infundibularis and 32
nucleus solitarius and 30
paraventricular nucleus and 32
periaqueductal gray and 32
preoptic region and 32
reticular formation and 32
septum and 32
supraoptic nucleus and 32
synthesis of 12, **12**
Adrenaline-containing cells and fibres 30–32, **31**
Adrenergic cell groups C1–C2 13, 30–32

Adrenocorticotropic hormone (ACTH), *see* Corticotropin
Affective behaviour 10
Agonistic behaviour 187
Aldosterone 109
Alzheimer's disease 10, 26, 172
Amino acid neurotransmitters 2, 6, 44–52, **45**
Analgesia
  acupuncture 96
  stimulation induced 96
  stress induced 96
Androgen-hormone-concentrating neurons 183–185, **184**
Angiotension I 108, **109**
Angiotensin II 6, 108–113, **109**, 116
  amygdaloid complex and 112
  area postrema and 109
  basal hypothalamus and 112
  dipsogenic effect of 110
  dorsomedial nucleus and 112
  hippocampus and 112
  intermediolateral column and 113
  locus coeruleus and 113
  medial preoptic area and 110
  median eminence and 112
  nucleus accumbens and 112
  organum vasculosum laminae terminalis and 109
  paraventricular nucleus and 110, 112
  periaqueductal gray and 113
  regulation of blood pressure and 113
  regulation of fluid balance and 113
  posterior pituitary and 110
  spinal anterior horn and 113
  spinal ganglion cells and 113
  spinal trigeminal nucleus and 113
  striatum and 112
  subfornical organ and 109
  substantia gelatinosa spinalis and 113
  supraoptic nucleus and 110, 112
Angiotensin II–containing cells and fibres 108–113, **111**
Angiotensin – converting enzyme 108
Angiotensinogen 108, **109**
Area postrema 109
Arginine-vasopressin (AVP), *see* vasopressin
Aromatic amino acid decerboxylese (AADC) 12, 14, 41
Arousal 4, 96
Aspartate (Aspartic acid) 2, 5, **45**, 48–51
Avian pancreatic polyptide (APP) **53**, 77, 79

Basal ganglia 130–133
Biogenic amines 11
Blood-brain barrier 4
Brachium conjunctivum 124
Bulbospinal serotoninergic pathways 38

Cardiovascular function 39, 192
Catecholaminergic cell groups 13
Catecholamines 11, **12**
  biosynthesis of 11, **12**
Cerebellar cortex 195, 158–162, **159**
  basket cells in 161
  climbing fibres in 160
  Golgi cells in 124, 161
  granule cells in 160
  mossy fibres in 160
  noradrenergic afferents of 162
  parallel fibres in 160
  Purkinje cells in 123, 158
  serotonergic afferents of 162
  stellate cells in 161
Cerebellar serotoninergic pathway 38
Cerebral cortex 154–156, **155**, **156**, **157**
Cerebral blood flow 39
Cerebral microcirculation 25
Cerebrovascular disorders 39
Chemical heterogeneity of neuronal populations 123
Chemical mediator 1
Chemical messengers 5
Chemical neuroanatomy 5, 194
Chemical (neuro)transmission
  non-synaptic 168–172, **169**
  synaptic 3, **169**
Chemo-architectonic subdivision 119, 195
Cholecystokinin (CCK) 2, 6, 18, 33, **53**, 60–65, 116
  amygdaloid complex and 62, 63
  analgesia and 63
  anterior commissure and 62
  anterior olfactory nucleus and 62
  area postrema and 63
  area tegmentalis ventralis and 62, 63
  bed nucleus of the stria terminalis and 63
  cerebral cortex and 62
  cingulate cortex and 62
  cuneate nucleus and 63
  diagonal band of Broca and 62
  dorsomedial hypothalamus and 62
  entorhinal cortex and 62
  fornix and 62
  hippocampal formation and 62
  inferior colliculus and 63
  interpeduncular nucleus and 63
  lateral geniculate body and 64
  lateral lemniscus and 64
  lateral olfactory tract and 62
  lateral parabrachial nucleus and 63, 64
  lateral septal nucleus and 62
  medial forebrain bundle and 62, 63, 64
  medial preoptic nucleus and 63
  medial septal nucleus and 62
  median eminence and 63, 64
  mesolimbic dopaminergic system and 63
  neocortex and 62
  nucleus accumbens and 62, 63
  nucleus of Edinger-Westphal and 65
  nucleus raphes dorsalis and 63
  nucleus raphes magnus and 63
  nucleus raphes obscurus and 64
  nucleus raphes pallidus and 64
  nucleus reticularis magnocellularis and 64
  nucleus solitarius and 63
  nucleus spinalis of the trigeminal nerve and 64
  olfactory tubercle and 63
  parabigeminal nucleus and 63
  paraventricular nucleus and 63
  periamygdaloid cortex and 62
  periaqueductal gray and 63
  piriform cortex and 62
  pituitary and 63, 64

prepiriform cortex and 62
release of oxytocin and 63
release of vasopressin and 63
schizophrenia and 63
spinal ganglion cells and 64
spinal ventral horn and 64
stria terminalis and 62
striatum and 62
superficial dorsal horn and 64
superior olivary complex and 64, 65
supramamillary nuclei and 63
supraoptic nucleus and 63
ventromedial nucleus and 63, 64
Cholecystokinin–containing cells and fibres 60–65, **61**
Choline 7
Choline acetyltransferase (Chat) 7, 10
Cholinergic cell groups Ch1–Ch6, 9, 10
Cholinergic cell groups and pathways 7–11, **8**
Chorea of Huntington 11, 55, 100, 176
Circumventricular organs 109, 110
Coexistence (of neuromediators) 4, 48, 52, 68, 103, 104, 112, 121–123, 132
Control
cardiovascular 192
respiratory 191
Core of the central nervous system 177–188, **179**, **184**, **190**
Colocalization (of neuromediators) *see* coexistence
Corticostriatal projection **131**, 132
Corticotropin 2, 6, **88**, 88–97
A1, A2, A4, A5, A8 catecholaminergic groups and 91
amygdaloid complex and 89
anterior zone of hypothalamus and 89
area tegmentalis ventralis and 91
bed nucleus of stria terminalis and 89
dorsomedial nucleus and 89
fasciculus longitudinalis dorsalis of Schütz and 91
infundibular nucleus and 89
lateral habenular nucleus and 91
lateral septal nucleus and 89
locus coeruleus and 91
medial preoptic area and 89
median eminence and 89
mediobasal hypothalamus and 89
neurohypophysis and 89
nucleus anterior dorsalis thalami and 91
nucleus cuneiformis and 91
nucleus reticularis gigantocellularis and 91
nucleus solitarius and 91
organum vasculosum laminae terminalis and 89
parabrachial nuclei and 91
paraventricular nucleus and 89
periaqueductal gray and 91
periventricular hypothalamic zone and 89
raphe nuclei and 91
spinal central gray and 91
ventral amygdalofugal pathway and 89
Corticotropin–containing cells and pathways 89–91, **90**
Corticotropin-like intermediate lobe peptide (CLIP) 87–89, **88**
Corticotropin-releasing factor (CRF) 6, **71**, 70–73
A1, A5, A7 noradrenergic groups and 70, 73
ACTH-immunoreactive fibres and 73
amygdala and 70
bed nucleus of stria terminalis and 70
cerebral cortex and 73
cingulate cortex and 73
dorsal raphe nucleus and 70, 73
hippocampus and 73
intermediolateral column and 73
Kölliker-Fuse nucleus and 73
lateral hypothalamic area and 70
lateral parabrachial nucleus and 70
lateral preoptic area and 70
lateral septal nucleus and 70
lateral tegmental area and 70
lateral tegmental nucleus and 70
locus coereleus and 70
medial forebrain bundle and 70
medial septal nucleus and 70
medial vestibular nucleus and 70
median eminence and 70
micturition reflex centre and 73
nucleus accumbens and 73
nucleus arcuatus and 73
nucleus cuneatus externus and 73
nucleus of the diagonal band and 70
nucleus of Edinger-Westphal and 70
nucleus paraventricularis and 70
nucleus preopticus medialis and 70
nucleus prepositus hypoglossi and 73
nucleus solitarius and 70
organum vasculosum laminae terminalis and 73
periaqueductal gray and 70
prefrontal cortex and 73
regulation of autonomic nervous system and 73
reticular formation and 70, 73
rhombencephalic periventricular gray and 73
spinal central gray and 73
stria terminalis and 70
substantia innominata and 70
ventral amygdalofugal pathway and 70
Corticotropin-releasing factor-containing cell groups and fibres 70–73, **72**
Cyclic AMP 3, 59
Cysteine-sulfinic acid decarboxylase (CSADCase) 51
Cytoarchitecture (cytoarchitectonics) 120, 194

Dale's principle 1, 4, 5
Diencephalo-septal dopaminergic projection 18
Dopamine 1, 6, 12, **12**, 14–19, 33
amygdaloid complex and 16, 17
anterior olfactory nucleus and 17
anterior cingulate cortex and 17
bed nucleus of stria terminalis and 17
bulbus olfactorius and 17
caudatus-putamen complex and 16
cell groups A8–A15 and 14–19
central canal and 18
dorsal vagal nucleus and 14
entorhinal cortex and 17
fasciculus longitudinalis dorsalis of Schütz and 19
fasciculus retroflexus and 18
frontal lobe and 17
globus pallidus and 16
hypothalamo-hypophysial portal system and 18
infundibular nucleus and 14

Dopamine
intermediolateral cell column and 18
lateral habenular nucleus and 18
lateral hypothalamic area and 16
lateral parabrachial nucleus and 18
lateral septal nucleus and 17, 18
locus coeruleus and 18
mamillothalamic tract and 14
medial forebrain bundle and 17
medial frontal cortex and 18
median eminence and 18
neostriatum and 16
neurointermediate lobe and 18
nucleus accumbens and 16, 18
nucleus raphes dorsalis and 14
olfactory tubercle and 16, 17
oxytocin release and 19
paraventricular thalamic nucleus and 19
paraventricular hypothalamic nucleus and 14
Parkinson's disease and 17
periglomerular cells and 14
piriform cortex and 17
prepiriform cortex and 17
prolactin secreting cells in pituitary and 18
putamen and 16
schizophrenia and 17
septum and 16
spinal dorsal horn and 18
spinal dorsolateral funiculus and 18
substantia nigra, pars compacta and 14, 16, 17
subthalamic nucleus and 18
supraoptic nucleus and 19
synthesis of 12, **12**
vasopressin release and 19
ventral tegmental area and 14, 16, 17, 18
Dopamine-containing cells and fibres 14–19, **15**
Dopamine β-hydroxylase (DBH) 12, 14
Dopaminergic cell groups A8–A15 14–19
Dorsal ascending serotoninergic pathway 37, 38
Dorsal noradrenergic bundle 20–24
Dorsal periventricular (catecholaminergic) bundle 19, 20, 23, 24, 26
Dorsal tegmental pathway of Shute and Lewis 9, 56
Dynorphin A 104, **105**
Dynorphin B 104, **105**
Dynorphins 6, 104, 104–108, **105**
A1, A2, A5 noradrenergic cell groups and 107
amygdaloid complex and 107
anterior hypothalamic nucleus and 107
area postrena and 104
bed nucleus of anterior commissure and 107
bed nucleus of stria terminalis and 104, 107
caudate-putamen complex and 107
cingulate cortex and 107
cochlear nuclei and 107
cuneiform nucleus and 107
dorsomedial nucleus and 107
entopeduncular nucleus and 107
griseum centrale mesencephali and 107, 108
hippocampus and 107
hypothalamo-hypophysial projection and 104
inferior colliculus and 107
lateral hypothalamic area and 104
lateral preoptic area and 104
lateral septal nucleus and 107
medial preoptic nucleus and 107
medial pretectal area and 107
median eminence and 107
motor cranial nerve nuclei and 108
neural lobe of pituitary and 104, 107
neuroendocrine control and 108
nociception and 108
nucleus accumbens and 107, 108
nuclei of the diagonal band and 104
nuclei of the lateral lemniscus and 107
nucleus cuneatus medialis and 107
nucleus infundibularis and 104, 107
nucleus paraventricularis and 104, 107
nucleus princeps of the trigeminal nerve and 108
nucleus solitarius and 104, 108
nucleus spinalis of the trigeminal nerve and 104, 108
nucleus supraopticus and 104, 107
olfactory bulb and 107
pain control and 108
parabrachial nuclei and 107
prefrontal cortex and 107
raphe nuclei and 108
reticular formation and 107, 108
spinal central gray and 108
spinal dorsal horn and 108
spinal ganglion cells and 108
striatum and 104
strio-pallidonigral projection and 108
substantia nigra and 104, 107
suprachiasmatic nucleus and 104, 107
ventral pallidum and 107, 108
ventromedial nucleus and 107
vestribular nuclei and 107
Dynorphin-containing cells and fibres 104–108, **106**

Emotional state 10
Endogenous depression 41
β-Endorphin (β-End) 2, 6, 87–89, **88**, 93, 94
A2 noradrenergic cell group and 94
amygdaloid complex and 94
analgesia 96
area tegmentalis ventralis and 94
autonomic regulations and 96
bed nucleus of stria terminalis and 94
cuneiform nucleus and 94
hypothalamic periventricular area and 94
lateral parabrachial nucleus and 94
lateral preoptic area and 94
lateral septal nucleus and 94
locus coeruleus and 94
medial preoptic area and 94
median eminence and 94
mediobasal hypothalamus and 93
neuroendocrine functions and 96
nucleus accumbens and 94
nucleus raphes dorsalis and 94
nucleus raphes magnus and 94
nucleus solitarius and 94
opiate receptors and 93
paraventricular nucleus and 94
periaqueductal gray and 94
periventricular thalamic nuclei and 94
reticular formation and 94
Enkephalins 6, 33, **97**, 97–104
A5 noradrenergic cell group and 102

amygdala and 98
anterior hypothalamic nucleus and 98
area postrema and 103
bed nucleus of stria terminalis and 98
chorea of Huntington and 100
cingulate cortex and 98
colliculus superior and 103
cortex cerebelli and 103
diagonal band of Broca and 103
dorsal cochlear nucleus and 103
dorsal motor nucleus of vagus and 103
dorsal tegmental nucleus and 101
dorsomedial nucleus and 100
entorhinal cortex and 98
globus pallidus and 98
hippocampus and 98,103
interpenduncular nucleus and 101
lateral habenular nucleus and 103
lateral hypothalamic area and 101
lateral geniculate body and 103
lateral septal nucleus and 98, 101
locus coeruleus and 103
magnocellular dorsal nucleus and 101
medial habenular nucleus and 103
medial preoptic nucleus and 100
medial septal nucleus and 98
median eminence and 100
motor cranial nerve nuclei and 103
neocortex and 98
neostriatum and 98
nucleus accumbens and 100
nucleus cuneiformis and 101
nucleus paraventricularis and 100
nucleus solitarius and 102, 103
nucleus supraopticus and 100
neuroendocrine control and 100
olfactory bulb and 98
olfactory tubercle and 103
olivocochlear bundle and 103, 104
organum vasculosum laminae terminalis and 103
pain control and 101
parabrachial nuclei and 102, 103
Parkinson's disease and 100
perforant path and 98
periaqueductal gray and 101
periventricular preoptic nucleus and 103
posterior pituitary and 100
raphe nuclei and 101, 102, 103
reticular formation and 102, 103
spinal dorsal horn and 102
spinal dorsolateral funiculus and 102
spinal trigeminal nucleus and 102
spinal ventral horn and 103
spinothalamic projection neurons and 103
stria terminalis and 98
strionigral projection and 100
striopallidal projection and 98
subcoeruleus area and 103
subiculum and 98
substantia nigra and 100, 103
subthalamic nucleus and 98
suprachiasmatic nucleus and 103
sympathetic preganglionic nucleus and 103
thalamic nuclei and 103
ventral amygdalofugal pathway and 98, 103
ventral pallidum and 100
ventromedial nucleus and 100
vestibular nuclei and 103
Enkephalin-containing cells and fibres 97–104, **99**
Ependymal cells 40
Estrogen-hormone-concentrating neurons 183, **184**
Excitatory postsynaptic potential(s) (EPSP) 3, 74
Exocytosis 1, 3
Extracellular space 3, 128
Extrapyramidal (motor) system 16, 132, 133

Fluid homeostasis 110
Formaldehyde-induced-fluorescence (FIF) technique 5

GABA (γ-aminobutyric acid) 1, 2, 6, 33, 44–48, **45**
amygdala and 48
basket cells in cerebellar cortex and 47
caudal hypothalamus and 47
caudatus-putamen complex and 45, 47
entorhinal cortex and 47
globus pallidus and 45
Golgi cells in cerebellar cortex and 45
granule cells in olfactory bulb and 47
hippocampus and 47
intralaminar thalamic nuclei and 47
intrinsic elements of hippocampus and 47
intrinsic elements of neocortex and 47
locomotor activity and 48
medial habenular nucleus and 47
medial septal nucleus and 47
neocortex and 47
nucleus accumbens and 47, 48
nucleus of the diagonal band and 47
nucleus raphes dorsalis and 45
nucleus reticularis thalami and 45
olfactory tubercle and 47
periglomerular cells in olfactory bulb and 47
pontomesencephalic tegmental region and 47
Purkinje cells and 44
release of anterior pituitary hormones and 47
release of hypophysiotropic peptides and 47
spinal interneurons and 44
stellate cells in cerebellar cortex and 44
strionigral projection and 47
striopallidal projection and 47
subthalamopallidal pathway and 47
substantia nigra, pars reticulata and 45, 47
superior colliculus and 45, 47
synthesis of 44, **45**
ventral pallidum and 48
ventral tegmental area and 47
ventrolateral thalamic nuclei and 47
GABA-containing cells and fibres 44–48, **46**
GABA-T-containing neurons 47
GABA-transaminase (GABA-T) 47
Gastrin 60
Glial cells 3, 197
Globus pallidus 123
Glutamate (Glutamic acid) 2, 6, 44, **45**, 48–51
auditory nerve and 50
bed nucleus of stria terminals and 50
caudate-putamen complex and 50
cerebellar cortex and 50
cerebellorubral fibres and 51
cerebellothalamic fibres and 51

Glutamate (Glutamic acid)
climbing fibres and 50
cochlear nuclei and 50
cornu ammonis and 50
corticopontine fibres and 50
corticospinal fibres and 50
corticostriatal fibres and 48
corticotectal fibres and 48
corticothalamic fibres and 48
cuneate nucleus and 50
fascia dentata and 50
fornix and 50
inferior olive and 50
lateral olfactory tract and 50
lateral septal nucleus and 50
mamillary body and 50
mediobasal hypothalamus and 50
neocortical pyramidal cells and 48
nucleus accumbens and 50
nucleus basalis and 50
nucleus of the diagonal band and 50
olfactory bulb and 50
parallel fibres and 50
perforant path and 50
prepiriform cortex and 50
red nucleus and 50
spinal dorsal roots and 50
subiculum and 50
substantia nigra and 50
trigeminal nerve and 50
Glutamate-(and Aspartate-) containing cells and fibres 48–51, **49**
Glutamic acid decarboxylase (GAD) 44
Glycine 2, 6, **45**, 51
caudate-putamen complex and 51
corticohypothalamic tract and 51
lateral hypothalamic area and 51
local circuit neurons and 51
prefrontal cortex and 51
substantia nigra and 51
ventromedial nucleus and 51
Golgi technique 195
Growth hormone release inhibiting hormone, *see* Somatostatin
Gut-brain peptides 52–70, **53**

Histamine 1, 6, 14, **41**, 41–44
amygdaloid complex and 43
anterior olfactory nucleus and 43
cerebral cortex and 43
caudate-putamen complex and 43
cochlear nuclei and 43
diagonal band of Broca and 43
dorsal nucleus of vagus nerve and 43
fasciculus longitudinalis dorsalis of Schütz and 43
inferior colliculus and 43
lateral geniculate body and 43
lateral lemniscus and 43
lateral parabrachial nucleus and 43
lateral septal nucleus and 43
mamillary body and 43
medial habenular nucleus and 44
medial septal nucleus and 43
medial forebrain bundle and 43
median eminence and 44
metencephalic central gray and 43
motor nucleus of VII and 43
neurons in caudal hypothalamus and 41
nucleus of the diagonal band and 43
nucleus of the trapezoid body and 43
nucleus raphes dorsalis and 43
nucleus solitarius and 43
nucleus vestibularis medialis and 43
paraventricular nucleus and 44
periaqueductal (central) gray and 43
periventricular thalamic nucleus and 43
pontine nucleus and 43
posterior hypothalamic area and 43
posterior pituitary and 44
prepiriform cortex and 43
stria terminalis and 43
subfornical organ and 44
suprachiasmatic nucleus and 44
supraependymal plexus and 44
supraoptic nucleus and 44
synthesis of 41
ventral amygdalofugal pathway and 43
ventromedial nucleus and 43
Histamine-containing cells and fibres 41–44, **42**
Histidine 41, **41**
Histidine-decarboxylase (HDC) 41
Hodology 195
Homeostasis 187
Hormone-architectonics 5
Huntington's disease, *see* chorea of Huntington
5-Hydroxyindolacetic acid (5-HIAA) 41
5-Hydroxytryptamine, *see* Serotonine
5-Hydroxytryptophan 12, **13**
Hypertension 113
Hypothalamo-hypophyseal portal system 2
Hypophysiotropic peptides (factors) 2, 70–83, **71**
Hypothalamo-hypophysial neurosecretory system 84
Hypothalamo-spinal dopaminergic projection 18

Immunohistochemistry 5
Incerto-hypothalamic dopaminergic projection 18
Indolamine 11
Interpeduncular nucleus 120, **142**, **143**, 144
Ionic channel(s) 1, 2, 3

L-dihydroxyphenylalanine (DOPA) 12, **12**
Learning 4, 95
Leucine-enkephalin (L-ENK) **97**, 97, 104
Limbic system 16, 17, 153, 181–183
β-Lipotropin (β-LPH) 87–89, **88**, 91, 92
γ-Lipotropin (γ-LPH) 87–89, **88**
Locomotor activity 17
Locus coeruleus 19–26
afferents 25
Alzheimer's disease and 172
efferents 20–26
functional significance 25, 26
Parkinson's disease and 175, **176**
Longitudinal catecholamine bundle 20, 24, 26, **127**, 128, 189
Luteinizing hormone releasing hormone (LHRH) 2, 6, **71**, 74–77
amygdala and 76
anterior hypothalamic area and 74
area postrema and 77
area tegmentalis ventralis and 76

bed nucleus of stria terminalis and 74, 76
ependyma and 76
fasciculus retroflexus and 76
gonadotrophs and 74
hippocampus and 76
indusium griseum and 76
infundibular nucleus and 74, 77
interpeduncular nucleus and 76
lamina terminalis and 74
lateral hypothalamic area and 76
medial forebrain bundle and 76
medial habenular nucleus and 76
medial mamillary nucleus and 76
medial preoptic nucleus and 74, 76
medial septal nucleus and 74, 76
median eminence and 74, 76
mediobasal hypothalamus and 74
nervus terminalis and 74
nucleus of the diagonal band and 74, 76
olfactory bulb and 76
olfactory tubercle and 76
organum vasculosum laminae terminalis and 76
periaqueductal gray and 76
posterior pituitry and 74
preoptico-infundibular projection and 74
reproduction and 77
reticular formation and 77
sexual behaviour and 77
stria medullaris and 76
stria terminalis and 76
subfornical organ and 77
superior colliculus and 74
suprachiasmatic nucleus and 74
supraoptic nucleus and 74
tubero-infundibular tract and 74
Luteinizing hormone-releasing hormone-containing cells and fibres 74–77, **75**

Medial forebrain bundle 125
Medial lemniscus 124
Medial longitudinal fasciculus 124
α-Melanocyte-stimulating hormone (α-MSH) 2, 6, 87–89, **88**, 92
caudate-putamen complex and 92
entorhinal cortex and 92
hippocampal formation and 92
intermediate hypothalamic zone and 92
lateral hypothalamic area and 92
mediobasal hypothalamus and 92
neocortex and 92
spinal cord and 92
zona incerta and 92
α-Melanocyte-stimulating hormone-containing cells and fibres **90**, 92
β-Melanocyte-stimulating hormone (β-MSH) 87–89, **88**, 93
γ-Melanocyte-stimulating hormone (γ-MSH) 87–89, **88**, 93
A1 and A8 catecholaminergic groups and 93
amygdaloid complex and 93
arcuate nucleus and 93
area tegmentalis ventralis and 93
Barrington's nucleus and 93
hypothalamic periventricular zone and 93
lateral parabrachial nucleus and 93
lateral preoptic area and 93
locus coeruleus and 93
medial preoptic area and 93
mediobasal hypothalamus and 93
nucleus accumbens and 93
nucleus solitarius and 93
paraventricular nucleus and 93
periaqueductal gray and 93
reticular formation and 93
septum and 93
stria terminalis and 93
Memory 10, 86, 95
Mesencephalic locomotor region 192
Mesocortical dopaminergic projection 17
Mesolimbic dopaminergic system (projection) 16, 17, 63, 67
Mesostriatal dopaminergic projection 16
Mesotelencephalic dopaminergic system 16
Methionine-enkephalin (M-ENK) 97, **97**
Microhomeostatic units 197
Monoamines 6, 11–14
Motilin 123, 160
Motivational state(drive) 4, 10, 95, 193

α-Neo-endorphin 104, **105**
β-Neo-endorphin 104, **105**
Neuroactive principles 5, 6
Neurochemicals 5
Neurocrine synaptic transmission 196
Neurohypophysial peptides 6, **84**, 84–87
Neuromediator profiles of cell masses 118, 119
Neuromediator profiles of fibre systems 124–126
Neuromediators 5, 114
Neuromediator specified neuronal populations 115–120
Neuromelanin 13
Neuromodulator(s) 3, 4, 5
Neuropeptides, *see* Peptides
Neurophysins 6, 84
Neuroregulators 5
Neurosecretory neurons 2
Neurotensin (NT) 2, 6, **53**, 65–70
amygdala and 67
anterior hypothalamic area and 68
anterior pituitory and 68
bed nucleus of stria terminalis and 67
blood pressure and 65
caudal nucleus of the trigeminal nerve and 69
caudate-putamen complex and 67
cingulate cortex and 65
dorsal motor vagus nucleus and 69
dorsomedial nucleus and 68
globus pallidus and 67
gastric secretion and 65
hypothermia and 65
infundibular nucleus and 68
lateral hypothalamic area and 68
lateral preoptic area and 68
locus coeruleus and 99
medial forebrain bundle and 67
medial preoptic nucleus and 68
medial septal nucleus and 67
median eminence and 68, 69, 70
midline nuclei of thalamus and 67
neocortex and 65
nucleus accumbens and 67, 69
nucleus of the diagonal band and 67, 69

Neurotensin (NT)
nucleus raphes dorsalis and 67, 69
nucleus raphes magnus and 65
nucleus raphes pontis and 69
nucleus solitarius and 69
olfactory bulb and 67
olfactory tubercle and 67
pain control and 68
parabrachial nuclei and 67, 69
paraventricular nucleus and 68
periventricular preoptic area and 68
periaqueductal gray and 69
posterior pituitary and 68
prepiriform cortex and 67
release of anterior pituitary hormones and 65
reticular formation and 69
schizophrenia and 67
sensitivity to pain and 65, 67
spinal central gray and 69
stria terminalis and 67
substantia gelatinosa spinalis and 69
substantia nigra and 69
thermoregulation and 68
ventral premamillary nucleus and 68
ventral tegmental area and 67, 69
ventromedial nucleus and 68
Neurotensin-containing cells and fibres 65–70, **66**
Neurotransmitter(s)
classical 5, 121
conventional 4
excitatory 3
general 1
genuine 5
inhibitory 2
putative 5
Nicotinamine adenine dinucleotide phosphate (NADPH)-diaphorase 77, 79
Nigrostriatal dopaminergic system 14
Nigrotectal projection **131**, 132
Nigrotegmental projection **131**, 132
Nissl staining 194
Non-synaptic chemical transmission 168–172, **169**
Noradrenaline 1, 6, 12, **12**, 33
Alzheimer's disease and 26
amygdala and 22, 23, 28
ansa peduncularis and 22
anterior commissure and 23
anterior hypothalamic area and 28
anterior olfactory nucleus and 22, 23
anterior perforated space and 22
anterior thalamic nuclei 22
area postrema and 19
attention function and 26
basal telencephalon and 23
bed nucleus of stria terminalis and 22, 26, 28
cardiovascular reactions and 25
cell groups A1–A7 and 19
central tegmental tract and 24
cerebellum and 23
cerebral microcirculation and 25
cingulum bundle and 23
coeruleocortical projection and 24
coeruleospinal fibres and 24
cochlear nuclei and 24
control of anterior pituitary hormones 30
diagonal band of Broca and 23
dorsal hypothalamic area and 28
dorsal motor vagus nucleus and 24
dorsal supraoptic decussation and 23
dorsomedial nucleus and 23, 26, 28
external capsule and 23
fasciculus longitudinalis dorsalis of Schütz and 23, 24
fasciculus retroflexus and 22
fornix and 23
inferior colliculus and 22
infundibular nucleus and 26, 28
interpeduncular nucleus and 22
hippocampal formation (hippocampus) and 22, 23
Kölliker-Fuse nucleus and 20
lateral funiculus of spinal cord and 24
lateral geniculate body and 22
lateral hypothalamic area and 28
lateral habenular nucleus and 22
lateral rhombencephalic tegmentum and 20
lateral septal nucleus and 28
lateral thalamic nuclei and 22
locus coeruleus (complex) and 19–26
mamillothalamic tract and 22
medial forebrain bundle and 22, 23, 26
medial geniculate body and 22
medial habenular nucleus and 22
medial parabrachial nucleus and 20
medial preoptic nucleus and 28, 29
medial septal nucleus and 22, 23
median eminence and 26, 28, 30
mesencephalic reticular formation and 26
neocortex and 22, 23
nuclei of the lateral lemniscus and 24
nucleus caudatus and 28
nucleus infundibularis and 28
nucleus of the diagonal band and 22, 23, 26
nucleus paraventricularis thalami and 28
nucleus raphes dorsalis and 22
nucleus solitarius and 24
olfactory bulb and 22, 23
olfactory tubercle and 22, 23
parasympathetic preganglionic neurons and 24
paraventricular nucleus and 23, 26, 28, 30
Parkinson's disease and 26
periaqueductal (central) gray and 20, 23, 26
periventricular nucleus and 28
piriform cortex and 28
pontine nuclei and 24
posterior commissure and 23
posterior hypothalamic area and 28
preoptic area and 26
principal sensory trigeminal nucleus and 24
schizophrenia and 26
septum and 26
spinal dorsal horn and 24
spinal intermediate gray and 24
spinal trigeminal nucleus and 24
spinal ventral horn and 24
stressful stimuli and 25
stria medullaris and 22
stria terminalis and 23
subcoeruleus area and 19, 23, 24
superior cerebellar peduncle and 23
superior colliculus and 22
supracallosal stria and 23
supraoptic nucleus and 23, 26, 28, 30

synthesis of 12, **12**
vasopressinergic neurons and 28
ventral amygdalofugal pathway and 22, 28
ventral thalamic nuclei and 22
ventromedial nucleus and 28
Noradrenaline-containing cells and fibres 19–30, **21**, **27**
Noradrenergic cell groups 19, 20
Nucleus accumbens 153, 192
Nucleus basalis of Meynert **172**, **173**, 175
Nucleus centralis superior 33
Nucleus cuneiformis 135
Nucleus raphes dorsalis 33–35, 135, 136
Nucleus raphes magnus 33–35, 135
Nucleus raphes obscurus 33–35
Nucleus raphes pallidus 33–35
Nucleus raphes pontis 33–35
Nucleus solitarius 130, 140
Nucleus tegmenti pedunculopontinus 132, 192
Nucleus ventralis posteromedialis thalami 130

Olfactory bulb 123, 162–167, **163**
afferents, centrifugal 166
Blanes cells of 166
efferents of 164
external plexiform layer of 162
glomerular layer of 162
glomeruli in 164
Golgi cells in 166
granular layer of 162
granule cells in 165
horizontal cells in 166
internal plexiform layer of 162
larger superficial interneurons in 167
mitral cell layer of 162
mitral cells in 164
olfactory nerve fibres in 162–164
periglomerular cells in 123, 162, 165
tufted cells in 164
Van Gehuchten cell in 167
vertical cells of Cajal in 166
Open multineuromediator channels 128
Opioid peptides 95, 104, 108
Opioid (opiate) receptors 95, 97, 100, 101, 104
Organum vasculosum laminae terminalis 109
Oxytoxin (Oxt) 6, **84**, 84–87, 116
lactation and 86
mammary glands 84
uterus and 84
Oxytoxin-containing cells and fibres 84–87, **85**

Pain, descending control system 40, 96, 101, 102, 108, 133–137, **134**, 193
Pallidonigral neurons 132
Pallidothalamic projection 124
Parabrachial nuclei 130, 140–144, **141**
Paracore lateral 189–193, **190**
Paracore median 189–193, **190**
Paracrine chemical transmission 197
Paracrine secretion **169**, 169–172, 187–189
Paralysis agitans, *see* Parkinson's disease,
Paraventricular nucleus 1, 146–148, **147**
Paraventriculo-hypophysial pathway 84
Parkinsonism-dementia complex of Guam 11, 175
Parkinson's disease 11, 17, 26, 55, 100, 175, **174**, **176**
Peptide(s) (neuropeptides),
anterior pituitary 2
general 2
gut-brain 2, 6, 5, 2–70, **53**
hypohysiotropic 6, 70–83, **71**
neurohypophyseal 6, **84**, 84–87
opioid 95, 104
transmitters 2
Peptidergic neurotransmitter 2, 3
Periaqueductal gray 133–136
afferents of 135
efferents of 135
Peripheral autonomic system 3
Phenylethanolamine-N-methyltransferase (PNMT) 12, 30–32
Pneumotaxic centre 191
Pontomedullary locomotor strip 192
Preoptic region 148–151, **150**
Pro-dynorphin 104, **105**
Pro-enkephalin 97, **97**
Prolactin 18, 82
Prolactin-inhibiting factor 82
Pro-neo-endorphin-dynorphin, *see* pro-dynorphin
Pro-opiomelanocortin (POMC) 2, 87–89, **88**, 94–97
corticotrophin releasing factor and 97
derivatives 6, 87–97, **88**
infundibular pool of cells and 95
solitarius pool of cells and 95
Propriobulbar noradrenergic connections 28
Pyramidal tract 126, 132

Receptors 1, 3, 4
excitatory 2
inhibitory 2
opposite 2
Renin 108
Renin-angiotensin system 108
Reproductive behaviour 187
Respiratory centres 191

Salivary glands 122
Schizophrenia 17, 26, 63, 67
Second messengers 3
Senile dementia of the Alzheimer type (SDAT) 10, 172–175, **173**
Sensory cortex 130
Serotonin 6, 11, **13**, 33–41
amygdala (amygdaloid complex) and 36, 37
ansa peduncularis and 36
anterior hypothalamic nucleus and 37
anterior olfactory nucleus and 37
anterior thalamic nucleus complex and 36
area tegmentalis ventralis and 36
caudatus-putamen complex and 36, 37
cerebellum and 38
cerebrovascular disorders and 39
cingulate cortex and 37
cingulum bundle and 37
corpus geniculatum laterale and 36
corpus mamillare and 36
diagonal band of Broca and 37
dorsal longitudinal fasciculus of Schütz and 37
dorsal tegmental nucleus and 38
dorsomedial nucleus and 37
endogenous depression and 41
entorhinal cortex and 37
external capsule and 37
fornix and 37

Serotonin
globus pallidus and 37
interpeduncular nucleus and 36
habenulo-interpeduncular tract 37
hippocampus and 37
inferior colliculus and 38
infundibular nucleus and 37
lateral hypothalamic area and 37
lateral septal nucleus and 37
lateral preoptic nucleus and 37
locus coeruleus and 35, 38, 40
mamillothalamic tract and 36
medial forebrain bundle and 36, 38
medial habenular nucleus and 36, 37
medial preoptic nucleus and 37
medial septal nucleus and 37
medial thalamic nucleus and 36
midline nuclei of thalamus and 36
neocortex and 37
nucleus accumbens and 37
nucleus ambiguus and 38
nucleus dorsalis nervi vagi and 38
nucleus intermediolateralis and 39
nucleus lateralis dorsalis thalami and 36
nucleus motorius nervi trigemini and 38
nucleus nervi abducentis and 38
nuclei parabrachialis and 38
nucleus parafascicularis thalami and 36
nucleus praepositus hypoglossi and 38
nucleus sensorius principalis of the trigeminal nerve and 38
nucleus solitarius and 38
nucleus spinalis of the trigeminal nerve and 38
nucleus tegmenti dorsalis and 38
olfactory bulb and 37
organum vasculosum of the lamina terminalis and 37
pain control and 40
parabrachial nuclei and 35
pedunculus cerebellaris medius and 38
periaqueductal (central) gray and 35, 38
posterior hypothalamic area and 36, 38
raphespinal projection and 40
regulation of cardiovascular function and 40
regulation of cerebral blood flow and 39
regulation of sleep and 40
reticular formation and 38
spinal dorsal horn and 38
spinal intermediate gray and 38
spinal ventral horn and 38, 39
stria medullaris and 36, 37
stria olfactory medialis and 37
stria terminalis and 37
subcoeruleus area and 35
subcommissural organ and 38
subfornical organ and 37
subiculum and 37
substantia nigra and 36, 40
superior colliculus and 38
supracallosal striae and 37
suprachiasmatic nucleus and 37
supraependymal plexus and 39, 41
synthesis of 12, **13**
ventral amygdalofugal pathway and 36
ventral thalamic nuclear complex and 36
ventromedial nucleus and 37
Serotonin-containing cells and fibres 33–41, **34**
Serotoninergic cell groups B1-B9 13, 33–36
Serotoninergic projections to rhombencephalic centres 38
Sexual behaviour 187
Shifting aptitude disorders 17
Sleep 40
Somatostatin (SST) 2, 6, 33, **71**, 77–82
abducens nucleus and 81
amygdala and 79, 80
anterior hypothalamic nucleus and 79
anterior olfactory nucleus and 79
bed nucleus of stria terminalis and 79, 80
brachium of the inferior colliculus and 81
cingulate cortex and 77
cochlear nuclei and 81
dorsal tegmental nucleus and 80, 81
facial nucleus and 81
fasciculus dorsolateralis of Lissauer and 81
fasciculus longitudinalis dorsalis of Schütz and 80, 81
fasciculus retroflexus and 80
fornix and 80
frontal cortex and 79
hippocampal formation and 77, 79, 80
hypoglossal nucleus and 81
inferior colliculus and 80, 81
infundibular nucleus and 79
intermediolateral nucleus and 82
interpeduncular nucleus and 80, 81
lateral lemniscus and 81
lateral mamillary nucleus and 80
lateral parabrachial nucleus and 80
lateral septal nucleus and 79, 80
limbic striatum and 79
medial forebrain bundle and 80
medial geniculate body and 80
medial habenular nucleus and 80
medial parabrachial nucleus and 81
medial pretectal area and 81
median eminence and 79
mesencephalic tegmentum and 80
neocortex and 77
neostriatum and 77
nuclei of the lateral lemniscus and 80, 81
nucleus accumbens and 79
nucleus ambiguus and 81
nucleus cuneatus lateralis and 81
nucleus cumatus medialis and 81
nucleus of Darkschewitsch and 81
nucleus entopeduncularis and 79
nucleus fastigii and 81
nucleus gracilis and 81
nucleus praepositus hypoglossi and 80, 81
nucleus raphes magnus and 81
nucleus raphes pontis and 81
nucleus solitarius and 80, 82
nucleus spinalis of the trigeminal nerve and 80
nucleus of the trapezoid body and 80, 81
oculomotor nucleus and 81
olfactory tubercle and 79, 80
Onuf's nucleus X complex and 82
organum vasculosum laminae terminalis and 79
paraventricular nucleus and 79, 81
periaqueductal gray and 81
periventricular anterior hypothalamic zone and 79

periventricular preoptic zone and 79
piriform cortex and 77
pontine nuclei and 81
posterior pituitary and 79
reticular formation and 80, 81
respiration and 82
septal region and 80
spinal central gray and 80, 82
spinal ganglion cells and 81
spinal trigeminal nucleus and 81
spinal ventral column and 82
spinal zona intermedia and 82
stria medullaris and 80
stria terminalis and 79, 80
substantia gelatinosa spinalis and 80, 81, 82
substantia nigra and 81
superior central nucleus and 81
superior colliculus and 81
suprachiasmatic nucleus and 79
supramamillary nuclei and 80
trochlear nucleus and 81
ventral amygdalofugal pathway and 80
ventral premamillary nucleus and 80
ventral tegmental area and 81
ventromedial nucleus and 80
vestibular nuclear complex and 81
Somatotostatin-containing cells and fibres 77–82, **78**
Somatotropin release-inhibiting factor, *see* somatostatin
Spinothalamic tract 124
Steroid hormones 5
Striatum (caudato-putamen complex) 130–133, **131**, 151–154, **152**
afferents of **131**, 132
chemoarchitecture of 151–154
efferents of **131**, 132
large aspiny cells in 133
limbic portions of 153
matrix compartment of 120, 151
medium-sized, aspiny cells in 133
medium-sized, densely spiny cells in 132
non-limbic (somatosensory) portions of 153
striosomes in 151–154, 158, 197
Subfornical organ 109
Substance P (SP) 2, 6, 33, 52–57, **53**
amygdaloid complex and 55, 56
anterior hypothalamic nucleus and 55, 56
area hypothalamica lateralis and 56
area tegmentalis ventralis and 56
baroceptive information and 52
bed nucleus of stria terminalis and 56
caudate-putamen complex and 55
central superior nucleus and 55
chemoreceptive information and 52
chorea of Huntington and 55
corpus geniculatum laterale and 56
cuneate fasciculus and 56
dorsal tegmental pathway and 56
entopeduncular nucleus and 56
fasciculus longitudinalis dorsalis of Schütz and 56
frontal cortex and 55, 56
globus pallidus 55
habenulointerpeduncular tract and 55
interpeduncular nucleus and 55
locus coeruleus and 56
lateral habenular nucleus and 55, 56
lateral hypothalamic region and 55
lateral lemniscus and 56
lateral septal nucleus and 55
medial forebrain bundle and 55, 56
medial habenular nucleus and 55, 56
medial lemniscus and 56
median eminence and 56
neocortex and 56
neural lobe of pituitary 57
nucleus ambiguus and 56
nucleus anterior dorsalis thalami and 56
nucleus anterior hypothalami and 57
nucleus arcuatus and 56
nucleus dorsomedialis and 56, 57
nucleus of Edinger-Westphal and 52
nucleus facialis and 56
nucleus infundibularis and 56
nucleus interpeduncularis and 56
nucleus laterodorsalis tegmenti and 55
nucleus mesencephalicus nervi trigemini and 56
nucleus olivaris superior and 56
nucleus preopticus lateralis and 57
nucleus preopticus medialis and 56, 57
nucleus preopticus periventricularis and 56
nucleus raphes magnus and 52
nucleus sensorius principalis nervi trigemini and 56
nucleus solitarius and 52, 55
nucleus supraopticus and 56
nucleus ventralis thalami and 56
nucleus ventromedialis and 57
olfactory bulb and 56
pain perception and 52
parabrachial nuclei and 56
paraventricular nucleus and 56
Parkinson's disease and 55
periaqueductal gray and 52, 56
periventricular zone of hypothalamus and 56
periventricular zone of rhombencephalon and 56
pontine nuclei and 56
reticular formation and 55, 56
spinal dorsal horn and 52
spinal ganglia and 52
spinal nucleus of the trigeminal nerve and 52
stria terminalis and 56
strionigral projection and 55
striopallidal projection and 55
substantia nigra and 55
subthalamic nucleus and 56
superior colliculus and 56
trigeminal ganglion and 52
tuberculum olfactorium and 56
Substance P-containing cells and fibres 52–57, **54**
Substantia gelatinosa (spinalis) 135–137
Substantia nigra 132, 144–145, **145**, **172**, **174**, 175
Suprachiasmatic nucleus 119, 148, **149**
Supra-ependymal histaminergic plexus 44
Supra-ependymal serotinergic plexus 39, 40
Supraoptic nucleus 2
Supraoptico-hypophysial pathway 84
Survival
of individual 187
of species 187
Sweat glands 4
Synapse(s) 1–3, 168, **169**
Synaptic transmission 3, 196

Tanycyte(s) 39

Tardive dyskinesia 175
Taurine 6, **45**, 51–52, 160
  basket cells in cerebellar cortex and 51
  cerebellar nuclei and 51
  Golgi cells in cerebellar cortex and 51
  granule cells in cerebellar cortex and 51
  Purkinje cells and 51
  stellate cells in cerebellar cortex and 51
Tectoreticular pathway 132
Tectospinal pathway 132
Terminal fields 138
  complementarity of 156, 157
  congruity of 156–158
Terminal plexus 138
Thyroïd-stimulating hormone (TSH), *see* thyrotropin
Thyrotropin 82
Thyrotropin releasing hormone (TRH) 2, 6, 33, **71**, 82–83
  bed nucleus of stria terminalis and 83
  dorsomedial nucleus and 83
  infundibular nucleus and 83
  lateral hypothalamus and 83
  lateral parabrachial nucleus and 83
  lateral septal nucleus and 83
  median eminence and 82, 83
  α-motoneurons and 83
  motor nuclei of cranial nerves and 83
  nucleus accumbens and 83
  nucleus raphes magnus and 83
  nucleus raphes pallidus and 83
  nucleus solitarius and 83
  organum vasculosum laminae terminalis and 83
  paraventricular nucleus and 83
  perifornical area and 83
  periventricular hypothalamic area and 83
  preoptic area and 83
  reticular formation and 83
  septal area and 83
  spinal gray and 83
  suprachiasmatic nucleus and 83
  ventromedial nucleus and 83
  zona incerta and 83
Tryptophan 12, **13**
Tryptophan hydroxylase 12
Tubero-infundibular dopaminergic projection 18
Tyrosine 12, **12**
Tyrosine hydroxylase (TH) 12, 14

Vas deferens 122
Vasoactive intestinal polypeptide 2, 4, 6, 11, 33, **53**, 57–60
  amygdaloid complex and 59, 60
  anterior hypothalamic area and 59
  anterior olfactory nucleus and 59
  bed nucleus of stria terminalis and 59, 60
  central spinal gray and 60
  cingulate cortex and 59
  caudate-putamen complex and 59
  corpus callosum and 57
  dorsomedial nucleus and 59
  entorhinal cortex and 59
  fasciculus dorsolateralis of Lissauer and 60
  glycogenolysis in cerebral cortex and 59
  hippocampal formation and 59
  lateral geniculate body and 60
  lateral parabrachial nucleus and 60
  lateral septal nucleus and 59
  medial forebrain bundle and 59
  median eminence and 60
  neocortex and 57, 59
  nodose ganglion and 60
  nucleus accumbens and 59
  nucleus solitarius and 60
  olfactory bulb and 59
  olfactory tubercle and 59
  paraventricular nucleus and 59
  periaqueductal (central) gray and 59
  premamillary nuclei and 59
  preoptic area and 59
  prepiriform cortex and 59
  pretectum and 60
  prolactin release and 60
  somatostatin release and 60
  spinal ganglion cells and 60
  stria terminalis and 59
  substantia gelatinosa and 60
  superior colliculus and 59
  suprachiasmatic nucleus and 59
  ventral pallidum and 59
  ventromedial nucleus and 59
Vasoactive intestinal polypeptide-containing cells and fibres 57–60, **58**
Vasopressin (AVP) 6, **84**, 84–87, 116
  accessory nuclei and 84
  amygdala and 86
  antidiuresis and 84
  bed nucleus of stria terminalis and 86
  control of blood pressure and 86
  diagonal band of Broca and 86
  dorsal motor vagus nucleus and 86
  dorsomedial hypothalamic nucleus and 86
  hippocampus and 87
  hypothalamo-hypophysial pathway and 84
  lateral habenular nucleus and 86
  lateral hypothalamus and 87
  lateral septal nucleus and 86
  learning processes and 86
  locus coeruleus and 86
  median eminence and 84
  mediodorsal thalamic nucleus and 87
  memory and 86
  mesencephalic central gray and 86
  neocortex and 87
  neurohypophysis and 84
  nociception and 86
  nucleus of the diagonal band and 86
  nucleus paraventricularis and 84
  nucleus solitarius and 86
  nucleus supraopticus and 84
  organum vasculosum laminae terminalis and 86
  parabrachial nuclei and 86
  periventricular thalamic nucleus and 86
  regulation of ACTH-secretion and 86
  spinal gray matter and 86
  suprachiasmatic nucleus and 86
  thermoregulation and 86
Vasopressin-containing cells and fibres 86
Ventral ascending serotoninergic pathway 84–87, **85**
Ventral noradrenergic pathway 36–37
Visceroceptive inputs to hypothalamus 30
Viscerosensory "vago-cortical" projection **129**, 130